T0384049

Contents

Preface

Over the years, wastewater has been evolving from *waste* to valuable resource. Recovery of water, nutrients, and/or energy are becoming essential elements of wastewater treatment design. The second edition of this book has been completely updated to reflect current advances in design, resource recovery practices, and research. A new chapter (Chapter 14) has been added on *resource recovery* and sustainability, along with real-life examples from all around the world. A new section on microbiology on the *Coronavirus* (SARS-CoV-2) is presented along with its implications in wastewater. Another highlight of this edition is chapter 15, which provides a *culminating design experience* of both urban and rural wastewater treatment systems.

This book is designed for a course on wastewater treatment and engineering for senior-level or early graduate-level students. The book covers the fundamental concepts of wastewater treatment followed by the engineering design of unit processes for the treatment of municipal wastewater and resource recovery. The students should have background knowledge of environmental chemistry and fluid mechanics. One important characteristic of this book is that each design concept is explained with the help of an underlying fundamental theory, followed by a mathematical model or formulation. Worked-out problems are used to demonstrate the use of the mathematical formulations and apply them in design.

Chapter 1 starts with a history of wastewater treatment, followed by current practices, emerging concerns, future directions in resource recovery, and pertinent regulations that have shaped the objectives and directions of this important area of engineering and research. Chapters 2 and 3 describe the fundamental concepts of reaction kinetics, reactor design, and wastewater microbiology. A new section on the *Coronavirus* (SARS-CoV-2) is provided along with its effects and implications in wastewater. Chapter 4 introduces natural purification processes and the dissolved oxygen sag curve. The concept of simple mass balances is introduced in this chapter. Chapters 5–10 describe the unit processes in primary and secondary treatment, in detail. Mass balance is used to develop design equations for biological treatment processes. A separate chapter (Chapter 11) is provided for anaerobic treatment, which is becoming more and more important due to the *energy production* potential from methane gas generation. Chapter 12 describes solids processing and disposal, together with pertinent regulations. A number of worked-out problems are used to demonstrate the calculation of mass and volume of sludge, perform mass balance of solids, and calculate process efficiency. Chapter 13 describes advanced and tertiary treatment processes. Recent advances in nitrogen and phosphorus removal are provided, followed by processes for solids removal. *Chapters 14 and 15 are new additions.* Chapter 14 presents sustainable design principles for the recovery of water, energy, and nutrients from wastewater. Examples of resource recovery and water reuse practices from around the world are provided. *Chapter 15 provides a culminating design experience* based on all the concepts explained in the previous chapters. Using the fundamental principles together with parameters used in practice, the overall design of a wastewater treatment system, urban and rural, is explained with the help of example problems.

The layout of the book is similar to the manner in which the first author, Dr. Riffat, has taught this course at George Washington University for the last 28 years. The course is taken by senior-level and graduate students of the Civil and Environmental Engineering Department. The material is covered in one semester consisting of 14 weeks. At the end of the course, the student should have an understanding of the fundamental concepts of wastewater treatment, and be able to design the unit processes for the treatment of municipal wastewater with an overall goal of resource recovery and sustainability.

Acknowledgments

I would like to thank my husband, Wahid Sajjad, for all his advice and support, which made it possible for me to complete the second edition. My daughter, Mehran, and my sister-in-law, Farhana Sazzad Mita, provided tremendous support at the family level, to make things easier for me during the coronavirus pandemic; and my son, Roshan, for his constant encouragement of my academic endeavors.

It was a pleasure to work with my co-author and former student, Taqsim Husnain, on the additions, modifications, and new chapters. He has also diligently and beautifully prepared all the diagrams and illustrations for this book. I would like to thank my former doctoral students Sebnem Aynur and Kannitha Krongthamchat, for their contributions to Chapters 11 and 12.

Finally, I would like to thank my extended family for all their love and inspiration.

Rumana Riffat

Writing a book is more rewarding than I could have ever imagined. And it has been possible for the enormous opportunity and support from my teacher and doctoral adviser, Dr. Rumana Riffat. I would like to express my deepest appreciation to her for providing constant mentorship and inspiration, and for always believing in me.

A very special thanks to my beloved wife, Nazia Salam. From reading the early draft to checking the calculations, she played an important role in getting this book completed. She stood by me during every struggle and all my successes with love, patience, and encouragement. My children, Rania and Zaheen, were very excited to advise me on the cover selection. Thank you so much, my dear family.

Taqsim Husnain

List of symbols

α	Oxygen transfer correction factor
β	Salinity–surface tension correction factor
C	Concentration
C_d	Drag coefficient
DO	Dissolved oxygen
d_p	Diameter of particle
F	Fouling factor
F_g	Force due to gravity
F_b	Force due to buoyancy
F_D	Drag force
Φ	Shape factor of particle
g	Acceleration due to gravity
K_s	Half saturation coefficient
k_1	BOD rate constant
k_2	Reaeration rate constant
k_d	Endogenous decay coefficient
k_t	Reaction rate coefficient
L_t	Oxygen equivalent of organic matter remaining at time t
M_o	Mass of oxygen
μ	Specific growth rate of biomass
μ_{max}	Maximum specific growth rate of biomass
μ_w	dynamic viscosity of water
ρ_p	Density of particle
ρ_w	Density of water
P_x	Biomass wasted
Q	Flowrate
R	Recycle ratio
R_e	Reynolds number
r_d	Rate of decay
r_g	Growth rate of biomass
r_{max}	Maximum biomass production rate
r_o	Rate of oxygen uptake
r_{su}	Rate of substrate utilization

S	Substrate concentration
S_t	Substrate concentration at time t
t	Time
T	Temperature
θ	Hydraulic retention time
θ_c	Solids retention time
V	Volume of reactor
V_L	Volumetric loading rate
v_t	Terminal settling velocity
X	Biomass concentration
Y	Biomass yield coefficient

List of abbreviations

AC	Alternating current
AEBR	Anaerobic expanded bed reactor
AMBR	Anaerobic migrating blanket reactor
AOTR	Actual oxygen transfer rate
APD	Acid phase digestion
AS	Activated sludge
ASBR	Anaerobic sequencing batch reactor
ATA	Anaerobic toxicity assay
ATAD	Autothermal thermophilic aerobic digestion
AWTP	Advanced wastewater treatment plant
bCOD	Biodegradable chemical oxygen demand
BOD	Biochemical oxygen demand
BOD_5	5 d Biochemical oxygen demand
BMP	Biochemical methane potential
BNR	Biological nutrient removal
BPR	Biological phosphorus removal
CEPT	Chemically enhanced primary treatment
CHP	Combined heat and power
COD	Chemical oxygen demand
CSTR	Continuous flow stirred tank reactor
DAF	Dissolved air flotation
DC	Direct current
DD	Dual digestion
DNA	Deoxyribonucleic acid
DO	Dissolved oxygen
DPR	Direct portable reuse
E. coli	Escherichia Coli
EEH	Enhanced enzymic hydrolysis
EGSB	Expanded granular sludge bed
EPA	Environmental Protection Agency
EU	European Union
FBBR	Fluidized bed bio-reactor
FC	Fecal coliform

GAC	Granular activated carbon
HRT	Hydraulic retention time
IFAS	Integrated fixed-film activated sludge
MAD	Mesophilic anaerobic digestion
MBBR	Moving bed biofilm reactor
MBR	Membrane biological reactor
MCFC	Molten carbon fuel cell
MF	Microfiltration
MGD	Million gallons per day
MLE	Modified Lutzack-Ettinger
MLSS	Mixed liquor suspended solids
MLVSS	Mixed liquor volatile suspended solids
MPN	Most probable number
N	Nitrogen
NH_3	Ammonia
NF	Nanofiltration
NOD	Nitrogenous oxygen demand
OLR	Organic loading rate
PAC	Powdered activated carbon
PAFC	Phosphoric acid fuel cell
PAO	Phosphorus accumulating organisms
PFR	Plug flow reactor
PFRP	Processes to further reduce pathogens
PSRP	Processes to significantly reduce pathogens
POTW	Publicly owned treatment works
PVC	Polyvinyl chloride
RAS	Return activated sludge
RBC	Rotating biological contactor
RNA	Ribonucleic acid
RO	Reverse osmosis
SBR	Sequencing batch reactor
sCOD	Soluble chemical oxygen demand
SDNR	Specific denitrification rate
SHARON	Single reactor system for high ammonium removal over nitrite
SOFC	Solid-oxide fuel cell
SOR	Surface overflow rate
SOTR	Oxygen transfer rate at standard temperature and pressure
SRT	Solids retention time
STP	Standard temperature and pressure
TC	Total coliform
TF	Trickling filter
ThOD	Theoretical oxygen demand
TOC	Total organic carbon
TPAD	Temperature phased anaerobic digestion
TS	Total solids

TSS	Total suspended solids
UASB	Upflow anaerobic sludge blanket
UF	Ultrafiltration
UV	Ultraviolet
VFA	Volatile fatty acid
VS	Volatile solids
VSS	Volatile suspended solids
WAS	Waste activated sludge
WRF	Water reclamation facility
WWTP	Wastewater treatment plant

About the authors

Dr. Rumana Riffat is Professor of the Civil and Environmental Engineering Department at George Washington University, in Washington, DC. She obtained her graduate degrees in Civil and Environmental Engineering from Iowa State University, in Ames, Iowa; and her Bachelor's degree in Civil Engineering from Bangladesh University of Engineering and Technology, in Dhaka, Bangladesh. She has been involved in teaching and research for the last 28 years.

Dr. Riffat's research interests are in wastewater treatment, specifically nutrient removal and anaerobic treatment of wastewater and biosolids. She and her research group have conducted extensive research on processes to further reduce pathogens, such as dual digestion, temperature-phased digestion, and thermal hydrolysis pretreatment options. Her nutrient removal research has focused on partial denitrification with anammox, determination of kinetics, and evaluation of various carbon sources for denitrification. Dr. Riffat is currently involved in several research projects with EPA, Water Research Foundation, and DC Water and Sewer Authority at Blue Plains Advanced Wastewater Treatment Plant, in Washington, DC, among others.

Dr. Riffat received the *Distinguished Teacher Award* from the School of Engineering and Applied Science of George Washington University in 2011. She received the *GW Service Excellence Award* for Sustainability team in 2012. She received the *George Bradley Gascoigne Wastewater Treatment Plant Operational Improvement Medal* of Water Environment Federation in 2010. She is a member of several professional organizations, including the American Society of Civil Engineers (ASCE), the Water Environment Federation (WEF), and the International Water Association (IWA). She is a registered Professional Engineer (PE) of the District of Columbia.

Dr. Taqsim Husnain is Assistant Professor in the Department of Engineering at William Jewell College in Liberty, Missouri. He earned a Ph.D. in Environmental Engineering from George Washington University in Washington, DC, an M.S. in Civil Engineering from Virginia Tech, an

M.Eng. in Transportation Engineering from the Asian Institute of Technology, Thailand, and a B.S. in Civil Engineering from Bangladesh University of Engineering and Technology, Dhaka, Bangladesh.

Dr. Husnain has been instrumental in the development of the new Civil Engineering program at William Jewell College. He has been involved in curriculum development and, design and setup of the environmental engineering and fluid mechanics laboratories. He has developed and taught a number of civil engineering courses including introduction to civil engineering, sustainability and environmental engineering, fluid mechanics for civil engineers, and water distribution and treatment design. Dr. Husnain's area of expertise is in sustainable water and wastewater treatment processes and wastewater reuse applications. He has conducted research in sustainable reuse of wastewater with advanced membrane processes and temperature-phased anaerobic digestion. He is currently involved with undergraduate researchers to investigate the potential application of temperature-driven membrane distillation for wastewater treatment plants.

Dr. Husnain is an ASCE ExCEEd fellow, and a member of several professional organizations, including the American Society of Civil Engineers (ASCE), American Water Works Association (AWWA), and the American Society of Engineering Education (ASEE). He is a registered Professional Engineer in the state of Missouri and an Envision Sustainability Professional (ENV SP).

Chapter 1

Sustainable wastewater treatment and engineering

1.1 INTRODUCTION AND HISTORY

The science and engineering of wastewater treatment has evolved significantly over the last century. As the population of the world has increased, our sources of clean water have decreased. This has shifted our focus towards pollution reduction and resource recovery. Disposal of wastes and wastewater without treatment in lands and water bodies is no longer an option. An increasing body of scientific knowledge relating waterborne microorganisms and constituents to the health of the population and the environment have spurred the development of new engineered technologies for the treatment of wastewater and potential reuse.

The term *wastewater* includes liquid wastes and wastes transported in water from households, commercial establishments, and industries, as well as storm water and other surface runoff. Wastewater may contain high concentrations of organic and inorganic pollutants, pathogenic microorganisms, as well as toxic chemicals. If the wastewater is discharged without treatment to a stream or river, it will result in severe pollution of the aquatic environment. The decline in water quality will render the stream water unusable for future drinking water purposes. *Sustainable wastewater engineering* involves the application of the principles of science and engineering for the treatment of wastewater, to remove and/or reduce the pollutants to an acceptable level prior to discharge to a water body or other environment, without compromising the self-purification capacity of that environment. The treatment, disposal, and beneficial reuse of the generated solids and other by-products are an integral part of the total process.

If we look back in time, wastewater engineering has progressed from collection and open dumping, to collection and disposal without treatment, to collection and treatment before disposal, all the way to collection and treatment prior to reuse. Evidence of waste collection in the streets, and then use of water to wash them through open sewers have been found in the ancient Roman Empire. In the early 1800s, the construction of sewers was started in London, in Great Britain. In 1843, the first sewer system of Hamburg, Germany, was officially designed by a British engineer Lindley (Anon, 2011).

DOI: 10.1201/9781003134374-1

1

In the United States (US), in seventeenth-century Colonial America, household wastewater management consisted of a privy (toilet) with an outlet constructed at ground level that discharged outside to a cesspool or a sewer. With low population densities, privies and cesspools constructed in this way did not cause many problems (Duffy, 1968). But as the population increased, the need for an engineered system for wastewater management in large cities became more evident. Scientists and public health officials started to understand the relationship between disease outbreaks and contamination of drinking water from wastewater. Nuisance caused by odors, outbreaks of diseases, e.g. cholera, and other public health concerns prompted the design of a comprehensive sewer system for the city of Chicago in the 1850s. At that time, the sewer system was used to transport the untreated wastewater outside of the residential community to a stream or river. Dilution of the wastewater with the stream water was the primary means of pollutant reduction. These were called water-carriage sewer systems.

Public health concerns in the 1850s also resulted in the planning and development of a water-carriage sewer system for the city of London. A cholera epidemic struck London in 1848 and again in 1854, causing a total of more than 25,000 deaths (Burian et al., 2000). Dr. John Snow was the first doctor at that time who established a connection between the cholera outbreak and a contaminated water supply at the Broad Street public well. In addition, he showed statistically that cholera victims had drawn their drinking water from a sewage-contaminated part of the river Thames, while those who remained healthy drew water from an uncontaminated part of the river. These findings together with the discoveries by Pasteur and Koch prompted the British parliament to pass an act in 1855, to improve the waste management system of the city of London. This led to the development of a comprehensive water-carriage sewer system for London, designed by Joseph Bazalgette (Hey and Waggy, 1979).

Toward the beginning of the twentieth century, sewage treatment plants mainly used settling tanks (primary treatment) to remove suspended particles from the wastewater before discharge to streams and rivers. In the early 1900s, about one million people in the US were served by 60 such treatment plants. During that time, the first trickling filter was constructed in Wisconsin, Madison to provide biological (secondary) treatment to wastewater. The Imhoff tank was developed by German engineer Karl Imhoff in 1906 for solids separation and further treatment. The first activated sludge process was constructed in San Marcos, Texas, in 1916 (Burian et al., 2000). Advances in sludge digestion and gas production were also being accomplished by researchers and utilities. From the mid-1900s to the present time, we have seen the development of various types of biological and biochemical processes for the removal of pollutants from wastewater. The earlier objectives were mainly to reduce the total suspended solids (TSS), biochemical oxygen demand (BOD), and pathogens. Primary and secondary biological treatment was considered sufficient for the production of treated wastewater of acceptable standards. With industrialization and scientific advances, chemical and toxic compounds have been detected in municipal wastewater. This has resulted in the need for additional treatment beyond the secondary, giving rise to tertiary treatment. Tertiary or advanced treatment can be physical, chemical or biological, or a combination of these processes.

1.2 CURRENT PRACTICE

The primary treatment in most municipal wastewater treatment plants consists of preliminary and primary stages. It typically includes screens, grit chambers, comminutors, and primary clarifiers, depending on the flow rates. Larger plants use chemically enhanced primary clarification for higher solids removal efficiency. The primary treatment is followed by a secondary treatment. The secondary treatment consists of a biological process followed by a secondary clarifier. If the secondary effluent meets the regulatory standards for BOD and TSS, it is discharged to receiving waters following disinfection. The solids and sludge collected from the various units undergo further processing and treatment before disposal. Various options are available for sludge processing. A conventional wastewater treatment plant is illustrated in Figure 1.1.

More than half of the municipal wastewater treatment plants in the US are capable of providing at least secondary treatment. About 92% of the total flow is treated by plants with a capacity of 0.044 m^3/s (1 million gallons per day or MGD) or larger (Metcalf and Eddy et al., 2013). In the last two decades, nutrient removal has become increasingly more important in

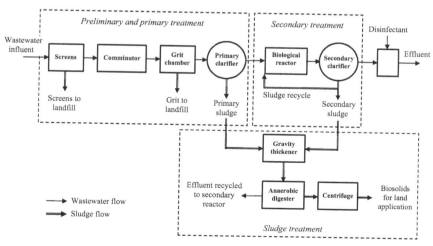

Figure 1.1 Flow diagram of a conventional wastewater treatment plant.

parts of the US, as well as in Europe and Asia. Eutrophication caused by excessive nitrogen and phosphorus in wastewater discharges has disrupted the aquatic life in receiving water bodies with a subsequent decline in water quality. Wastewater treatment plants in affected areas and watersheds have to provide additional nutrient removal prior to discharge. Biological nutrient removal is incorporated as part of the secondary treatment, or as tertiary treatment. Nutrient removal is no longer considered an advanced treatment option. An example of this is the Chesapeake Bay watershed in the eastern US, and the municipal wastewater treatment plants within the watershed. Most of the plants use biological nitrification–denitrification together with BOD removal, and/or chemical precipitation for removal of phosphorus. The use of granular media filtration as a tertiary treatment for the reduction of TSS is also quite common. Table 1.1 presents the pollutants commonly found in municipal wastewater and the physical, chemical, and biological processes used to remove or reduce their concentrations.

Table 1.1 Common wastewater pollutants and the processes used to reduce/ remove them

Pollutants	Unit processes
Suspended solids	Coarse screens, fine screens
	Grit chamber
	Clarification
	Filtration
	Chemically enhanced clarification
Colloidal and dissolved solids	Chemical precipitation
	Membrane filtration
	Ion exchange
	Activated carbon adsorption
Biodegradable organics	Suspended growth processes (aerobic and anaerobic)
	Attached growth processes (aerobic and anaerobic)
	Ponds and lagoons
	Membrane bioreactors
Pathogens	Chlorination
	Ozonation
	Ultraviolet disinfection
Nutrients	
Nitrogen	Biological nitrification–denitrification (suspended and fixed film variations)
	Air stripping
	Breakpoint chlorination
Phosphorus	Biological phosphorus removal
	Chemical precipitation
Volatile organic compounds	Activated carbon adsorption
	Air stripping

Source: Adapted from Metcalf and Eddy et al. (2013), and Peavy et al. (1985).

1.3 EMERGING ISSUES

The following are areas of importance and concern for municipal wastewater treatment plants:

- Rising energy costs for the operation of treatment plants.
- Disposal of biosolids in a sustainable manner.
- Performance and reliability of treatment plants in the digital age.
- Presence of endocrine-disrupting compounds (EDC) in wastewater.
- Presence of toxic chemicals in wastewater from household products.
- More stringent discharge limits due to continued degradation of water bodies.
- Scarcity of freshwater sources.
- The need to upgrade aging infrastructure and treatment plants.
- The need for adequate mathematical models and software for process analysis and control.

1.4 FUTURE DIRECTIONS

Based on the emerging issues, wastewater engineering and research should be focused on the following areas in the future.

- **Energy generation** – Typically, wastewater treatment plants have high energy requirements for plant operation. They are big consumers of energy or electricity. Wastewater plants can generate significant amounts of methane gas from the anaerobic digestion of the sludge. The gas can in turn be used to heat the digesters, as well as generate power that can be used by the plant or sold to nearby industries. With rising energy costs, this should be the future direction of operation of wastewater treatment plants. For sustainable operation, treatment plants need to evolve into energy producers from energy consumers.

  Blue Plains Advanced Wastewater Treatment Plant (AWTP) of DC Water has implemented the thermal hydrolysis process (THP) together with anaerobic digestion to produce Class A biosolids and generate energy. This is the first THP facility in North America and the largest in the world. The THP process pressure-cooks the waste solids to generate combined heat and power, generating a net 10 MW of electricity (DC Water, 2019a and 2019b). The project has resulted in a 30% reduction in energy purchased from the grid, a 41% reduction in greenhouse gas production, and a 50% reduction in biosolids shipping costs (CDM Smith, 2021).

  Another example of energy production using anaerobic digestion is the East Bay Municipal Utility District, California, which cogenerates electricity and thermal energy onsite from waste methane.

This resulted in an annual reduction in energy costs from $4.6 million to $2.9 million (EBMUD, 2021). Other examples are the Encina Wastewater Authority and Point Loma Wastewater Treatment Plant in California, among others. The West Point Treatment Plant in Kings County, Washington, uses the methane gas generated from anaerobic digestion to run generators that produce electricity. They are able to produce 1.5–2 MW of electricity, which is sold to the local utility company after meeting plant demands (WPTP, 2021).

- **Beneficial reuse of biosolids** – The cost of processing and disposal of the biosolids produced at a wastewater treatment plant can amount to almost 50% of the total capital and operation costs. Future direction should be more toward producing a product that can be reused in a beneficial manner, such as fuel or fertilizer. One example is the production of Class A biosolids by Blue Plains AWTP and marketed under the product name *Bloom*. This is applied on land as a soil blender, fertilizer, etc., and nutrients are recycled back to the soil (CDM Smith, 2021). Another example is the Encina Wastewater Authority which produces biosolids pellets that are sold to a cement-manufacturing facility as an alternative fuel (EWPCF, 2021).
- **Wastewater reuse** – As freshwater resources become more scarce, the need for recycled wastewater is increasing. Future directions should include increased research on water quality and safety of recycled wastewater, as well as public education for direct potable reuse. The island nation of Singapore uses reclaimed water that is produced from a multiple-barrier wastewater treatment process. The wastewater is first treated by conventional treatment, followed by microfiltration/ ultrafiltration, reverse osmosis, and finally by ultraviolet disinfection. NEWater is the brand name given to the reclaimed water in Singapore (NEWater, 2021). The water is used for non-potable use and indirect potable use for reservoir recharge. There are five NEWater plants currently supplying up to 40% of Singapore's water needs. The goal is to meet up to 55% of the future water demand by 2060 (PUB, 2021).
- **Fundamental research** – With an increasing aging population and an increase in the use of pharmaceutical products, a vast number of new and emerging contaminants have found their way into wastewater treatment plants. Of concern are the endocrine-disrupting compounds (EDCs) which have caused the feminization of fish in the waters of Maryland, among others. A number of these compounds pass through the treatment plant unchanged and end up in streams and rivers, having various consequences on aquatic life. Fundamental research is necessary to determine the characteristics of these compounds of concern and to develop methods for treatment and removal.
- **Mathematical modeling** – Wastewater engineering is still in its infancy when compared to other engineering disciplines, with regard to the

development and availability of process models for design and control of treatment operations. There are a few models developed by IWA (International Water Association), and Biowin®, among others. In order for this body of science and engineering to have a significant positive impact on the planet's scarce water resources, the future direction should be focused on attracting brilliant scientific minds to develop adequate and versatile process models.

1.5 REGULATORY REQUIREMENTS

Regulatory requirements have played a significant role in the development and application of wastewater treatment processes. Emerging research and subsequent regulations have shifted current goals, or added new goals to the treatment process from time to time. This has resulted in the innovation of new engineered processes. In the following sections, the development of regulations and standards pertaining to wastewater in the US, European Union, and the United Kingdom (UK) will be discussed in detail.

1.5.1 United States regulations

The *Federal Water Pollution Control Act* (FWPCA) of 1948 was the first legislation enacted by the Federal government to address urban wastewater management issues (Public Law, 1948). The act provided for comprehensive programs for eliminating or reducing the pollution of interstate waters and tributaries, and research and technical assistance for improving the sanitary condition of surface and ground waters. Major amendments to the FWPCA were enacted in 1961, 1966, 1970, 1972, 1977, and 1987. The 1966 amendments, titled the Clean Water Restoration Act, strongly addressed the issue of protecting water quality (Public Law, 1966). It provided for authorization of a comprehensive study of the effects of pollution, including sedimentation, in the estuaries and estuarine zones of the US on fish and wildlife, sport and commercial fishing, recreation, water supply and power, and other specified uses. The legislation established a set of water quality standards. Protecting public health was the primary goal, but additional goals of protecting aquatic life and the aesthetics of water resources were included.

The Federal Water Pollution Control Act Amendments of 1972 stipulated broad national objectives to restore and maintain the chemical, physical, and biological integrity of the Nation's waters (Public Law, 1972). This became known as the *Clean Water Act* (CWA) together with subsequent amendments in 1977. The CWA established the basic structure for regulating discharges of pollutants into the nation's waters and regulating quality standards for surface waters. New regulations were established for industrial and agricultural polluters. The CWA authorized the Environmental

Protection Agency (EPA) to establish the National Pollutant Discharge Elimination System (NPDES) permit program. All municipal, industrial, and other facilities that discharged their wastewater to surface waters were required to obtain an NPDES permit from EPA, which specified *technology-based* effluent standards for specific pollutants.

The CWA also authorized significant federal funding for research and construction grants, with the ambitious goal of eliminating all water pollution by 1985. All Publicly Owned Treatment Works (POTWs) were required to meet the minimum standards for secondary treatment.

In 1973, the US EPA published its definition of minimum standards for *secondary treatment*. This was amended in 1985 to include percent removal requirements for treatment plants served by separate sewer systems. The standards were amended again in 1989 to clarify percent removal requirements during dry periods for treatment facilities served by combined sewers. The secondary treatment standards are provided in Table 1.2. The standards are published in the Code of Federal Regulations (40 CFR, Part 133.102). Three important effluent parameters are included: BOD_5 (5 d BOD), TSS, and pH. $CBOD_5$ or carbonaceous BOD_5 may be substituted for BOD_5 at the option of the permitting authority. Special considerations and alternative requirements are permitted for POTWs that receive industrial flows or use waste stabilization ponds and trickling filters (US EPA, 2022).

The CWA was amended in 1987 to emphasize identification and regulation of toxic compounds in sludge, as well as authorize penalties for permit violations. This amendment was known as the Water Quality Act. The Act established funding for States to develop and implement, on a watershed basis, nonpoint source management and control programs. A significant amendment of the CWA was made in 2000 (Section 303(d) of the CWA), which required the establishment of a Total Maximum Daily Load (TMDL) or amount of a pollutant that a water body could receive without compromising water quality standards (Preisner et al., 2020).

Table 1.2 Secondary treatment standards as defined by US EPA (2012)

Effluent parameter	Average 30-d concentration	Average 7-d concentration
BOD_5	30 mg/L	45 mg/L
TSS	30 mg/L	45 mg/L
Removal	85% BOD_5 and TSS	
pH	Within 6.0–9.0 at all times	
$CBOD_5$	25 mg/L	40 mg/L

Source: US EPA, 2012.

Note: Treatment facilities using stabilization ponds and trickling filters are allowed to have higher average 30-d and average 7-d concentrations of 45 mg/L and 65 mg/L of BOD_5 and TSS, as long as the water quality of the receiving body is not adversely affected. Exceptions are also permitted for facilities with combined sewers, etc. The $CBOD_5$ may be substituted for BOD_5 at the discretion of the permitting authority.

The use and disposal of treated sludge or biosolids are regulated under 40 CFR, Part 503 (US EPA, 1994). The regulation was promulgated in 1993 to regulate the use and disposal of biosolids from municipal wastewater treatment plants, and establish limits for contaminants (e.g. metals), pathogens, and vector attraction. The regulations are applicable to all treatment plants that use land application for final disposal of biosolids. The regulations are self-implementing, i.e. permits are not required by the plants. But failure to conform to the regulations are considered to be violations of the law. The frequency of monitoring and reporting requirements are provided in detail. The Part 503 Rule defines two types of biosolids, *Class A* and *Class B*, based on the level of pathogen reduction, metal concentrations, and vector attraction reduction. Class A biosolids can be applied to land without any restrictions. Sludge stabilization requirements and pathogen reduction alternatives are specified in the law. Additional details of the Part 503 Rule are provided in Section 12.9.

1.5.2 European Union regulations

The European Union (EU) has established a number of policies or directives that address the quality of surface and groundwaters. Water supply and sanitation are the responsibility of each member nation in the EU. However, the EU directives serve as a baseline for individual nations to form their own legislation.

There are three major EU directives. They are:

- The Urban Waste Water Treatment Directive (91/271/EEC) of 1991 pertaining to discharges of municipal and some industrial wastewaters;
- The Drinking Water Directive (98/83/EC) of 1998 pertaining to potable water;
- The Water Framework Directive (2000/60/EC) of 2000 pertaining to the management of surface water and groundwater resources.

The Urban Waste Water Treatment Directive was aimed at protecting the environment from adverse effects due to the collection, treatment, and discharge of wastewater from municipal and some industrial treatment facilities (Europa, 2012). The major elements of the Directive were: (i) Depending on the population size and designated location, all built-up areas were required to have urban wastewater collection and treatment systems by the year 1998, 2000, or 2005 (new members by 2015); (ii) the level of treatment had to be primary, secondary or tertiary, depending on the sensitivity of the receiving water (van Riesen, 2004). Member states had to establish lists of sensitive areas. Primary treatment was deemed sufficient for less sensitive areas. The Directive was amended by the Commission Directive 98/15/EC in 1998. The discharge standards for normal areas are provided in Table 1.3. Discharge requirements for nitrogen and phosphorus in sensitive areas are provided in Table 1.4.

Table 1.3 EU standards for effluent discharge in normal areas

Parameter	Concentration, mg/L	Minimum reduction, %
BOD_5	25	70–90
COD	125	75
TSS	35	90

Source: Adapted from van Riesen (2004).

Table 1.4 EU standards for nutrient discharge in sensitive areas

Parameter	Concentration, max. annual mean	Minimum reduction, %
Total Phosphorus	2 mg/L as P (10,000–100,000 PE) 1 mg/L as P (>100,000 PE)	80
Total Nitrogen	15 mg/L as N (10,000–100,000 PE) 10 mg/L as N (>100,000 PE)	70–80

Source: Adapted from van Riesen (2004).

Note: PE indicates population equivalent.

The European Commission (EC) has published a number of reports on the implementation of the directive. The last report was published in 2020. The report noted that collection and treatment of wastewater had improved, with compliance rates of 95% for collection, 88% for secondary treatment, and 86% for removal of nitrogen and phosphorus (European Commission, 2020). However, there were countries where more infrastructure needed to be built or improved. Finance and planning were the main challenges facing these countries. According to an earlier report of the European Commission (2004), the directive represented the most cost-intensive European legislation in the environmental sector. The EC estimated that 152 billion Euros were invested in wastewater treatment from 1990 to 2010. The EU provided support of about five billion Euros per year for the implementation of the directive. According to the 2020 report, the member countries would have to invest an additional 229 billion Euros to ensure long-term compliance.

A review and public feedback were initiated in July 2020, as part of an impact assessment process for potential revision of the Urban Waste Water Treatment Directive (European Commission, 2021).

1.5.3 United Kingdom regulations

An example of the adoption of the Urban Waste Water Treatment Directive is discussed in terms of the UK. In the UK, the Urban Waste Water Treatment Regulations of 1994 were enacted based on the Urban Waste Water Treatment Directive of the EU. These were later amended in 2003. The regulations set the standards for the collection and treatment of wastewater. The law stipulated that a sewerage system be provided for all urban areas above a specified size, and that the collected sewage should receive at least secondary (biological) treatment before it is discharged to the environment. Uncontrolled discharges from the sewerage systems are only allowed under storm conditions. The law identified sensitive areas, e.g. eutrophic waters. Larger treatment plants have to reduce their nutrient loads prior to discharge to eutrophic waters. The regulation also banned the disposal of sludge to sea, by the end of 1998 (DEFRA, 2012).

The Department of Environment, Food and Rural Affairs (DEFRA) is responsible for policy on the implementation of the regulations in England, the Department of the Environment in Northern Ireland, the Scottish Government in Scotland, and the Welsh Government in Wales. Their environmental regulators, (the Environment Agency for England, Northern Ireland Environment Agency, Scottish Environment Protection Agency, and Environment Agency Wales), are responsible for monitoring discharges from treatment plants for compliance with the legislation's treatment standards (DEFRA, 2012). The regulatory limits set for the UK are the same as those provided in Tables 1.3 and 1.4, according to the updated guidance of 2019 (Environment Agency, 2019).

REFERENCES

Anon. 2011. The History of Wastewater Treatment. http://www.cityoflewisville.com/wcmsite/publishing.nsf/AttachmentsByTitle/Wastewater+Treatment+History/$FILE/The+History+of+Wastewater+Treatment3.pdf

Burian, S. J., Nix, S. J., Pitt, R. E. and Durrans, S. R. 2000. Urban Wastewater Management in the United States: Past, Present and Future. *Journal of Urban Technology*, 7, 3: 33–62.

CDM Smith. 2021. Driving Net-Zero at DC Water. https://cdmsmith.com/en/Client-Solutions/Projects/DC-Water-Driving-Net-Zero

DC Water. 2019a. DC Water's Thermal Hydrolysis and Anaerobic Digester Project. https://www.dcwater.com/sites/default/files/documents/BioenergyFacility.pdf

DC Water. 2019b. *Blue Plains Advanced Wastewater Treatment Plant Brochure*. DC Water and Sewer Authority, Washington, DC.

DEFRA. 2012. http://www.defra.gov.uk/environment/quality/water/sewage/sewage-treatment/

Duffy, J. 1968. *A History of Public Health in New York City: 1625–1866*. Russell Sage Foundation, New York.

EBMUD. 2021. East Bay Municipal District Website, http://www.ebmud.com/about-ebmud, (accessed August 11, 2021).

Environment Agency. 2019. Waste Water Treatment Works – Treatment Monitoring and Compliance Limits. https://www.gov.uk/government/publications/waste-water-treatment-works-treatment-monitoring-and-compliance-limits/waste-water-treatment-works-treatment-monitoring-and-compliance-limits#lut-for-bod-and-cod

Europa. 2012. Summaries of EU Legislation. http://europa.eu/legislation_summaries/environment/water_protection_management/l28008_en.htm

European Commission. 2004. European Commission Report. p. 108. http://eur-lex. europa.eu/LexUriServ/site/en/com/2004/com2004_0248en01.pdf

European Commission. 2020. 10th Implementation Report. https://ec.europa.eu/ environment/water/water-urbanwaste/implementation/implementationreports_ en.htm

European Commission. 2021. https://ec.europa.eu/environment/water/water-urbanwaste/index_en.html

EWPCF. 2021. Encina Water Authority Website, https://www.encinajpa.com/environment-resource-management (accessed on 11 August 2021).

Hey, D. L. and Waggy, W. H. 1979. Planning for Water Quality: 1776 to 1976. *ASCE Journal of the Water Resources Planning and Management Division*, 105, (March), 121–131.

Metcalf and Eddy, Tchobanoglous, G., Stensel, H., Tcuchihashi, R., and Burton, F. 2013. *Wastewater Engineering: Treatment and Resource Recovery*. 5th edn. McGraw-Hill, Inc., New York, NY, USA.

NEWater. 2021. http://en.wikipedia.org/wiki/NEWater

Peavy, H. S., Rowe, D. R., and Tchobanoglous, G. 1985. *Environmental Engineering*. McGraw- Hill, Inc., New York, NY, USA.

Preisner, M., Neverova-Dziopak, E. and Kowalewski, Z. 2020. An Analytical Review of Different Approaches to Wastewater Discharge Standards with Particular Emphasis on Nutrients. *Environmental Management*. 66: 694–708.

PUB. 2021. NEWater. PUB, Singapore's National Water Agency. https://www.pub. gov.sg/watersupply/fournationaltaps/newater

Public Law. 1948. P.L. 845, Ch 758, 33 U.S.C. 1251–1376.

Public Law. 1966. P.L. 89–753, 33 U.S.C. 466.

Public Law. 1972. P.L. 92–500, 33 U.S.C. 1251, 1311–1317.

US EPA. 1994. A Plain English guide to the EPA Part 503, Biosolids Rule. EPA/832/R-93/003. Washington, DC.

US EPA. 2012. EPA NPDES Permit Writers' Manual. Chapter 5, section 5.2. http:// www.epa.gov/npdes/pubs/chapt_05.pdf

US EPA. 2022. Municipal Wastewater. https://www.epa.gov/npdes/municipal-wastewater

van Riesen, S. 2004. European Wastewater Standards. Presented at the Wastewater Forum at IFAT Conference, China.

WPTP. 2021. West Point Treatment Plant Website, http://www.kingcounty.gov/environment/wtd/About/System/West.aspx (accessed 11 August 2021).

Chapter 2

Reaction kinetics and chemical reactors

2.1 REACTION KINETICS

A variety of chemical and biochemical reactions take place in the environment that are of importance to environmental engineers and scientists. These include reactions between various elements of the air, water, and soil, as well as with microorganisms. A number of these reactions are dependent on time, temperature, pressure, and/or concentration, e.g. biodegradation of organic matter, bacterial growth and decay, chemical disinfection, etc.

Reaction kinetics can be defined as the study of the effects of temperature, pressure, and concentration of reactants and products on the rate of a chemical reaction (Henry and Heinke, 1996). Reactions that occur within a single phase (solid, liquid, or gaseous) are called *homogeneous reactions*, e.g. nitrification in wastewater. Reactions that involve two or more phases are called *heterogeneous reactions*, e.g. gas adsorption on activated carbon.

The *rate of reaction*, r_i, is used to describe the rate of formation of a product, or rate of disappearance of a reactant. For homogeneous reactions, r_i is calculated as the moles or mass produced or consumed, per unit volume per unit time.

Let us consider the following homogeneous reaction:

$$a\text{A} + b\text{B} \rightarrow c\text{C} \tag{2.1}$$

where
  C = product
  A, B = reactants
  a, b, c = stoichiometric coefficients

The rate equation for the above reaction is

$$r_\text{A} = -k[\text{A}]^\alpha [\text{B}]^\beta = k[\text{C}]^\gamma \tag{2.2}$$

DOI: 10.1201/9781003134374-2

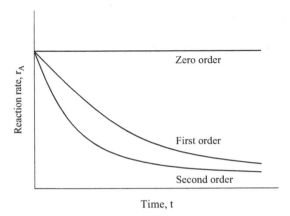

Figure 2.1 Variation of reaction rate with time.

where
  α, β, γ = empirically determined exponents
  [A], [B], [C] = molar concentrations of A, B, and C
  k = reaction rate constant

The *order of a reaction* is the sum of the empirically determined exponents, e.g. the order is $(\alpha + \beta)$ with respect to the reactants A and B, while the order is γ with respect to the product C. The order of a reaction can be a whole number (e.g. 0, 1, 2) or a fraction. Figure 2.1 illustrates the variation of reaction rate r_A with time for zero, first, and second order reactions. For a homogeneous, irreversible, elementary reaction that occurs in a single step, the empirically determined exponents are equal to the stoichiometric coefficients. In that case, Equation (2.2) becomes

$$r_A = -k[A]^a[B]^b = k[C]^c \tag{2.3}$$

2.2 HOW TO FIND THE ORDER OF A REACTION

Consider the following irreversible elementary reaction where reactant A is converted to a product C.

$$A \rightarrow C \tag{2.4}$$

The rate equation can be written as

$$r_A = -k[A]^{\alpha}$$

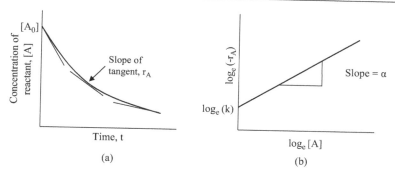

Figure 2.2 (a) Concentration of A versus time plot, (b) logarithmic plot of reaction rate versus concentration of A.

or

$$\log_e(-r_A) = \log_e(k) + \alpha \log_e[A] \tag{2.5}$$

where
α = order of the reaction (e.g. 0, 1, 2, etc.)
k = reaction rate constant

An experiment is conducted where the above reaction is allowed to proceed. The concentrations of A ($[A]$) at various time intervals (t) are measured. Plot $[A]$ versus t, as shown in Figure 2.2(a). Calculate the slope (r_A) of the tangent at various points along the curve. Plot $\log_e(-r_A)$ versus $\log_e[A]$, as shown in Figure 2.2(b). A best-fit line is drawn to represent Equation (2.5). The slope of the best-fit line is equal to the order of the reaction.

Example 2.1: The data shown in Table 2.1 were obtained from a batch experiment for the reaction A → P. Determine the order of the reaction.

SOLUTION

An Excel spreadsheet is used to calculate the values, shown in Table 2.2.
Figure 2.3(a) is a plot of concentration versus time. The section of the curve between each time interval is assumed to be a straight line, and the rates are calculated from the slope of that section. So, $r_A = dA/dt = (100 - 74)/(0 - 10) = -2.59$ for the first interval and so on. Figure 2.3(b) is a plot of $\ln(-r_A)$ versus $\ln(A)$. The slope of the best-fit line is 0.935, which can be rounded to 1. So the reaction is first order.

Table 2.1 Concentration data from a batch experiment (for Example 2.1)

Time (min)	0	10	20	40	60	80	100
A (mg/L)	100	74	55	30	17	9	5

Table 2.2 Spreadsheet calculation (for Example 2.1)

t, min	A, mg/L	$log_e(A)$	r_A	$log_e(-r_A)$
0	100	4.61		
10	74	4.31	−2.59	0.95
20	55	4.01	−1.92	0.65
40	30	3.41	−1.24	0.21
60	17	2.81	−0.68	−0.39
80	9	2.21	−0.37	−0.99
100	5	1.61	−0.20	−1.59

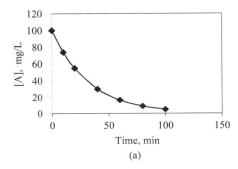

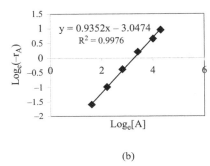

Figure 2.3 Plot of (a) concentration versus time, and (b) $log_e(-r_A)$ versus $log_e(A)$ (for Example 2.1).

2.3 ZERO ORDER REACTION

A zero order reaction proceeds at a rate that is independent of the concentration of the reactants or products. Consider the following irreversible elementary reaction where reactant A is converted to product C.

$$A \rightarrow C \tag{2.6}$$

If this reaction is zero order, the rate expression can be written as

$$r_A = -k \tag{2.7}$$

or

$$\frac{d[A]}{dt} = -k \tag{2.8}$$

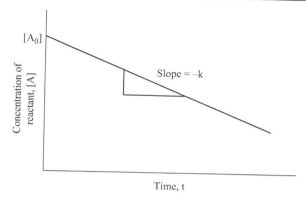

Figure 2.4 Concentration versus time plot for zero order reaction.

where

$$\frac{d[A]}{dt} = \text{rate of change of concentration of A with time}$$

k = reaction rate constant, time^{-1}

Integrate Equation (2.8) between initial value and values after time t,

$$\int_{[A_0]}^{[A_t]} d[A] = -k \int_0^t dt$$

$$\text{or,} [A_t] - [A_0] = -kt$$

$$[A_t] = [A_0] - kt \tag{2.9}$$

where

$[A_0]$ = initial concentration of reactant A at time zero, mg/L
$[A_t]$ = concentration of A after time t, mg/L

In order to determine the rate constant k for zero order kinetics (Equation 2.9), an experiment is conducted where the concentration of A is measured at regular intervals of time. The concentration of A versus time is plotted. A best-fit line is drawn through the data points as shown in Figure 2.4. The slope represents the rate constant k, and the intercept represents $[A_0]$.

2.4 FIRST ORDER REACTION

Consider the irreversible elementary reaction represented by Equation (2.6). If the reaction is first order with respect to the concentration of A, the rate expression becomes

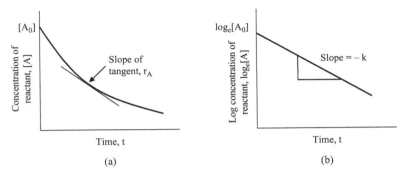

Figure 2.5 Plots of concentration versus time for a first order reaction on (a) arithmetic scale and (b) semi-logarithmic scale.

$$\frac{d[A]}{dt} = -k[A] \tag{2.10}$$

Integrate Equation (2.10) between initial value and values after time t,

$$\int_{[A_0]}^{[A_t]} \frac{d[A]}{[A]} = -k\int_0^t dt$$

$$\text{or}, \log_e \frac{[A_0]}{[A_t]} = kt$$

$$\text{or}, [A_t] = [A_0]e^{-kt} \tag{2.11}$$

An experimental procedure similar to the previous one is followed, in order to determine the rate constant k, for first order kinetics. The concentration of A versus time is plotted. For a first order reaction, a curve is obtained similar to Figure 2.5(a). The slope of the tangent at any point on the curve represents Equation (2.10). A plot of $\log_e[A]$ versus time should yield a straight line, as shown in Figure 2.5(b). The slope of the best-fit line is equal to the rate constant k.

2.5 SECOND ORDER REACTION

Let us consider the irreversible elementary reaction represented by Equation (2.6). If the reaction is second order with respect to the concentration of A, the rate expression becomes

$$\frac{d[A]}{dt} = -k[A]^2 \tag{2.12}$$

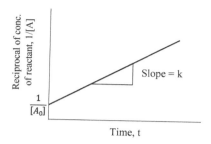

Figure 2.6 Plot of 1/[A] versus time for a second order reaction.

Integrate Equation (2.12) between initial value and values after time t,

$$\int_{[A_0]}^{[A_t]} \frac{d[A]}{[A]^2} = -k \int_0^t dt$$

$$\text{or, } \frac{1}{[A_t]} - \frac{1}{[A_0]} = kt \tag{2.13}$$

An experimental procedure similar to the previous one is followed, in order to determine the rate constant k, for second order kinetics. Values of 1/[A] versus time are plotted, as shown in Figure 2.6. The slope of the best-fit line provides the value of k.

2.6 REACTORS

A reactor is a tank or vessel where chemical, biological, or biochemical reactions take place, usually in a liquid medium. Reactions can also take place in a solid or gaseous medium, or in a combination of mediums. Chemical reactors are used in coagulation–flocculation, lime softening, taste and odor control, disinfection, and other unit processes that involve chemical reactions, in a water treatment plant. Reactors used in wastewater treatment plants involve mostly biochemical and biological reactions, e.g. activated sludge reactor, membrane bioreactor, etc.

There are three types of ideal reactors. They are (i) Batch reactor, (ii) Plug Flow Reactor (PFR), and (iii) Continuous-Flow Stirred Tank Reactor (CSTR). The hydraulics and conversion efficiencies of these reactors can be determined using mathematical models. Models developed for ideal reactors can be further modified to represent real-life processes and flow conditions for reactors used at treatment plants. In the following sections, basic design equations for ideal reactors will be discussed.

2.6.1 Conversion of a reactant

The conversion or removal of a reactant is calculated as

$$f = \frac{[A_o] - [A_t]}{[A_o]} = 1 - \frac{[A_t]}{[A_o]} \tag{2.14}$$

where
   f = conversion or removal efficiency
   $[A_o]$ = initial concentration of reactant A at time zero, mg/L
   $[A_t]$ = concentration of A after time t, mg/L

2.6.2 Detention time in a reactor

The theoretical detention time or residence time of the fluid particles in a reactor is given by

$$t = \frac{V}{Q} \tag{2.15}$$

where
   t = detention time in reactor
   V = volume of reactor
   Q = volumetric flow rate, volume/time

The actual detention time in a reactor can be determined by adding a tracer or dye to the influent during the steady state flow, and measuring the concentration of the tracer in the effluent over a period of time. The tracer concentration in the effluent is plotted versus time on a graph paper, and the centroid of the resulting curve is located as the actual detention time, as illustrated in Figure 2.7. The actual detention time is usually less than the theoretical detention time calculated using Equation (2.15). This can be due to back-mixing and short-circuiting of fluid in the reactor.

2.7 BATCH REACTOR

In a batch reactor, reactant(s) are added to the reactor and mixed for a requisite amount of time for the reactions to occur, as shown in Figure 2.8(a). At the end of the reaction time, the contents are removed from the reactor. One characteristic of the batch reactor is that all fluid particles have the same residence time in the reactor. Homogeneous mixing

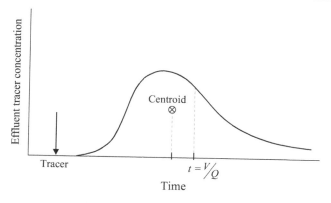

Figure 2.7 Effluent tracer profile for calculation of detention time in a reactor.

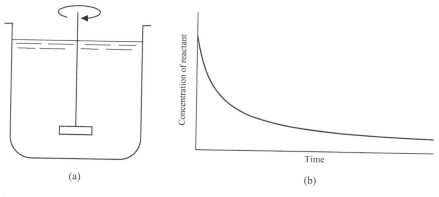

Figure 2.8 (a) Batch reactor, (b) concentration profile for a batch reactor with time.

is assumed so that the composition of the mixture is the same through-out the reactor. The concentration varies with time as the reaction pro-ceeds. Figure 2.8(b) illustrates the variation of reactant concentration with time.

Batch reactors are generally used for bench-scale experiments and liquid phase reactions. They are useful in determining the effects of variables on a reaction process. A number of experiments can be conducted at the same time in batch reactors, thus facilitating the study of process variables. They are used extensively in pharmaceutical and other industries.

Batch reactors are not suitable for gas phase reactions, or large-scale com-mercial applications. Labor costs and materials handling costs can run high, due to the time and effort involved in filling, emptying, and cleaning the reactors.

Table 2.3 Design equations for batch reactor, PFR, and CSTR

Order	Rate of reaction	Batch/PFR	CSTR
0	$-k$	$kt = [A_0] - [A_t]$	$kt = [A_0] - [A_t]$
1	$-k[A]$	$kt = \log_e\left(\dfrac{[A_o]}{[A_t]}\right)$	$kt = \left(\dfrac{[A_o]}{[A_t]}\right) - 1$
2	$-k[A]^2$	$kt = \dfrac{1}{[A_o]}\left(\dfrac{[A_o]}{[A_t]} - 1\right)$	$kt = \dfrac{1}{[A_t]}\left(\dfrac{[A_o]}{[A_t]} - 1\right)$

Example 2.2: Consider a first order reaction taking place in a batch reactor. Develop an expression for the detention time in the reactor.

SOLUTION

For a first order reaction, $r_A = -k\,[A]$
Substitute the expression for r_A in Equation (2.17),

$$-k[A] = \frac{d[A]}{dt}$$

This is similar to Equation 2.10. Upon integration between limits we obtain

$$\log_e \frac{[A_o]}{[A_t]} = kt$$

$$\text{or}, t = \frac{1}{k}\log_e \frac{[A_o]}{[A_t]}$$

2.7.1 Design equation

Consider the following mass/material balance of a reactant A converted to a product C in a reactor:

$$\left(\text{Rate of input}\right) = \left(\text{Rate of output}\right) + \left(\text{Rate of accumulation}\right) \\ - \left(\text{Rate of consumption}\right) \tag{2.16}$$

For a batch reactor, time period for reaction begins just after the reactor is filled, and ends just before contents are emptied. So, rate of input = 0, and rate of output = 0. Equation 2.16 becomes

Rate of consumption = Rate of accumulation

So, the design equation is written as

$$r_A = \frac{d[A]}{dt} \tag{2.17}$$

where

r_A = rate of consumption of limiting reactant A, concentration/time
[A] = concentration of limiting reactant A

When the order of the reaction is known, an expression for r_A can be substituted into the left-hand side of Equation (2.17), and the resulting differential equation can be integrated to obtain the design expression. Table 2.3 presents the design equations for zero, first, and second order reactions in a batch reactor.

2.8 PLUG FLOW REACTOR

In a PFR, fluid particles flow through the tank and are discharged in the same sequence as they entered. The fluid particles move through the reactor tube as plugs moving parallel to the tube axis, as illustrated in Figure 2.9(a). There is no longitudinal mixing of fluid, though there may be some lateral mixing. All fluid elements have the same residence time in the reactor. Figure 2.9(b) presents the concentration gradient from reactor inlet to outlet. This is due to the conversion of reactant as it flows through the reactor. The velocity profile at any given cross section is flat, as there is no back mixing or axial diffusion. As a result, the concentration of reactant across any vertical cross section is the same, as illustrated in Figure 2.9(c).

The PFR is suitable for gas phase reactions that take place at high pressure and temperature. An insulating jacket can be placed around the reactor to maintain the desired temperature. There are no moving parts inside the reactor. The average reaction rate is usually higher in a PFR as compared to a CSTR of similar volume, for the same feed composition and reaction temperature. The PFR makes more efficient use of reactor volume, which makes it suitable for processes that require large volumes. With sufficiently high recycle rates, the behavior of the PFR becomes similar to that of a CSTR.

2.8.1 Design equation

Consider the differential section dx (Figure 2.9(a)) with a differential volume dV in the reactor. A mass/material balance on limiting reactant A in the differential volume is

$$\left(\text{Rate of input}\right) = \left(\text{Rate of output}\right) + \left(\text{Rate of accumulation}\right)$$
$$- \left(\text{Rate of consumption}\right)$$

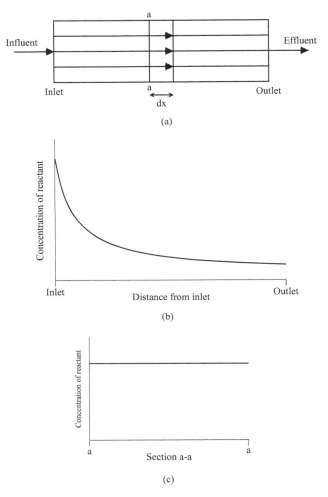

Figure 2.9 (a) Flow through a PFR, (b) variation of reactant concentration in a PFR, (c) Longitudinal distribution of reactant concentration for section a-a in the PFR.

For steady state conditions, the design equation is written as

$$r_A = \frac{d[A]}{dt} \tag{2.18}$$

which is the same as the design equation for a batch reactor. When the order of the reaction is known, an expression for r_A can be substituted into the left-hand side of the above equation, and the resulting differential equation can be integrated to obtain the design expression. Table 2.3 presents the design equations for zero, first, and second order reactions in a PFR.

Example 2.3: A reaction takes place in a PFR, where reactant A is converted to product P. The rate equation is

$$r_A = -0.38 \, [A] \text{ mol/L s}$$

Determine the volume of PFR required for 95% conversion of A. The initial concentration of A is 0.25 mol/L and the volumetric flow rate is 5 m³/s.

SOLUTION

The given reaction is first order with $k = 0.38 \text{ s}^{-1}$, $[A_o] = 0.25$ mol/L.
   With 95% conversion, $[A] = (1 - 0.95) \, [A_o] = 0.05 \times 0.25 = 0.0125$ mol/L
   From Table 2.1, first order design equation for a PFR is

$$kt = \ln\left(\frac{[A_o]}{[A_t]}\right)$$

or, $0.38t = \ln(0.25/0.0125)$

or, $t = 7.88 \text{ s}$

Volume of PFR, $V = Qt = (5 \text{ m}^3/\text{s})(7.88 \text{ s}) = 39.42 \text{ m}^3$

2.9 CONTINUOUS-FLOW STIRRED TANK REACTOR

CSTR or completely mixed reactors are used mainly for liquid phase reactions at low or atmospheric pressures. In this reactor, the reactant flows continuously into the reactor, the product effluent flows out continuously, and the reactor contents are mixed on a continuous basis, as shown in Figure 2.10(a). This type of reactor is also called back-mix reactor or completely mixed reactor.

The basic assumption for an ideal CSTR is that the reactor contents are completely mixed and homogeneous throughout. When a reactant $[A_o]$ enters the reactor, it is subjected to instantaneous and complete mixing, resulting in an immediate reduction to the final effluent concentration $[A_t]$. The effluent composition and temperature are the same as those of the reactor contents. This remains the same over time, as shown in Figure 2.10(b).

A tracer molecule in the influent has an equal probability of being located anywhere in the reactor after a small time interval, within the limit of complete mixing (Hill, 1977). Thus all fluid elements in the reactor have an equal probability of leaving the reactor with the effluent in the next time increment. As a result, there is a broad distribution of residence times for various fluid particles as illustrated in Figure 2.10(c).

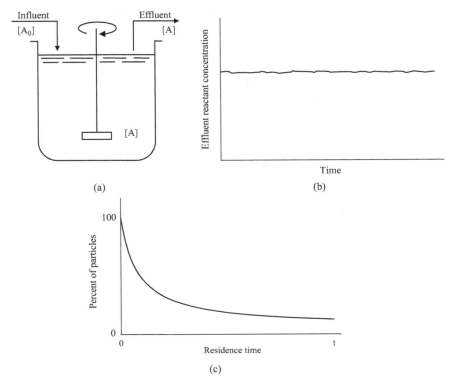

Figure 2.10 (a) CSTR, (b) effluent concentration variation for a CSTR, (c) residence time distribution of fluid particles in a CSTR.

Lower conversion of reactant is achieved in a CSTR as compared to a PFR, at the same operating temperature and feed composition. This is mainly due to the variation of particle residence times within the reactor, and the inability to achieve complete mixing. As a result, a CSTR of larger volume is required to achieve the same conversion as a PFR.

2.9.1 Design equation

A mass/material balance can be written for the limiting reactant A, assuming homogeneous conditions throughout the reactor:

$$\left(\text{Rate of input}\right) = \left(\text{Rate of output}\right) + \left(\text{Rate of accumulation}\right)$$
$$- \left(\text{Rate of consumption}\right)$$

At steady state conditions, the rate of accumulation = 0. So the design equation can be written as

$$r_A = \frac{[A_t] - [A_o]}{t} \tag{2.19}$$

where all the terms have the same meanings as defined in the previous sections. When the order of the reaction is known, an expression for r_A can be substituted into the left-hand side of the above equation, to obtain the design expression. Table 2.1 presents the design equations for zero, first, and second order reactions in a CSTR.

> Example 2.4: A chemical reaction takes place in a CSTR, where A is converted to product P. The initial concentration of A is 45 mg/L. After 5 min, the concentration of A is measured as 36 mg/L.
>
> a. Calculate the rate coefficient assuming that the reaction is first order.
> b. Calculate the rate coefficient assuming that the reaction is second order.
>
> SOLUTION
>
> a. For first order reaction, use the design equation from Table 2.1
>
> $$kt = \left(\frac{[A_o]}{[A_t]} \right) - 1$$
>
> $$\text{Therefore, } k = \frac{1}{5\,\text{min}} \left(\frac{45\,\text{mg/L}}{36\,\text{mg/L}} - 1 \right) = 0.05\,\text{min}^{-1}$$
>
> b. For second order reaction, use the design equation from Table 2.1
>
> $$kt = \frac{1}{[A_t]} \left(\frac{[A_o]}{[A_t]} - 1 \right)$$
>
> $$\text{Therefore, } k = \frac{1}{5\,\text{min} \times 36\,\text{mg/L}} \left(\frac{45\,\text{mg/L}}{36\,\text{mg/L}} - 1 \right) = 0.0014\,(\text{mg/L min})^{-1}$$

2.10 REACTORS IN SERIES

One method of increasing the removal efficiency of a process is to use a number of reactors in series. This is usually applicable for CSTRs, though a combination of CSTR and PFR can also be used (Reynolds and Richards, 1996).

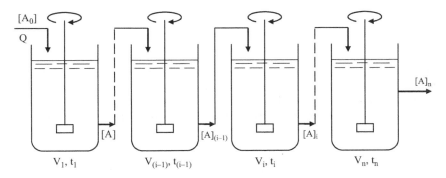

Figure 2.11 Series of CSTRs.

When a series or cascade of CSTRs are used, the effluent from one reactor serves as the influent to the next reactor, as shown in Figure 2.11. There is a stepwise decrease in the composition of reactant and temperature as the flow travels from one reactor to the next one. Assuming that the conditions in any individual reactor in the series are not influenced by downstream conditions, and conditions of the inlet stream and those prevailing in the reactor are the only variables that influence reactor performance (Hill, 1977), the following design equation can be written for steady state conditions:

$$r_{A_i} = \frac{[A]_i - [A]_{(i-1)}}{t_i} \tag{2.20}$$

where
   r_{Ai} = rate of consumption of A in the ith reactor
   t_i = detention time in ith reactor
   $[A]_i$ = concentration of A in effluent from the ith reactor
   $[A]_{(i-1)}$ = concentration of A in effluent from the $(i - 1)$th reactor
            = concentration of A in influent to the ith reactor

The detention time in the ith reactor is given by

$$t_i = \frac{V_i}{Q} \tag{2.21}$$

where
   V_i = volume of the ith reactor
   Q = volumetric flow rate into reactor

In a series of n reactors, the overall conversion is given by

$$f = \frac{[A_o] - [A_n]}{[A_o]} \tag{2.22}$$

where

[A_o] = concentration of A in influent to the 1st reactor
[A_n] = concentration of A in effluent from the nth reactor

Conversion in individual reactors can be calculated from influent and effluent reactant concentrations of that reactor.

Example 2.5: Consider the same first order chemical reaction from Example 2.4. Two reactors are used in series, a CSTR followed by a PFR for product formation. The detention time in the first reactor (CSTR) is 5 min. The two reactors are operated at the same temperature and have the same volume. What will be the effluent concentration of A from the PFR? What is the conversion efficiency?

SOLUTION

The two reactors are operated at the same temperature.
Therefore, $k_{CSTR} = k_{PFR}$
The two reactors have the same volume, and if the flow rate is the same, detention time, $t_{CSTR} = t_{PFR}$

The reactor system is illustrated in Figure 2.12.
From Example 2.4, A_o = 45 mg/L, A_1 = 36 mg/L, k_{CSTR} = 0.05 min^{-1}
Therefore, k_{PFR} = 0.05 min^{-1}
Use design equation for PFR for first order reaction from Table 2.1

$$kt = \ln\left(\frac{A_1}{A_2}\right)$$

or $\quad 0.05\,\text{min}^{-1} \times 5\,\text{min} = \ln\left(\frac{36\,\text{mg}/\text{L}}{A_2}\right)$

or $\quad A_2 = 28.04\,\text{mg}/\text{L}$

Overall conversion efficiency $= \dfrac{45 - 28.04}{45} \times 100\% = 37.7\%$

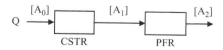

Figure 2.12 Reactor system (for Example 2.5).

Example 2.6: In Example 2.5, if another CSTR was used as the second reactor instead of the PFR, what would be the effluent concentration of A? Calculate the conversion efficiency. Determine the concentration of reactant A in the first and second reactors.

SOLUTION

Use design equation for CSTR for first order reaction, for reactor #2
$A_1 = 36$ mg/L, $k_{CSTR} = 0.05$ min^{-1}, $t = 5$ min

$$kt = \left(\frac{A_1}{A_2} \right) - 1$$

or $\;\; 0.05\,\text{min}^{-1} \times 5\,\text{min} = \left(\dfrac{36\,\text{mg/L}}{A_2} - 1 \right)$

or $\;\; A_2 = 28.8\,\text{mg/L}$

Overall conversion efficiency $= \dfrac{45 - 28.8}{45} \times 100\% = \mathbf{36\%}$

For a CSTR, concentration in effluent = concentration in reactor

Concentration of A in reactor #1 $= A_1 = \mathbf{36\,mg/L}$

Concentration of A in reactor #2 $= A_2 = \mathbf{28.8\,mg/L}$

2.11 SEMI-BATCH OR SEMI-FLOW REACTORS

Reactors used in actual treatment plants and processes may be operated somewhere in between ideal reactor modes (Metcalf and Eddy et al., 2013). Reactor operation can be semi-batch or semi-flow. A few examples are given below:

    a. A reactor where all the reactants are added at the same time as a batch, but the products are discharged continuously.
    b. A reactor where the reactants are added at different time intervals.
    c. A reactor where the products are removed at different time intervals.
    d. A batch reactor partially filled with one reactant, with progressive addition of other reactants until the reaction is completed.

PROBLEMS

2.1   It was observed from an experimental study that the rate of a chemical reaction did not depend on the concentration of the reactant, but was influenced by the concentration of the product. What is the order of the reaction with respect to the reactant?

2.2   Draw the curves for reaction rate versus time for zero, first, and second order reactions. Write down the rate expressions for each curve.

2.3   A denitrification experiment was conducted by a graduate student in the environmental engineering laboratory at George Washington University, where nitrate (NO_3) was converted to nitrite and nitrogen gas. The concentration of nitrate was measured at regular time intervals. The data is shown in Table 2.4. Determine the rate of the reaction.

Table 2.4   Concentration of nitrate at regular time intervals (for Problem 2.3)

Time, h	NO_3, mg/L
0.0	30.0
0.5	23.3
1.0	19.0
1.5	15.3
2.0	11.0
2.5	8.3
3.0	7.0
3.5	6.3
4.0	5.7
4.5	5.3
7.75	4.7

2.4   What is a Plug Flow Reactor? What are the advantages and disadvantages of using a PFR?

2.5   Illustrate graphically the variation of reactant concentration with time in (i) a PFR, and (ii) a CSTR.

2.6   What is a CSTR? Mention two advantages and two disadvantages of a CSTR.

2.7   A wastewater is treated in a reactor vessel. A first order reaction takes place with respect to the organic matter in the wastewater. The rate constant is determined to be 0.23 d^{-1}. The initial concentration of organic matter is 150 mg/L, and it is desired to achieve 90% conversion. The flow rate of the wastewater is 500 m^3/d.

   a. Calculate the detention time and volume of PFR required to achieve this conversion.
   b. Calculate the volume of CSTR required to achieve the same conversion.
   c. Which option seems better to you and why?

2.8   A laboratory analysis is carried out in batch reactors. The initial concentration of reactant was 0.25 mol/L, and 85% conversion was achieved in 20 min. It was assumed that the reaction was zero order with respect to the reactant.

    a. Calculate the zero order rate coefficient.

    b. After further experimentation, it was discovered that the rate was first order and not zero order. Calculate the correct rate coefficient.

2.9 The ammonia in wastewater is to be converted to nitrate in a bioreactor. The initial concentration of ammonia is 145 mg/L. It is desired to achieve 90% conversion in a 50 m³ reactor. The design engineer is trying to select between a PFR and CSTR mode for the operation of the reactor. Which mode of operation will allow the engineer to process a larger volume of wastewater within a shorter period of time?

2.10 An industrial wastewater is treated in a CSTR. The conversion of reactant A to product C is governed by the following rate equation:

$$r_A = -1.2[A] \, mg/L.h$$

    a. The volume of the reactor is 60 m³. What is the volumetric flow rate of the wastewater, corresponding to a conversion efficiency of 95%?

    b. If only 90% conversion efficiency is desired, can we use a smaller reactor volume to handle the same flow rate? What would be the volume?

2.11 A dairy wastewater is treated in a series of CSTRs. The initial concentration of complex organics in the wastewater is 1500 mg/L. The first order rate coefficient is 0.45 d⁻¹. The detention time in each reactor is 1.5 d. If two reactors are used in series, calculate the final effluent concentration of the organic matter. What is the conversion efficiency?

2.12 Using the data from problems 2–11, calculate the overall efficiency if three reactors are used in series. Would it be feasible to use three reactors?

2.13 It is desired to increase the conversion efficiency of a chemical process. Would you use multiple reactors in series or in parallel to achieve this? Why?

2.14 An existing wastewater treatment process is experiencing a very high volume of wastewater due to an increase in population. The design engineer is planning to increase the capacity of the process. Should the engineer use multiple reactors in series or in parallel to achieve this? Why?

2.15 A pharmaceutical wastewater is treated at a flow rate Q in a series of two CSTRs. The initial concentration of BOD_5 is 1200 mg/L. BOD conversion is assumed to be first order. For the first reactor, the first order rate coefficient is 0.48 d⁻¹, and detention time is 1.5 d. The second reactor is *half* the volume of the first reactor. Reactor 1 is operated at 20°C and Reactor 2 is operated at 30°C. Arrhenius coefficient θ is 1.04.

a. Calculate the final effluent BOD_5 concentration from the second reactor.
b. What is the concentration of BOD_5 in Reactor 1?
c. What is the overall conversion efficiency of BOD_5?

2.16 An industrial wastewater is treated in a series of 2 reactors, a CSTR followed by a PFR. The flow rate is Q. The initial concentration of BOD_5 is 5200 mg/L. BOD conversion is assumed to be first order. For the first reactor, the first order rate coefficient is 0.48 d^{-1}, and detention time is 2 d. The first reactor is twice the volume of the second reactor. Reactor 1 is operated at 20°C and Reactor 2 is operated at 30°C. Arrhenius coefficient θ is 1.04.

a. Calculate the final effluent BOD_5 concentration from the second reactor.
b. What is the overall conversion efficiency of BOD_5?

REFERENCES

Henry, J. G., and Heinke, G. W. 1996. *Environmental Science and Engineering*. 2nd edn., Prentice Hall, Inc., Upper Saddle River, NJ, USA.

Hill, C. G. Jr. 1977. *An Introduction to Chemical Engineering Kinetics & Reactor Design*. John Wiley & Sons, New York, NY, USA.

Metcalf and Eddy, Tchobanoglous, G., Stensel, H., Tcuchihashi, R., and Burton, F. 2013. *Wastewater Engineering: Treatment and Resource Recovery*. 5th edn. McGraw-Hill, Inc., New York, NY, USA.

Reynolds, T. D. and Richards, A. R. 1996. *Unit Operations and Processes in Environmental Engineering*. 2nd edn. PWS Publishing Company, Boston, MA, USA.

Chapter 3

Wastewater microbiology

3.1 INTRODUCTION

Wastewater contains a wide variety of microorganisms, some of which are pathogens, while others play a significant role in the degradation of organic matter. Bacteria, protozoa, and other microorganisms play an active role in the conversion of biodegradable organic matter to simpler end products that result in the stabilization of the waste. This is a continuous process occurring in streams and rivers as a natural purification process. This is described in more detail in Chapter 4. These natural purification processes are enhanced and accelerated in engineered biological treatment systems at wastewater treatment plants. For efficient removal of organic matter and other pollutants, it is essential to have a thorough understanding of the nature, growth kinetics, and process requirements of the microorganisms involved and utilized in the biological treatment processes. This chapter will provide an overview of the major groups of microorganisms used in the biological treatment of wastewater.

The three major domains of living organisms are the Bacteria, the Archaea, and the Eukarya. This is according to the Universal Phylogenetic (Evolutionary) Tree, which was derived from comparative sequencing of 16S or 18S ribosomal RNA (Madigan et al., 2021). Based on cell structure, all living organisms are divided into two types: *Prokaryotic and eukaryotic.* The major structural difference between prokaryotes and eukaryotes is their nuclear structure. The eukaryotic nucleus is surrounded by a nuclear membrane, contains deoxyribonucleic acid (DNA) molecules, and undergoes division by mitosis. On the other hand, the prokaryotic nuclear region is not surrounded by a membrane, contains a single DNA molecule whose division is non-mitotic. The prokaryotes include bacteria, blue-green algae (cyanobacter), and archaea. Figures 3.1(a) and (b) show typical cell structures of prokaryotes and eukaryotes, respectively. The archaea are separated from bacteria due to their DNA composition and unique cellular chemistry. Examples of archaea are the methane-producers, e.g. *Methanococcus, Methanosarcina.* The eukaryotes are much more complex and include plants and animals, as well as protozoa, fungi, and algae. Table 3.1 presents the

DOI: 10.1201/9781003134374-3

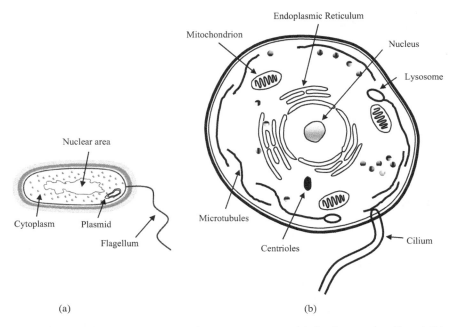

Figure 3.1 Typical cell structure of microorganisms: (a) Prokaryotic cell and (b) eukaryotic cell.

Table 3.1 General classification of organisms

Organisms	Eukaryotes	Prokaryotes
Macroorganisms	Animals	None known
	Plants	
Microorganisms	Algae	Archaea
	Fungi	Bacteria
	Protozoa	

classifications. Macroscopic animals include *Rotifers*, *Crustaceans*, etc. Rotifers act as polishers of effluent from wastewater treatment plants by consuming organic colloids, bacteria, and algae. The microorganisms are discussed in more detail in the following sections.

3.2 BACTERIA

Bacteria are unicellular prokaryotic microorganisms. They use soluble food. Bacteria usually reproduce by binary fission, although some species reproduce sexually or by budding. They are generally characterized by the shape, size, and structure of their cells. Bacteria can have one of three general

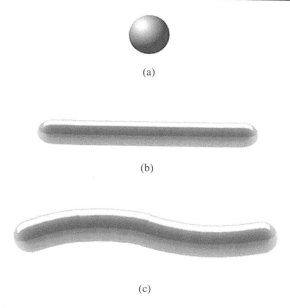

(a)

(b)

(c)

Figure 3.2 Different shapes of bacteria: (a) Coccus, (b) rod, and (c) spirillum.

shapes: (a) Spherical (*coccus*), (b) cylindrical or rod-shaped (*bacillus*), and (c) spiral-shaped (*spirillum*). Bacteria can range in size from 0.5 to 5.0 µm long and from 0.3 to 1.5 µm wide. Cocci are about 0.1 µm in diameter (Henry and Heinke, 1996). Figure 3.2 illustrates the different shapes of bacteria.

3.2.1 Cell composition and structure

A bacterial cell has about 80% water and 20% dry matter. Of the dry matter, 90% is organic and 10% is inorganic. The bacterial cell is generally expressed by the simple chemical formula: $C_5H_7O_2N$. This can be expanded to include sulfur and phosphorus. Figure 3.3 illustrates a typical bacterial cell. The *cell wall* is a rigid structure that provides shape to the cell and protects it from osmotic pressure. The wall is usually 0.2–0.3 µm thick and accounts for 10–50% of the dry weight of the cell. Inside the cell wall is the cytoplasmic membrane. The *cytoplasmic membrane* is a critical permeability barrier, which regulates the transport of food into the cell and of waste products out of the cell. The interior of the cell contains the cytoplasm, the nuclear area, and the polyribosomes. The *cytoplasm* is a colloidal suspension of proteins, carbohydrates, and other complex organic compounds. The cytoplasm contains RNA (ribonucleic acid) which causes the biosynthesis of proteins. The RNA, together with proteins, form densely packed particles called polyribosomes. *Polyribosomes* manufacture enzymes for each specific biochemical reaction. The nuclear area contains DNA. DNA contains all the

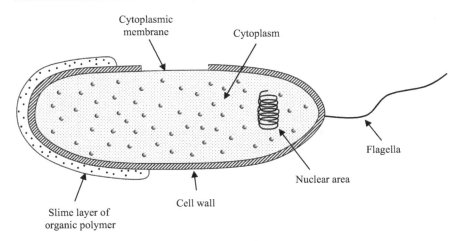

Figure 3.3 Diagram of a typical bacterial cell.

genetic information necessary for reproduction and is considered to be the blueprint of the cell. Some bacteria occasionally have *inclusions* consisting of excess nutrients that are stored for future use. The thickness of the inclusion or slime layer depends on the age of the cell.

3.2.2 Bacterial growth curve

A number of factors affect the growth and death of bacteria. These include the type of food or carbon source, abundance of food, nutrients, pH, temperature, presence or absence of oxygen, and toxic substances. Given the presence of optimal conditions, bacteria can grow in logarithmic proportions. A batch experiment with a limited amount of food or substrate can produce a bacterial growth curve similar to the one illustrated in Figure 3.4.

The bacterial growth curve exhibits four distinct phases, as shown in Figure 3.4. They are the following:

1. The first phase is called the *lag phase*. This represents the time needed by the bacteria to adjust to the new environment, and start producing enzymes necessary to degrade the substrate surrounding them. If the substrate is readily degradable, then the lag phase is short. If the substrate is not readily biodegradable, then it may take time for the bacteria to produce the necessary enzymes. This may result in a long lag phase, as illustrated in Figure 3.5(a), until the bacteria is acclimated to the substrate and then starts reproducing. If the bacteria are not able to synthesize the necessary enzymes, the substrate may be toxic and eventually result in the death of the cells, as illustrated in Figure 3.5(b).
2. After the lag phase comes the *logarithmic or exponential growth phase*. In this phase, the population doubles at regular intervals of time due to the abundance of the substrate, and optimal growth conditions.

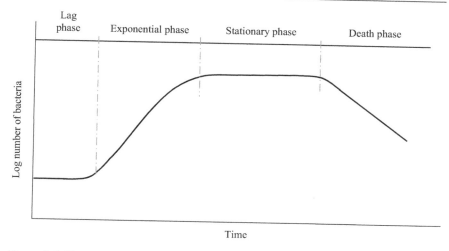

Figure 3.4 Typical bacterial growth curve from a batch study.

3. Eventually, as the substrate concentration decreases, the growth rate starts decreasing and the *stationary phase* is observed. During this phase, the growth rate equals the death rate resulting in a dynamic equilibrium at which there is no further increase in population. This phase corresponds to a very low substrate concentration.

4. The last phase is called the *endogenous decay or death phase* when one or more nutrients or the substrate is completely exhausted. Cell death and lysis release some soluble organics that are used by surviving bacteria for a while. The death rate keeps on increasing until all bacterial cells die off.

Most biological wastewater treatment processes are operated somewhere in between the stationary phase and the death phase. This is true for biological reactors operated as CSTRs since this corresponds to a very low substrate or BOD concentrations in the reactor and effluent.

3.2.3 Classification by carbon and energy requirement

All cells need a source of carbon and a source of energy to carry out cell synthesis. One of the goals of wastewater treatment is to convert both the carbon and energy of the wastewater into microbial cells, which can then be removed from water by settling or filtration. Bacterial cells can be divided into two broad groups according to carbon and energy sources:

   i. *Heterotrophic* – use organic compounds as both their carbon and energy source. A large number of wastewater bacteria are heterotrophic.
   ii. *Autotrophic* – use inorganic compounds as a carbon source (e.g. CO_2, HCO_3^-) and sunlight or inorganic compounds for energy. Two types

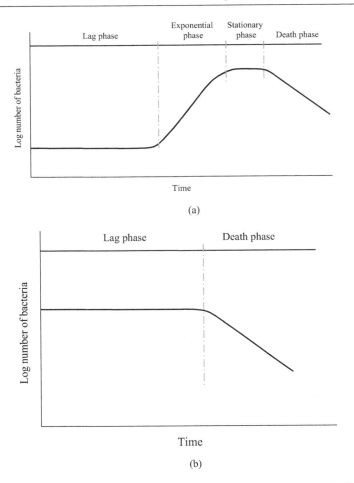

Figure 3.5 Bacterial growth curves exhibiting (a) acclimation, and (b) toxic response.

of autotrophs are of interest. They are (a) photoautotrophs that obtain energy from sunlight and carbon from CO_2, and (b) chemoautotrophs that obtain energy from the oxidation of inorganic compounds and carbon from CO_2. For example, nitrifying bacteria – *Nitrosomonas* and *Nitrobacter*.

The nitrifying bacteria are of great significance in wastewater treatment for nitrogen removal. The nitrifying bacteria carry out the two-step process of nitrification, which results in the conversion of ammonia to nitrites in the first step followed by conversion of nitrites to nitrates in the second step.

The nitrates are converted to nitrogen gas in a subsequent step called deni-trification. The nitrification reaction is given below:

$$NH_3 + O_2 \xrightarrow{\text{nitrosomonas}} NO_2^- + energy \tag{3.1}$$

$$NO_2^- + O_2 \xrightarrow{\text{nitrobacter}} NO_3^- + energy \tag{3.2}$$

3.2.4 Classification by oxygen requirement

Bacteria can be divided into the following groups based on their oxygen requirements:

i. *Aerobic* – requires oxygen for growth and survival.
ii. *Anaerobic* – grows in absence of oxygen. They cannot survive in presence of oxygen.
iii. *Facultative* – can grow both in the presence and absence of oxygen. For example, Denitrifying bacteria are facultative anaerobes that grow under anoxic conditions.

3.2.5 Classification by temperature

Certain groups of bacteria grow at specific ranges of temperatures. They are:

i. *Cryophilic or Psychrophilic* – grows at temperatures below 20°C, usually between 12°C and 18°C.
ii. *Mesophilic* – grows between 25°C and 40°C, optimum at 35°C.
iii. *Thermophilic* – grows between 50°C and 75°C, optimum at 55°C.

Growth is not limited to these temperature ranges only. Bacteria will grow at slower rates at other temperatures and can survive over a wide range of temperatures. Some can survive at temperatures as low as 0°C. If frozen rapidly, bacteria can be stored for a long time with insignificant death rates. Bacteria reproduce by binary fission as illustrated in Figure 3.6.

3.2.6 Bacteria of significance

Bacteria are the most important group of microorganisms in the environment. The largest population of microorganisms present in water and wastewater are bacteria. Some of them are pathogenic and cause diseases in humans and animals. Other groups of bacteria are important in biological wastewater treatment processes, natural purification processes in lakes and

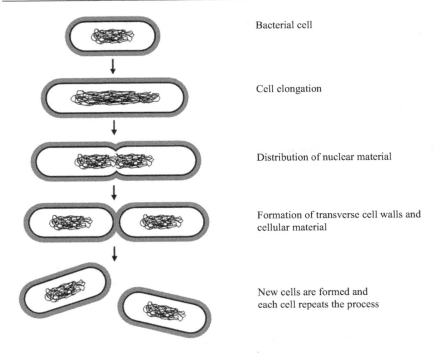

Bacterial cell

Cell elongation

Distribution of nuclear material

Formation of transverse cell walls and cellular material

New cells are formed and each cell repeats the process

Figure 3.6 Cell reproduction by binary fission

Source: Adapted from Henry and Heinke, 1996.

streams, and decomposition of organic matter in soil and landfills. Table 3.2 presents some significant groups of bacteria and their functions.

3.3 ARCHAEA

At the molecular level, both archaea and bacteria are structurally prokaryotic. But they are evolutionarily distinct from one another. Archaea was previously known as the *archaebacteria* (Madigan et al., 2021). Their cell wall, cell material, and RNA composition are different from bacteria. Some archaea are important in anaerobic processes, e.g. methanogens that produce methane gas from the degradation of organic matter under anaerobic conditions. These include *Methanobacterium*, *Methanosarcina*, and *Methanothrix* that are important in the anaerobic digestion of sludge. Some archaea exhibit highly specialized metabolic pathways and are found under extreme environmental conditions. One distinct group is the hyperthermophiles. They are obligate anaerobes and have a temperature optimum above 80°C. Examples are *Thermoproteus*, *Sulfolobus*, *Methanopyrus*. Extremely halophilic archaea are another diverse group that inhabits highly

Table 3.2 Important groups of bacteria in water and wastewater

Bacteria	Genus	Importance
Nitrifying bacteria	Nitrosomonas	Oxidizes ammonia to nitrites
	Nitrobacter	Oxidizes nitrites to nitrates
Denitrifying bacteria	Pseudomonas, Bacillus	Reduces nitrite and nitrate to nitrogen gas
Iron bacteria	Leptothrix, Crenothrix	Oxidizes ferrous iron to ferric iron
Sulfur bacteria	Thiobacillus	Oxidizes sulfur and iron, causes corrosion of iron sewer pipes
Photosynthetic bacteria	Chromatium, Chlorobium	Reduces sulfides to sulfur
Indicator bacteria	Escherichia, Enterobacter	Indicates fecal pollution
Pathogenic bacteria	Salmonella	Causes Salmonellosis
	Vibrio cholera	Causes Cholera
	Salmonella typhi	Causes Typhoid fever
	Legionella pneumophila	Causes Legionnaire's disease

Source: Adapted from Henry and Heinke (1996) and Metcalf and Eddy et al. (2013).

saline environments, such as solar salt evaporation ponds. Examples are *Halobacterium* and *Halococcus*.

3.4 PROTOZOA

Protozoa are mostly unicellular eukaryotes that lack cell walls. They can be free-living or parasitic. Most are aerobic heterotrophs, some are aerotolerant anaerobes, and a few are obligate anaerobes. They reproduce by binary fission. They can range in size from a few to several hundred micrometers. They are an order of magnitude larger than bacteria. Protozoa act as polishers of effluent from biological treatment processes, by feeding on bacteria, algae, and particulate organic matter. Some protozoa have hair-like strands called flagella, which provide motility by a whiplike action, e.g. *Giardia*. Some flagellated species feed on soluble organics. Free-swimming protozoa have cilia, which are used for propulsion and gathering of organic matter, e.g. *Paramecium*.

A number of protozoa are important in water and wastewater, as they cause enteric diseases in humans and animals. These include:

a. *Amoeba* – They move by extending their cytoplasm in search of food. These extensions are called pseudopods or false feet, illustrated in Figure 3.7. They are pathogenic and cause amoebic dysentery in humans.

b. *Giardia lamblia* – These are parasitic protozoa. They range in size from 8 to 18 μm long and from 5 to 15 μm wide (Hammer and

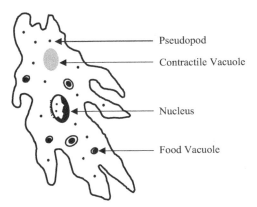

Figure 3.7 Amoeba.

Hammer, 2012). Inside a host body, the Giardia cyst releases a tropho-
zoite that feeds, grows, and reproduces causing a gastrointestinal disease called Giardiasis. *Giardiasis* causes cramps, diarrhea, and fatigue,
and can become severe. In a drinking water treatment plant, coagulation–flocculation followed by filtration and disinfection is required to
kill them. Drinking water treatment plants in the US have to achieve
99.9% removal of Giardia.

c. *Cryptosporidium* – It forms a thick-walled oocyst in the environment,
and can survive for long periods of time. The oocyst is spherical with
a diameter of 4–6 μm. Cryptosporidium oocysts are present in small
numbers in surface waters. When humans ingest it with drinking
water, the oocyst opens in the small intestine, releases sporozoites that
attach themselves to the walls of the intestine, and disrupts intestinal
functions causing Cryptosporidiosis. Cryptosporidiosis causes severe
diarrhea and can become life-threatening. Chlorination cannot kill the
oocysts. Based on the size, they can be removed by using enhanced
coagulation–flocculation processes, and ozone disinfection in drinking
water treatment processes.

3.5 ALGAE

Algae are autotrophic, photosynthetic eukaryotic plants. One exception is
the *blue-green algae* or *cyanobacter* which is prokaryotic, and produces toxins that are harmful to fish and birds, e.g. *Anabena*. Algae can be unicellular
or multicellular. Their size ranges from 5 μm to 100 μm or more, when they
are visible as a green slime on water surfaces, e.g. *Pediastrum*. They have
no roots, stems, or leaves. Multicellular colonies can grow in filaments or
simple masses of single cells that clump together. All algal cells are capable
of photosynthesis. The simplified reaction is given below:

$$CO_2 + PO_4^{3-} + NH_3 + H_2O \xrightarrow{\text{Sunlight}} \text{new cells} + O_2 + H_2O \qquad (3.3)$$

Algae use sunlight as their energy source and carbon dioxide or bicarbonates as their carbon source. The oxygen that is produced during photosynthesis replenishes the dissolved oxygen content of the water. They are often used in aerobic oxidation ponds since they can produce the oxygen necessary for aerobic bacteria. Algae are important primary producers in the aquatic food chain.

Excessive algae growths can cause taste and odor problems, clog water intakes at treatment plants, and shorten filter runs. Algae can grow very quickly to produce algal blooms when high concentrations of nutrients, such as nitrogen and phosphorus are available. This leads to a condition called *Eutrophication* of lakes, streams, and estuaries. The algal blooms form a green-colored mat on the water surface blocking the penetration of sunlight. This adversely affects other aquatic plants. At night during respiration, algae use up oxygen from the water and produce carbon dioxide according to the following simplified reaction:

$$\text{Algal cells} + O_2 \rightarrow CO_2 + H_2O \qquad (3.4)$$

This causes significant depletion of dissolved oxygen in the lake and can affect the fish population. Most game fish require at least 4 mg/L dissolved oxygen (DO) for survival. Other aquatic species are adversely affected below 2 mg/L DO. The excess dissolved oxygen produced during photosynthesis cannot be stored and is released into the atmosphere. Thus eutrophic lakes are characterized by unsightly green polluted waters, loss of species diversity, very low dissolved oxygen, and absence of game fish. One example is the deterioration and impaired waters of the Chesapeake Bay in the Eastern US. Agricultural runoff and other sources contribute nutrients that lead to Eutrophication. Control of these sources is necessary to limit algal growths. Different types of algae are illustrated in Figure 3.8.

3.6 FUNGI

Fungi are multicellular, non-photosynthetic, heterotrophic eukaryotes. Most are obligate or facultative aerobes. They can reproduce sexually or asexually by fission, budding, or spore formation. Fungi can grow under low nitrogen, low moisture, and low pH conditions. The optimum pH is about 5.6, but the range is between 2 and 9. They can also degrade cellulose, which makes them useful in composting processes. There are mainly three groups of fungi: Molds, yeasts, and mushrooms. Yeasts are used in baking, distilling, and brewing operations. Fungi are illustrated in Figure 3.9.

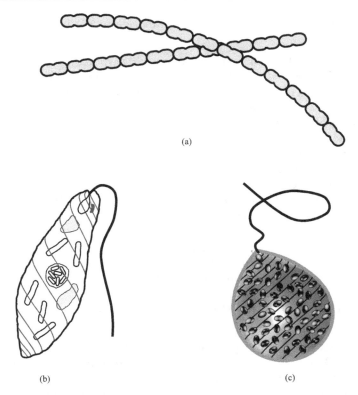

Figure 3.8 Different types of algae: (a) *Anabena*, (b) *Euglena*, and (c) *Lepocinclis*

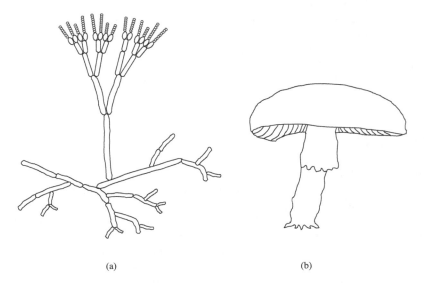

Figure 3.9 Types of fungi: (a) Mold and (b) mushroom.

3.7 VIRUS

A virus is a non-cellular genetic element that uses a host cell for its replication, and also has an extracellular state. In the extracellular state, it is called a *virion*. A virion is metabolically inert and does not carry out respiratory or biosynthesis functions. Viruses are obligate intracellular parasites. They are composed of a nucleic acid core that contains either a DNA or RNA, surrounded by a protein shell called the capsid. According to shape and structure, they can be polyhedral, helical, or combination T-even, as illustrated in Figure 3.10. Viruses are usually very small, ranging in size from a few nanometers to about 100 nm. Viruses are classified based on the host that they infect, e.g. animal viruses, plant viruses, bacterial viruses, or bacteriophages. Viruses of concern in wastewater are the ones excreted in large numbers in human feces. These include *poliovirus*, *hepatitis A virus*, and enteroviruses that cause diarrhea, among others. Drinking water treatment plants have to achieve 99.99% removal of viruses.

A virus cannot reproduce or replicate on its own. It can only replicate inside a host body. The various phases of the replication process of a bacteriophage are given below (Madigan et al., 2021) and illustrated in Figure 3.11.

    i. Attachment – adsorption of the virion to a susceptible host cell.
   ii. Penetration – injection of the virion or its nucleic acid into the cell.

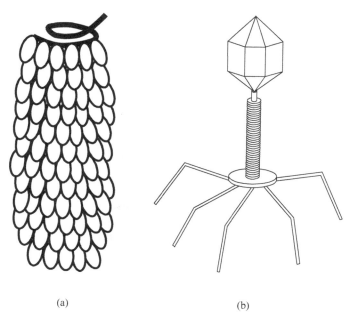

(a)   (b)

Figure 3.10 Viruses of different shapes, (a) helical, and (b) combination T-even.

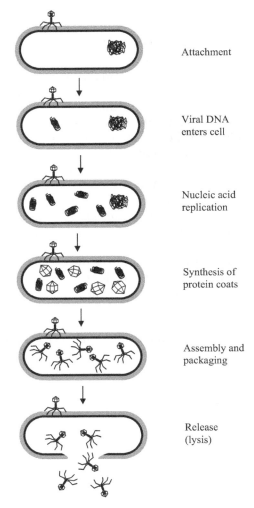

Attachment

Viral DNA
enters cell

Nucleic acid
replication

Synthesis of
protein coats

Assembly and
packaging

Release
(lysis)

Figure 3.11 Cell replication of a virus in a bacterial cell.

Source: Adapted from Madigan and Martinko, 2006.

   iii. Replication – of the virus nucleic acid. The virus alters the cell's metab-
      olism to synthesize new virus nucleic acids.
   iv. Synthesis – of protein subunits of the virus coat.
    v. Assembly and packaging – of protein subunits and nucleic acid into
      new virus particles.
   vi. Release – of mature new viruses from the cell by lysis as the cell breaks
      open.

For a virus infecting a bacteria, the whole replication process can be completed in 30–40 minutes.

3.8 MAJOR OUTBREAKS

A sudden increase in the number of cases of an infectious disease within a community or geographic area is defined as an outbreak or *epidemic*. When a disease spreads across several countries affecting a large number of people, it becomes a *pandemic*. Microorganisms, especially viruses, bacteria, and protozoa have been responsible for many outbreaks in Europe, the US, and other parts of the world. Smallpox outbreak in the 1600s, the cholera outbreak in the mid-1800s, and the typhoid outbreak in the early 1900s caused millions of deaths all over the world. Influenza outbreak, also known as the *Spanish flu* in 1918, was caused by the H1N1 virus and was responsible for over 25 million deaths in the first 6 months. Diphtheria epidemic by a bacterial infection in 1921, Polio outbreak by a viral infection in the early to mid-1900s, and Measles outbreak through a virus infection in the 1980s caused illnesses and deaths of millions of people.

In 1993, there was a devastating outbreak of cryptosporidiosis in Milwaukee, Wisconsin in the US. This resulted in around 400,000 people becoming sick from the protozoa *Cryptosporidium*, and a number of deaths. The outbreak became the impetus for a tremendous amount of research on the survival of the protozoa, and its efficient monitoring and removal techniques from drinking water systems. The concept of multiple barrier systems in treatment plants also gained more importance. During 2012–2017, health officials in 26 states reported 111 giardiasis outbreaks in the US, caused by the protozoa *Giardia lamblia*. The disease is transmitted through water exposure, person-to-person contact, and contaminated food (Conners et al., 2021).

3.8.1 SARS-CoV-2 (Coronavirus)

The severe acute respiratory syndrome coronavirus-2 (SARS-CoV-2), commonly known as coronavirus was first discovered in December 2019 in Wuhan, China. The coronavirus caused the respiratory illness known as COVID-19, which was responsible for more than 5.45 million deaths all over the world as of January 2022 (WHO, 2022). Symptoms of COVID-19 include coughing, shortness of breath/breathing difficulties, fever, chills, muscle pain, headache, sore throat, and loss of taste and smell (CDC, 2021). The basic structure of SARS-CoV-2 (Coronavirus) is illustrated in Figure 3.12.

The virus is mainly transmitted through respiratory droplets and direct contact, although other possibilities of transmission exist through feces, wastewater, and drinking water. Wang (2020) suggested that the virus was most prevalent in the respiratory tract, and could spread through fecal matter in the wastewater. Virus RNA has been found in fecal samples and semen

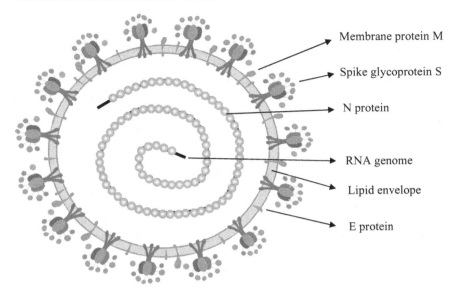

Membrane protein M

Spike glycoprotein S

N protein

RNA genome

Lipid envelope

E protein

Figure 3.12 Basic structure of SARS-CoV-2 (Coronavirus).

from infected individuals (Holshue et al., 2020). Other researchers are trying to identify the community transmission of the virus several days in advance, through testing of wastewater samples for SARS-CoV-2. In an effort to reduce the likelihood of transportation of the virus through water, the World Health Organization released a series of wastewater treatment and facility management suggestions. These treatment suggestions include using a waste stabilization pond with a 20 d retention time, which combined with sunlight, elevated pH levels, and biological activity is very effective in destroying pathogens. Filtration and disinfection should also be an effective measure in eliminating the virus using free chlorine at a concentration greater than 0.5 mg/L for at least 30 minutes at a pH higher than 8.0 with a chlorine residual remaining for the distribution system. In areas without centralized water treatment facilities or safe distribution systems, households should consider boiling, ultrafiltration or nano-membrane filtration, solar irradiation, and UV irradiation and carefully dosed free chlorine, which can be effective in removing or deactivating viruses before consuming the water (WHO and UNICEF, 2020).

The coronavirus is zoonotic; meaning it is derived from animals. Historically, many of the viruses trivial to animals have mutated and affected humans in significant ways. Mutations cause new variants to occur; some emerge and disappear, while others persist and spread. The H1N1 virus, which was the mutation of the bird flu virus, had contracted pigs and then affected humans. Although there are controversies related to the specific origin of SARS-CoV-2, it has shown a high correlation with Bat Coronavirus

RaG13 of low risk, which became significantly dangerous through mutation. Although many variants of the coronavirus were observed in many parts of the world, some variants are of major concern with regard to how they spread, and are resistant to treatment and vaccination. The Alpha (B.1.1.7), Beta (B.1.351), and Gamma (P.1) variants were initially detected in the UK, South Africa, and Brazil and eventually spread to other parts of the world. The Delta (B.617.2) variant which was initially identified in India, soon became the most prevalent one in mid-2021. It spread more easily and quickly than the other variants and led to an increased number of infections, hospitalizations, and deaths. In November 2021, the Omicron (B.1.1.529) variant was identified in South Africa, which spread quickly all over the world. Many other variants may eventually appear, and the effects of those variants on humans will be a continuing area of research for scientists and researchers.

Traditional vaccine development can take 15 years or more, starting with a lengthy discovery phase in which vaccines are designed and exploratory preclinical experiments are conducted. This is usually followed by a phase in which more formal preclinical experiments and toxicology studies are performed and production processes are developed (Krammer, 2020). To provide acquired immunity from SARS-CoV-2, there was an unprecedented collaboration between the global pharmaceutical industries and governments for expedited development of vaccines. At least 19 vaccines are currently authorized in different parts of the world, and 8.7 billion doses of vaccines have been administered worldwide as of January, 2022 (WHO, 2022). Although many of the vaccines are conventional inactivated vaccines, there are few developed with "next-generation" strategies with precise targeting of COVID-19 infections. The RNA vaccine, produced by Pfizer-BioNTech and Moderna, contains RNA which, when introduced into a tissue, acts as messenger RNA (mRNA) to cause the cells to build the foreign protein and stimulate an adaptive immune response, which teaches the body to identify and destroy the corresponding pathogen (Park et al., 2021). Adenovirus vector vaccines are examples of non-replicating viral vector vaccines, using an adenovirus shell containing DNA that encodes a SARS-CoV-2 protein (CDC, 2021). Although there are several challenges with the rapid development of such vaccines regarding the safety, efficacy, and dosing trials that generally require years, most researches show significant effectiveness of the vaccines and a direct correlation between vaccine dose and virus transmission, with more COVID-19-related hospitalizations and death in less vaccinated areas.

The *coronavirus* can be detected in fecal samples and wastewater several days before an infected person exhibits signs of infection. Due to this reason, wastewater monitoring and surveillance is becoming a tool for early detection of the virus in a city or community. One example is the early detection of the Delta variant in the city of Davis, California (Abbott, 2021). The CDC has allocated $33 million to 31 public health laboratories for wastewater testing for the coronavirus.

PROBLEMS

3.1   What are the differences between eukaryotes and prokaryotes? Explain with diagrams of their cells.
3.2   Draw a typical bacterial cell and label the different parts. Mention the functions of the cytoplasmic membrane, cytoplasm, and DNA.
3.3   Illustrate a typical bacterial growth curve and label the different phases.
3.4   What is the logarithmic growth phase? Develop a model to calculate the number of bacteria in the logarithmic growth phase assuming first order kinetics.
3.5   Explain the process of nitrification with the help of equations. What types of bacteria are involved in the process? Name them.
3.6   Write short notes on algae, protozoa, virus.
3.7   Conduct a literature review to find out about the last waterborne disease outbreak in your city/country. What type of microorganism was responsible, what were the reasons for the outbreak, and what measures were taken for control and future prevention?
3.8   What is Cryptosporidium? Why is chlorine disinfection unable to remove it from the drinking water supply?
3.9   What is Eutrophication?
3.10 What are the steps in the replication of a bacteriophage? Explain with the help of a diagram.
3.11 Arrange the following microorganisms according to size and predation from largest to smallest: Bacteria, virus, protozoa, crustaceans.
3.12 Is the *coronavirus* present in wastewater? From a public health perspective, how can we take advantage of its presence in wastewater?

REFERENCES

Abbott, B. 2021. Wastewater Helps Health Officials Spot Covid-19 Warning Signs. *The Wall Street Journal*, 25, 2021.
CDC. 2021. *Understanding Viral Vector COVID-19 Vaccines*. US Centers for Disease Control and Prevention (CDC), USA.
Conners E.E., Miller A.D., Balachandran N., Robinson B.M., and Benedict K.M. 2021. Giardiasis Outbreaks - United States, 2012–2017. *MMWR Morb Mortal Wkly Report*, 70:304–307.
Hammer, M. J. and Hammer, M. J. 2012. *Water and Wastewater Technology*. 7th edn. Pearson-Prentice Hall, Inc., Hoboken, NJ, USA.
Henry, J. G., and Heinke, G. W. 1996. *Environmental Science and Engineering*. 2nd edn., Prentice Hall, Inc., Upper Saddle River, NJ, USA.
Holshue ML, DeBolt C, Lindquist S, Lofy KH, Wiesman J, Bruce H, et al. 2020. First Case of 2019 Novel Coronavirus in the United States. *The New England Journal of Medicine*. 382 (10): 929–936.
Krammer, F. 2020. SARS-CoV-2 Vaccines in Development. *Nature*, 586:516–527.

Madigan, M., Bender, K. S., Buckley, D. H., Sattley, W. M., and Stahl, D. A. 2021. *Brock Biology of Microorganisms*. 16th edn. Pearson, Hoboken, NJ, USA.

Madigan, M. and Martinko, J. 2006. *Brock Biology of Microorganisms*. 11th edn. Prentice Hall, Inc., Hoboken, NJ, USA.

Metcalf and Eddy, Tchobanoglous, G., Stensel, H., Tcuchihashi, R., and Burton, F. 2013. *Wastewater Engineering: Treatment and Resource Recovery*. 5th edn. McGraw-Hill, Inc., New York, NY, USA.

Park, K.S., Sun, X., Aikins, M. E., and Moon, J. J. 2021. Non-viral COVID-19 Vaccine Delivery Systems. *Advanced Drug Delivery Reviews*. 169:137–151.

Wang, W., Xu, Y., Gao, R. et al. 2020. Detection of SARS-CoV-2 in Different Types of Clinical Specimens. *Journal of American Medical Association*, 323 (18):1843–1844.

WHO. 2022. *WHO Coronavirus (COVID-19) Dashboard*. World Health Organization. https://covid19.who.int/

WHO and UNICEF. 2020. *Water, Sanitation, Hygiene, and Waste-Management for the COVID-19 Virus*. World Health Organization, Apr. 2020, Doc No. WHO/ 2019-nCoV/IPC_WASH/2020.2.https://apps.who.int/iris/handle/10665/331499

Chapter 4

Natural purification processes

4.1 IMPURITIES IN WATER

A wide variety of pollutants are present in natural waters. These include sand, silt, clay, organic matter from decaying vegetation, and products of chemical conversions, among others. Natural purification processes are continually active in streams and rivers to reduce the levels of the pollutants to acceptable or negligible concentrations. These processes include dilution, sedimentation, filtration, heat transfer, and chemical and biological conversions. These natural purification mechanisms are slow and can restore the health of water bodies over a period of time, depending on the concentration of the pollutants.

As human and industrial activity has increased, so has the amount of pollutants discharged into the water bodies. Various types of industrial chemicals, fertilizers, and pesticides end up in water bodies. In most cases, natural purification processes become insufficient to reduce the levels of pollutants. As a result, the health of the water body becomes impaired. Environmental regulations are introduced and enforced in an effort to reduce the pollution of natural streams and rivers.

Engineered systems are used in wastewater treatment plants to reduce the pollutant concentrations to acceptable levels prior to discharge to streams and rivers. These systems are designed based on the principles of natural purification processes. The difference lies in the rates of reaction and conversion. In treatment plants, unit processes are designed to achieve conversions within a short period of time. For this reason, it is essential to understand the basic principles and kinetics of natural purification processes, as well as quantification of the pollution strength of wastewater.

4.2 DILUTION

Dilution is a process whereby the concentration of pollutants is reduced due to the mixing of a small volume of polluted water with a large body of water, e.g. a stream or river. This usually happens when a wastewater is

DOI: 10.1201/9781003134374-4

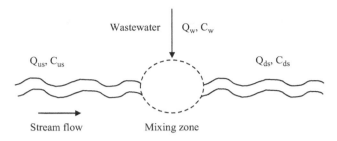

Figure 4.1 Stream flow with wastewater discharge.

discharged into a stream. If the stream has a low or negligible amount of pollutants, and its volume flow rate is much greater than the wastewater, dilution will take place and is reflected in downstream water characteristics. Low pollutant concentrations, adequate mixing, temperature, as well as hydraulic characteristics will dictate the success of dilution.

The principles of continuity and mass balance can be used to calculate the dilution capacity of a stream. Consider the following example, as illustrated in Figure 4.1.

A wastewater is discharged into a stream at a flow rate Q_w with a concentration C_w of a pollutant. Prior to discharge, the stream flow rate was Q_{us} with a concentration C_{us} of the pollutant. Assuming complete mixing at the point of discharge and no accumulation or chemical conversion, we can calculate the downstream flow rate Q_{ds} and concentration C_{ds}, of the mix after discharge.

From the principle of Continuity,

$$Q_w + Q_{us} = Q_{ds} \tag{4.1}$$

From the principle of mass balance,

$$\left(\text{Mass flow rate of pollutants}\right)_{in} = \left(\text{Mass flow rate of pollutants}\right)_{out} \tag{4.2}$$

$$Q_w.C_w + Q_{us}.C_{us} = Q_{ds}.C_{ds} \tag{4.3}$$

Example 4.1: A tanning industry discharges wastewater with ammonia into a stream as illustrated in Figure 4.1. Prior to discharge, the flow rate of the stream is 30 m³/s with an ammonia concentration of 0.2 mg/L. The flow rate of the industrial discharge is 1.3 m³/s with an ammonia concentration of 50 mg/L. Calculate the resultant flow rate and ammonia concentration downstream from the point of discharge.

SOLUTION

Calculate resultant flow rate using equation of continuity (Equation 4.1)

$$Q_{ds} = 30\,m^3/s + 1.3\,m^3/s$$

$$Q_{ds} = 31.3\,m^3/s$$

Write a mass balance between upstream and downstream points (Equation 4.2)

$$\left(\text{Mass flow rate of ammonia}\right)_{in} = \left(\text{Mass flow rate of ammonia}\right)_{out}$$

$$\left(30\,m^3/s \times 0.2\,mg/L\right) + \left(1.3\,m^3/s \times 50\,mg/L\right) = 31.3\,m^3/s \times C_{ds}$$

$$C_{ds} = 2.27\,mg/L$$

4.3 SEDIMENTATION

Sedimentation is a process that involves the removal of suspended solids from a water body by settling them out. The size of the solid particles plays a major role in the efficiency of sedimentation. Larger particles settle out quickly, whereas smaller particles may remain suspended for longer periods and eventually settle out. Stream characteristics, such as flow rates, bed depth, and roughness, also affect the rates of sedimentation. Excessive turbulence or flooding can cause resuspension of deposited solids. This can transfer solids deposits from one location to another.

4.4 MICROBIAL DEGRADATION

Wastewater discharged from municipal sources contains a large amount of organic matter. When untreated wastewater is discharged into streams and rivers, the organic matter is used as food by bacteria, protozoa, and other microorganisms in the water bodies. Aerobic microorganisms use oxygen during the aerobic oxidation of organic matter. This creates a substantial oxygen demand in the water body and can lower the dissolved oxygen concentrations significantly. The oxygen in water bodies is replenished by transfer from the atmosphere.

4.5 MEASUREMENT OF ORGANIC MATTER

A number of different methods can be used to measure the organic content of wastewater. The commonly used techniques include (i) Biochemical Oxygen Demand (BOD), (ii) Chemical Oxygen Demand (COD), and (iii) Total Organic Carbon (TOC).

BOD is the most widely used parameter for the measurement of the amount of biodegradable organic matter present in a wastewater. Standard BOD test results are obtained after 5 days. This is discussed in more detail in the following section.

COD is defined as the oxygen equivalent of organic matter that can be oxidized by a strong chemical oxidizer in an acidic medium. COD measures both biodegradable and non-biodegradable organic matter. The results can be obtained in a few hours.

The TOC test measures the total amount of organic carbon that can be oxidized to carbon dioxide in the presence of a catalyst. The test can be performed rapidly and results obtained in a short period of time.

4.5.1 Biochemical oxygen demand (BOD)

The BOD is used as a measure of the pollution potential of wastewater. It gives us an idea of the amount of biodegradable organic matter that is present in a wastewater. BOD is defined as the amount of oxygen utilized by a mixed population of microorganisms during aerobic oxidation of organic matter at a controlled temperature of 20°C for a specified time.

Theoretically, it would take an infinitely long time for the microorganisms to degrade all the organic matter present in the sample. The BOD value is time-dependent. Within a 20-day period, the oxidation of the carbonaceous organic matter is about 95% complete. In the wastewater industry, the BOD_5 in mg/L of O_2 is used as a standard value that is obtained from a BOD test conducted for 5 days. About 60–70% of the organic matter is oxidized after 5 days. A measure of the total amount of organic matter present in the sample is obtained from the ultimate BOD or BOD_{ult}.

If the wastewater contains proteins and other nitrogenous matter, the nitrifying bacteria will also exert a measurable demand after 6–7 days. The delay in the exhibition of the nitrogenous oxygen demand (NOD) is due to the slow growth rate of the nitrifying bacteria, as compared to the growth rate of the heterotrophic bacteria responsible for the exertion of the carbonaceous oxygen demand typically known as carbonaceous BOD or simply BOD. Figure 4.2 illustrates a typical BOD curve and NOD curve.

4.5.1.1 BOD kinetics

The rate at which organic matter is utilized by microorganisms can be assumed to be a first order reaction (Peavy et al., 1985). In other words,

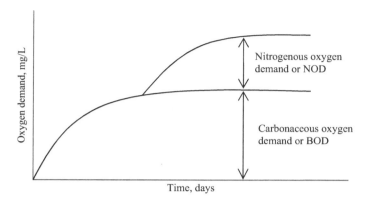

Figure 4.2 Typical BOD and NOD curves.

the rate at which organic matter is utilized is proportional to the amount of organic matter remaining. This can be expressed as

$$\frac{dL_t}{dt} = -kL_t \tag{4.4}$$

where
L_t = oxygen equivalent of organic matter remaining at time t, mg/L
k = reaction rate constant, d^{-1}

Equation 4.4 can be rearranged and integrated as

$$\int_{L_0}^{L_t} \frac{dL_t}{L_t} = -k\int_0^t dt \tag{4.5}$$

$$\ln \frac{L_t}{L_0} = -kt$$

$$L_t = L_0 e^{-kt} \tag{4.6}$$

where
L_0 = oxygen equivalent of total organic matter at time 0.

Figure 4.3 illustrates the relationship of organic matter remaining to the exertion of BOD. The amount of organic matter decays exponentially with time. Since L_0 is the oxygen equivalent of the total amount of organic matter, the amount of oxygen used in the degradation of organic matter, or the BOD, can be determined from the L_t value. Therefore,

$$BOD_t = L_0 - L_t \tag{4.7}$$

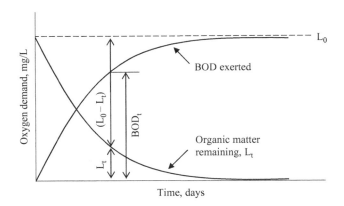

Figure 4.3 Organic matter remaining and BOD exertion curves.

Substituting the value of L_t from Equation (4.6) in Equation (4.7),

$$BOD_t = L_o - L_o e^{-kt} \qquad (4.8)$$

The BOD_{ult} of the wastewater approaches L_o in an asymptotic manner, indicating that the ultimate BOD is equal to the initial total amount of organic matter present in the sample, as shown in Figure 4.3. Equation (4.8) can be written as

$$BOD_t = BOD_{ult}\left(1 - e^{-kt}\right) \qquad (4.9)$$

Another form of the BOD equation can be written as follows when Equation 4.5 is simplified using log base 10:

$$BOD_t = BOD_{ult}\left(1 - 10^{-k't}\right) \qquad (4.10)$$

where k' = BOD rate constant (base 10) corresponding to Equation (4.10).
     In Equations (4.9) and (4.10), BOD_{ult} is a constant for a particular wastewater, regardless of time or temperature, since it corresponds to the total amount of organic matter initially present in the sample. Typical values of BOD_{ult} for municipal wastewater can range from 100 mg/L to 300 mg/L or more. The value of the BOD rate constant k (or k') represents the rate of the reaction and is temperature-dependent. Since microorganisms are more active at higher temperatures, the k value increases with temperature. The van't Hoff–Arrhenius model can be used to determine k, when k at 20°C is known.

$$k_T = k_{20}\,\theta^{(T-20)} \qquad (4.11)$$

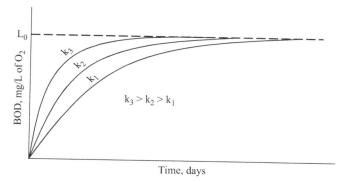

BOD, mg/L of O$_2$

L$_0$

k$_3$

k$_2$

k$_1$

k$_3$ > k$_2$ > k$_1$

Time, days

Figure 4.4 Variation of BOD curves with different rate constants.

where θ = Arrhenius coefficient; 1.047 is often used for BOD.

The value of k can vary from 0.1 to 0.4 or more, depending on the biodegradability of the organic matter. Sugars and simple carbohydrates that are easily degraded by microorganisms have a higher k value, as compared to complex compounds and fats that are difficult to degrade and have a lower k value. Figure 4.4 illustrates BOD curves for wastewaters with the same ultimate BOD, but with different rate constants.

Example 4.2: The BOD$_5$ of a municipal wastewater is 200 mg/L at 20°C. The amount of organic matter remaining in the sample after 5 days is equivalent to 151.93 mg/L of O$_2$. Calculate the BOD$_8$ of the sample at 30°C. Use θ = 1.047 as the Arrhenius coefficient for the BOD rate constant.

SOLUTION

Given, BOD$_5$ at 20°C = 200 mg/L, and L$_5$ = 151.93 mg/L, calculate L_o using Equation (4.7).

$$BOD_5 = L_o - L_5$$

$$200 = L_o - 151.93$$

Therefore, $BOD_{ult} = L_o = 200 + 151.93 = 351.93 \, mg / L$

Calculate k_{20} using Equation (4.9) with t = 5 d and $k = k_{20}$

$$BOD_5 = BOD_{ult}\left(1 - e^{-k.5}\right)$$

$$200 = 351.93\left(1 - e^{-k.5}\right)$$

Therefore, $k = k_{20} = 0.168\,\mathrm{d}^{-1}$

Calculate k_{30} using Equation (4.11) with $T = 30$, and $\theta = 1.047$

$$K_{30} = 0.168 \times (1.047)^{30-20} = 0.266\,\mathrm{d}^{-1}$$

Calculate BOD_8 at 30°C using Equation (4.9) with $t = 8$ d and $k = k_{30}$

$$BOD_8 = BOD_{ult}\left(1 - e^{-k.8}\right)$$

$$= 351.93\left(1 - e^{-0.266 \times 8}\right)$$

$$= 310.02\,\mathrm{mg/L}$$

4.5.1.2 Laboratory measurement

There are two types of tests that are used to determine the BOD of a wastewater sample in the laboratory. They are (i) Unseeded BOD test, and (ii) Seeded BOD test. The unseeded test is used for wastewater that has a sufficient population of microorganisms in it to exert a measurable oxygen demand for 5 days or more. The seeded test is used for wastewater that does not have enough microorganisms in it to exert a measurable demand during the test. Additional seed microorganisms are added to the sample.

Two criteria have to be satisfied for a valid BOD test (AWWA et al., 2017): (i) At least 2 mg/L of dissolved oxygen should be consumed by the microorganisms after 5 days, and (ii) final dissolved oxygen of the sample should not be less than 1 mg/L.

4.5.1.3 Unseeded BOD test

The BOD test is carried out in 300 mL BOD bottles according to Standard Methods (AWWA et al., 2017). A measured volume of wastewater is added to the bottle, together with dilution water. The dilution water is prepared by adding phosphate buffer, magnesium sulfate, calcium chloride, and ferric chloride. The water is then saturated with oxygen. The wastewater supplies the organic matter and microorganisms; and dilution water provides the oxygen and nutrients. An inhibitor such as 2-chloro-6 (trichloromethyl) pyridine can be added to the BOD bottle to prevent nitrification reactions. The bottle is incubated at 20°C for a specific time. Depletion of dissolved oxygen in the test bottle is measured daily, to determine the oxygen used by the microorganisms in degrading the organic matter. The BOD after time t days is calculated from the following equation:

$$\text{BOD}_t = \frac{D_1 - D_2}{P} \tag{4.12}$$

where

D_1 = Initial dissolved oxygen concentration in the BOD bottle, mg/L
D_2 = Final dissolved oxygen concentration in the BOD bottle after t days, mg/L

$$P = \frac{\text{volume of wastewater sample}}{\text{volume of BOD bottle}} = \frac{x\,\text{mL}}{300\,\text{mL}}$$

The volume of wastewater (x mL) that is added to the BOD bottle depends on the BOD of the wastewater, which is an unknown quantity. A range of BOD is assumed, based on which tests are conducted with a number of different x values. Using the two criteria mentioned above, and assuming the initial dissolved oxygen concentration to be close to the saturation concentration of 9.17 mg/L at 20°C, Equation (4.12) can be used to calculate a range of appropriate x values. Table 4.1 presents wastewater sample volumes to be used for different BOD values.

Table 4.1 Wastewater sample volumes for BOD tests

Wastewater sample (x mL)	Range of BOD (mg/L)
0.2	3000–12,000
0.5	1200–4800
1.0	600–2400
2.0	300–1200
5.0	120–480
10.0	60–240
20.0	30–120
50.0	12–48
100.0	6–24
300.0	2–8

Note: Values were calculated using Equation (4.12) and D_1 = 9.17 mg/L at 20°C.

Example 4.3: In a laboratory BOD test, 8 mL of wastewater with no dissolved oxygen is mixed with 292 mL of dilution water containing 8.9 mg/L dissolved oxygen in a BOD bottle. After 5 days of incubation, the dissolved oxygen content of the mixture is 3.4 mg/L. What is the BOD_5 of the wastewater?

SOLUTION

Calculate initial dissolved oxygen D_1 of the mixture from mass balance (Equation 4.3)

$$D_w \cdot V_w + D_d \cdot V_d = D_1 \cdot V_1$$

$$(0\,mg/L \times 8\,mL) + (8.9\,mg/L \times 292\,mL) = (D_1 \times 300\,mL)$$

$$D_1 = 8.66\,mg/L$$

Calculate BOD_5 using Equation (4.12)

$$BOD_5 = \frac{(8.66 - 3.4)\,mg/L}{\frac{8}{300}} = 197.25\,mg/L$$

4.5.1.4 Seeded BOD test

When wastewater does not have enough microbial population in it to exert a measurable oxygen depletion during the BOD test, seed microorganisms are added to the mixture. Activated sludge from an aeration basin or wastewater from a stabilization pond can be used to provide the seed. A general rule of thumb is to add a volume of seed wastewater such that 5–10% of the total BOD of the mixture results from the seed alone (Hammer and Hammer, 2012). The BOD of the seed is calculated separately and subtracted from that of the mixture. The following equation is used to calculate the BOD of a seeded wastewater:

$$BOD_t = \frac{(D_1 - D_2) - (B_1 - B_2)f}{P} \tag{4.13}$$

where
   D_1 = Initial dissolved oxygen concentration of diluted seeded wastewater mixture, mg/L
   D_2 = Final dissolved oxygen concentration of diluted seeded wastewater mixture, mg/L
   B_1 = Initial dissolved oxygen concentration of seed mixture from seed BOD test, mg/L
   B_2 = Final dissolved oxygen concentration of seed mixture from seed BOD test, mg/L
   f = Ratio of seed volume in seeded wastewater mixture to seed volume used in seed BOD test

$$= \frac{\text{mL of seed in } D_1}{\text{mL of seed in } B_1} = 0.05 - 0.10$$

$$P = \frac{\text{mL of wastewater sample in } D_1}{300 \text{ mL}}$$

Example 4.4: Determine the BOD_5 of a food processing wastewater. The data from the seeded BOD test are as follows:

Volume of wastewater sample in seeded mixture = 15 mL
Volume of seed in seeded mixture = 1 mL
Initial dissolved oxygen of seeded mixture = 8.9 mg/L
Final dissolved oxygen of seeded mixture = 4.5 mg/L

For BOD test for seed conducted separately:

Volume of seed = 10 mL
Initial dissolved oxygen = 8.7 mg/L
Final dissolved oxygen = 5.1 mg/L

SOLUTION

$$f = \frac{1 \text{ mL}}{10 \text{ mL}} = 0.1$$

$$P = \frac{15}{300}$$

$$BOD_5 = \frac{(8.9 - 4.5) - (8.7 - 5.1) \times 0.1}{\dfrac{15}{300}} = 80.8 \text{ mg}/L$$

4.5.1.5 Determination of k and L_o

The values of the constants k and L_o or BOD_{ult} can be determined from a series of BOD measurements. A number of techniques can be used, e.g. (i) Thomas' graphical method (Droste and Gehr, 2018; and Thomas, 1950); (ii) Least-squares method (Metcalf and Eddy et al., 2013; and Moore et al., 1950); and (iii) Fujimoto method (Fujimoto, 1961), among others.

4.5.1.6 Thomas' graphical method

A BOD test is conducted for 7–10 days, and daily measurements are taken. For each day, BOD and (time/BOD)$^{1/3}$ are calculated. A plot of (time/BOD)$^{1/3}$

versus time is made, and the best-fit line is drawn. The best-fit line has a slope (S) and an intercept (I). The slope and intercept values are used to calculate k (base 10) and L_o from the following relationships. The derivations for Equations (4.14) and (4.15) are provided elsewhere (Droste and Gehr, 2018; and Thomas, 1950).

$$k = 2.61\frac{S}{I} \qquad (4.14)$$

$$L_0 = \frac{1}{2.3kI^3} \qquad (4.15)$$

This is an approximate method. It is not valid for BOD > $0.9L_o$, or after 90% of the BOD has been exerted.

4.5.2 Theoretical oxygen demand

The theoretical oxygen demand (ThOD) for a compound or substance can be determined from the chemical oxidation reactions of that compound. If the substance is a complex of carbohydrates and proteins, then the *total ThOD* is the sum of the *carbonaceous ThOD* and the *nitrogenous ThOD*. The *carbonaceous ThOD* is equivalent to the BOD_{ult} of the substance. The calculations are illustrated in the following example.

Example 4.5: A wastewater contains 250 mg/L of glucose ($C_6H_{12}O_6$) and 60 mg/L of NH_3–N. Calculate the total ThOD for the wastewater. Atomic weights: C 12, H 1, O 16.

Under anaerobic conditions, glucose is converted to carbon dioxide and methane

$$C_6H_{12}O_6 \rightarrow CO_2 + CH_4$$

Methane undergoes further oxidation to carbon dioxide and water

$$CH_4 + O_2 \rightarrow CO_2 + H_2O$$

SOLUTION

Adding the two chemical reactions and balancing the resultant reaction,

$$C_6H_{12}O_6 + 6O_2 \rightarrow 6CO_2 + 6H_2O$$

According to this reaction, 6 moles of O_2 are required to completely oxidize each mole of glucose.

$$\text{Molecular wt of } C_6H_{12}O_6 = (12 \times 6) + (1 \times 12) + (16 \times 6) = 180\,\text{g/mol}$$

$$\text{Molecular wt of } O_2 = 16 \times 2 = 32\,\text{g/mol}$$

Carbonaceous ThOD

$$= 250\,\text{mg/L}\,C_6H_{12}O_6 \times \frac{6\,\text{mol}\,O_2}{\text{mol}\,C_6H_{12}O_6} \times \frac{32\,\text{g}\,O_2}{\text{mol}\,O_2} \times \frac{1000\,\text{mg}}{\text{g}}$$
$$\times \frac{1\,\text{mol}\,C_6H_{12}O_6}{180\,\text{g/mol}} \times \frac{\text{g}}{1000\,\text{mg}}$$

$$= 266.67\,\text{mg/L} = BOD_{ult}$$

Nitrification reaction (balanced): $NH_3\text{--}N + 2O_2 \rightarrow NO_3^-\text{--}N + H^+ + H_2O$

According to this reaction, 2 moles of O_2 are required to completely oxidize each mole of NH_3-N. Note, the concentration of NH_3 is given in terms of N.

Nitrogenous ThOD

$$= 60\,\text{mg / L}\,NH_3 - N \times \frac{2\,\text{mol}\,O_2}{\text{mol}\,NH_3 - N} \times \frac{32\,\text{g}\,O_2}{\text{mol}\,O_2} \times \frac{1000\,\text{mg}}{\text{g}}$$
$$\times \frac{1\,\text{mol}\,NH_3 - N}{14\,\text{g N/mol}} \times \frac{\text{g}}{1000\,\text{mg}}$$

$$= 274.29\,\text{mg / L}$$

$$\textbf{Total ThOD} = \text{Carbonaceous ThOD} + \text{Nitrogenous ThOD}$$

$$= 266.67\,\text{mg/L} + 274.29\,\text{mg/L} = 540.96\,\text{mg/L}$$

4.6 DISSOLVED OXYGEN BALANCE

One of the most important parameters for maintaining a healthy ecology of natural streams and rivers is the dissolved oxygen (DO) concentration. Most aquatic plants and animals require a minimum concentration of 2 mg/L DO to survive. Game fish and other higher life forms require 4 mg/L or more for survival (Peavy et al., 1985).

When a wastewater with a high BOD is discharged into a stream, dissolved oxygen from the water is used up by the microorganisms in the degradation of BOD or organic matter. This results in a drop in dissolved oxygen concentration of the stream. The amount of oxygen that can be dissolved in water at a given temperature is defined as its equilibrium or saturation concentration, or solubility. This can be calculated using Henry's law (Mihelcic and Zimmerman, 2010). Equilibrium concentrations of oxygen in water at various temperatures and salinity values are provided in Table A-2. The difference between the saturation DO (DO_{sat}) and the measured actual stream DO concentration (DO_{stream}) is called the *dissolved oxygen deficit* (D).

$$D = DO_{sat} - DO_{stream} \qquad (4.16)$$

At equilibrium DO_{sat} is constant, so the rate of change of deficit $\left(\dfrac{dD}{dt}\right)$ is proportional to the rate of change of DO of the stream. The rate at which dissolved oxygen decreases is also proportional to the rate at which BOD is exerted. Thus, we can obtain the following relationship (Peavy et al., 1985):

$$r_D = k_1 L_t \qquad (4.17)$$

where
  r_D = rate of change of deficit due to oxygen utilization
  k_1 = BOD rate constant
  L_t = organic matter remaining after time t

The natural process of replenishment of dissolved oxygen is called *reaeration*, which is the rate at which oxygen is resupplied from the atmosphere. The dissolved oxygen deficit is the driving force for reaeration. The rate of reaeration (r_R) increases as the concentration of dissolved oxygen decreases. The rate of reaeration is a first order reaction with respect to the oxygen deficit (D). This can be written as

$$r_R = -k_2 D \qquad (4.18)$$

where
  k_2 = reaeration rate constant

If algae are present in the water, they can replenish the dissolved oxygen in the water in presence of sunlight, as they produce oxygen during photosynthesis. Excessive algal growths sometimes outweigh these benefits, since they can lead to eutrophication as explained in Chapter 3. Excess dissolved oxygen cannot be stored in the water for future use, and usually escapes to the atmosphere due to turbulence and wind action. Also, in the absence of sunlight especially at night, algae use dissolved oxygen from the stream during respiration. This can lead to significant oxygen deficits in the stream.

4.6.1 Dissolved oxygen sag curve

When a wastewater with a significant amount of organic matter is discharged into a stream or river, the dissolved oxygen level decreases and drops to a minimum value. As reaeration slowly replenishes the dissolved oxygen, over time and with distance, the stream DO level comes back to pre-discharge concentration. This is illustrated in Figure 4.5 and is known as the dissolved oxygen sag curve. Streeter and Phelps developed one of the earliest models of the dissolved oxygen sag curve in 1925. Their basic model will be discussed here, which predicted changes in the deficit as a function of BOD exertion and stream reaeration.

According to the model, the rate of change in deficit is a function of oxygen depletion due to BOD exertion and stream reaeration. This can be expressed mathematically as

$$\frac{dD}{dt} = r_D + r_R \tag{4.19}$$

$$= k_1 L_t - k_2 D \tag{4.20}$$

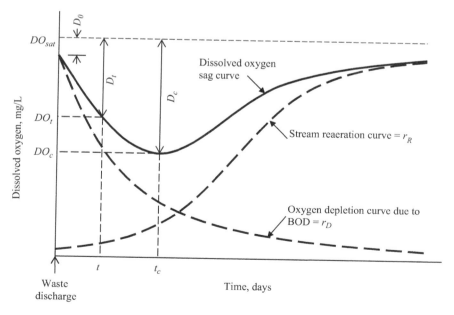

Figure 4.5 Dissolved oxygen sag curve.

Equation (4.20) can be written as a first order differential equation, integrated and solved using boundary conditions to obtain the following expression for the deficit at any time t (Peavy et al., 1985):

$$D_t = \frac{k_1 L_0}{k_2 - k_1}\left(e^{-k_1 t} - e^{-k_2 t}\right) + D_o e^{-k_2 t} \tag{4.21}$$

where

D_o = initial deficit, mg/L = DO_{sat} − $DO_{initial}$
L_o = BOD_{ult}, mg/L
t = time of travel in the stream from the point of discharge, d

If x is the distance traveled along the stream and v is the stream velocity, then

$$t = \frac{x}{v} \tag{4.22}$$

4.6.1.1 Critical points

The lowest point on the oxygen sag curve where the deficit is the greatest is called the *critical deficit* D_c (Figure 4.5). This point represents the maximum impact of the waste discharge on the dissolved oxygen content of the stream. If the BOD of the waste is too high, it may result in a deficit that causes *anaerobic* conditions in the stream, i.e. DO level goes to zero. The time taken to reach the critical deficit is called *critical time*, t_c, and the corresponding distance *critical distance* x_c. It is imperative to determine the deficit at the critical location. If standards are met at this location, they will be met at other locations too. Equation (4.21) is differentiated with respect to time, set to zero since D_c is maximum at t_c, and simplified to obtain

$$t_c = \frac{1}{k_2 - k_1}\ln\left\{\frac{k_2}{k_1}\left(1 - D_o\frac{k_2 - k_1}{k_1 L_o}\right)\right\} \tag{4.23}$$

An expression for critical deficit can be written in terms of critical time as

$$D_c = \frac{k_1}{k_2}L_o\, e^{-k_1 t_c} \tag{4.24}$$

Critical DO concentration, $DO_c = DO_{sat} - D_c$ (4.25)

DO Concentration at any time t, $DO_t = DO_{sat} - D_t$ (4.26)

Equations (4.21) to (4.26) can be used to determine the deficits and dissolved oxygen concentrations along a stream following waste discharge, and the oxygen sag curve can be produced. This is illustrated in the following example.

Example 4.6: An industrial process discharges its effluent into a stream. It is desired to determine the effects of the waste discharge on the dissolved oxygen concentration of the stream. k_1 at 20°C is 0.23 d^{-1}, and k_2 at 20°C is 0.43 d^{-1}. The characteristics of the stream and industrial wastewater are given in Table 4.2.

Table 4.2 Characteristics of the stream and industrial wastewater (for Example 4.6)

Characteristics	Stream	Industrial wastewater
Flow rate, m³/s	6.5	0.5
Temperature, °C	19.2	25
Dissolved oxygen, mg/L	8.2	0.5
BOD$_5$ at 20°C	3.0	200.0

a. Calculate the critical values and the distance from the point of discharge at which the critical values will occur. The stream velocity is 0.2 m/s.
b. Calculate the values and draw the dissolved oxygen sag curve for a 100 km reach of the stream from the point of discharge.

SOLUTION

Step 1: Determine the characteristics of the stream–wastewater mixture using mass balance.

Mix flow rate, $Q_m = 6.5 + 0.5 = 7.0 \, \text{m}^3/\text{s}$

Mix temp, $T_{m=} \dfrac{6.5 \times 19.2 + 0.5 \times 25.0}{6.5 + 0.5} = 19.61°C$

Mix DO, $DO_{m=} \dfrac{6.5 \times 8.2 + 0.5 \times 0.5}{6.5 + 0.5} = 7.65 \, \text{mg/L}$

Mix BOD$_5$, $BOD_{5m=} \dfrac{6.5 \times 3.0 + 0.5 \times 200.0}{6.5 + 0.5} = 17.07 \, \text{mg/L}$

DO_{sat} at T_m = 9.24 mg/L (using Table A-2, and interpolating between the DO_{sat} values for 19°C and 20°C)

$D_o = DO_{sat} - DO_m = 9.24 - 7.65 = 1.59 \, \text{mg/L}$

Calculate L_o or BOD_{ult} using Equation (4.9). Note: BOD_5 is always measured at 20°C, unless otherwise mentioned. So, use $k = k_{20}$ in Equation (4.9).

$$17.07 = L_o\left(1 - e^{-0.23 \times 5}\right)$$

or, $L_o = 24.98 \, \text{mg/L}$

Step 2. Determine reaction rate constants for mix temperature using Equation (4.11).

$$k_{1 \, \text{at} \, 19.61°C} = 0.23\left(1.047\right)^{(19.61-20)} = 0.226 \, \text{d}^{-1}$$

$$k_{2 \, \text{at} \, 19.61°C} = 0.43\left(1.016\right)^{(19.61-20)} = 0.427 \, \text{d}^{-1}$$

Step 3. Determine critical values using Equations (4.23) and (4.24).

$$t_c = \frac{1}{k_2 - k_1} \ln\left\{ \frac{k_2}{k_1}\left(1 - D_o \frac{k_2 - k_1}{k_1 L_o}\right) \right\}$$

$$= \frac{1}{0.427 - 0.226} \ln\left\{ \frac{0.427}{0.226}\left(1 - 1.59 \frac{0.427 - 0.226}{0.226 \times 24.98}\right) \right\}$$

$$= 2.87 \, \text{d}$$

$$D_c = \frac{k_1}{k_2} L_o \, e^{-k_1 t_c}$$

$$= \frac{0.226}{0.427} 24.98 \, e^{-(0.226)(2.87)}$$

$$= 6.91 \, \text{mg/L}$$

$$\text{DO}_c = \text{DO}_{\text{sat}} - D_c = 9.24 - 6.91 = 2.33 \, \text{mg/L}$$

Critical distance, $x_c = t_c \, v$

$$= \left(2.87 \, \text{d}\right) \times \left(0.2 \, \text{m/s} \times 86{,}400 \, \text{s/d} \times \text{km/1000 m}\right)$$

$$= \left(2.87 \, \text{d}\right) \times \left(17.28 \, \text{km/d}\right)$$

$$= 49.59 \, \text{km downstream from point of waste discharge}$$

Step 4. Determine the deficit at various points along the stream, e.g. 20, 40, 60, 80, and 100 km from the point of waste discharge. First, calculate the times corresponding to these distances.

$$\text{For } x = 20\,\text{km}, t_{20} = \frac{x\,\text{km}}{v\,\text{km/d}} = \frac{20\,\text{km}}{17.28\,\text{km/d}} = 1.16\,\text{d}$$

$$\text{For } x = 40\,\text{km}, t_{40} = \frac{40\,\text{km}}{17.28\,\text{km/d}} = 2.32\,\text{d}$$

$$\text{For } x = 60\,\text{km}, t_{60} = \frac{60\,\text{km}}{17.28\,\text{km/d}} = 3.48\,\text{d}$$

$$\text{For } x = 80\,\text{km}, t_{80} = \frac{80\,\text{km}}{17.28\,\text{km/d}} = 4.64\,\text{d}$$

$$\text{For } x = 100\,\text{km}, t_{100} = \frac{100\,\text{km}}{17.28\,\text{km/d}} = 5.80\,\text{d}$$

Next, calculate the deficits at these times using Equation (4.21).

$$D_t = \frac{k_1 L_o}{k_2 - k_1}\left(e^{-k_1 t} - e^{-k_2 t}\right) + D_o e^{-k_2 t}$$

$$D_{20} = \frac{0.226 \times 24.98}{0.427 - 0.226}\left(e^{-(0.226 \times 1.16)} - e^{-(0.427 \times 1.16)}\right) + 1.59\,e^{-(0.427 \times 1.16)}$$

$$= 5.46\,\text{mg/L}$$

$$D_{40} = 6.79\,\text{mg/L}$$

$$D_{60} = 6.80\,\text{mg/L}$$

$$D_{80} = 6.19\,\text{mg/L}$$

$$D_{100} = 5.35\,\text{mg/L}$$

Calculate the DO levels in the stream from the deficits.

$$\text{DO}_x = \text{DO}_{\text{sat}} - D_x$$

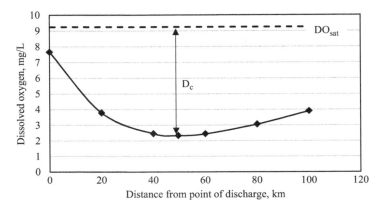

Figure 4.6 Dissolved oxygen sag curve (for Example 4.6).

$$DO_{20} = 9.24 - 5.46 = 3.78 \, \text{mg/L}$$

$$DO_{40} = 9.24 - 6.79 = 2.45 \, \text{mg/L}$$

$$DO_{60} = 9.24 - 6.80 = 2.44 \, \text{mg/L}$$

$$DO_{80} = 9.24 - 6.19 = 3.05 \, \text{mg/L}$$

$$DO_{100} = 9.24 - 5.35 = 3.89 \, \text{mg/L}$$

Using these values, the dissolved oxygen sag curve is drawn in Figure 4.6.

4.6.1.2 Limitations of the oxygen sag curve model

1. The model assumes only one source of BOD discharge into the stream. If there are multiple waste discharges along the stream, they have to be taken into account. The stream can be divided into segments consisting of a single source of waste discharge, and the model applied sequentially from the first to the last segment.
2. Oxygen demand due to nitrification or algal respiration is not taken into account.
3. The contribution of algae to reaeration is not considered.
4. Steady state conditions are assumed along the stream channel, resulting in the use of a single value of k_2. Streambed characteristics, slopes, impoundments, etc., are not considered.

Computer models have been developed in recent years based on the Streeter–Phelps model that have included the nitrification process, the diurnal effect of algal photosynthesis and respiration, as well as stream characteristics affecting reaeration rates.

PROBLEMS

4.1 A municipal wastewater is discharged into a stream as illustrated in Figure 4.1. Prior to discharge, the flow rate of the stream is 45 m³/s with a BOD_5 of 1.5 mg/L. Downstream from the point of discharge, the stream flow rate is 47.2 m³/s with a BOD_5 of 50 mg/L. Calculate the characteristics of the municipal discharge.

4.2 An industry discharges a wastewater that has 200,000 mg/L of dissolved solids into a stream. The wastewater flow rate is 100 L/min. The concentrations of dissolved solids upstream and downstream of the point of discharge are 50 mg/L and 1000 mg/L, respectively. Determine the upstream and downstream flow rates.

4.3 A stream flows through a small town where an industry discharges its effluent into the stream. Upstream characteristics prior to industrial discharge are as follows: flow rate 1000 m³/d, BOD_5 2.5 mg/L, nitrates 2.0 mg/L, and temperature 19°C. The industry discharges at 50 m³/d. According to regulatory requirements, the maximum allowable values in the stream following any discharge are BOD_5 30 mg/L, nitrates 4.0 mg/L, and temperature differential of 4°C from upstream conditions. Calculate the maximum allowable values for the industrial discharge.

4.4 For a BOD test, calculate the size of the sample expressed as a percent, if the 5-day BOD is estimated at 650 mg/L and total oxygen consumed in the BOD bottle is limited to 2 mg/L?

4.5 The BOD value of a wastewater was measured at 2 days and 8 days and found to be 125 and 225 mg/L, respectively. Determine the 5-day BOD value using the first order rate model.

4.6 For a BOD analysis, 30 mL of wastewater with a DO of zero mg/L, is mixed with 270 mL of dilution water with a DO of 9 mg/L. The sample is then placed in an incubator. The final DO is measured after 7 days. The final DO is measured at 3.0 mg/L. However, it is discovered that the incubator was set at 30°C. Assume a k_1 of 0.2 d⁻¹ (base e) at 20°C and $\theta = 1.05$. Determine the 5-day, 20°C BOD of the sample.

4.7 In a BOD test, the amount of organic matter remaining in the wastewater was measured at certain time intervals, instead of measuring the dissolved oxygen content. The amount of organic matter remaining after 4 and 9 days was measured as 52.86 mg/L and 8.11 mg/L, respectively. Calculate the ultimate BOD and BOD rate constant k for the wastewater.

4.8   An unseeded BOD test was conducted on a raw domestic wastewater sample. The wastewater portion added to each 300 mL BOD bottle was 8.0 mL. The dissolved oxygen values and incubation periods are listed in Table 4.3. Plot a BOD versus time curve and determine the 4-day BOD value.

Table 4.3 Dissolved oxygen values (for Problem 4.8)

Bottle number	Initial DO mg/L	Incubation days	Final DO mg/L
1	8.4	0	8.4
2	8.4	0	8.4
3	8.4	1	6.2
4	8.4	1	5.9
5	8.4	2	5.2
6	8.4	2	5.2
7	8.4	3	4.4
8	8.4	3	4.6
9	8.4	5	0.8
10	8.4	5	3.5

4.9   Compute the Ultimate Carbonaceous Oxygen Demand of a waste represented by the formula $C_9N_2H_6O_2$, and use the reaction below.

$$C_9N_2H_6O_2 + O_2 => CO_2 + NH_3$$

4.10 A BOD test was run on wastewater taken from the Blue Plains Advanced Wastewater Treatment Plant. The test was continued for 12 days at 20°C, and the dissolved oxygen content was measured every 2 days. A plot of $(t/BOD)^{1/3}$ versus t yielded a straight line. The equation of the straight line is

$$(t/BOD)^{1/3} = 0.015t + 0.18$$

   a. Calculate the BOD rate coefficient k and the ultimate BOD for this wastewater.
   b. Calculate the BOD_5 at 25°C.

4.11 If the BOD_5 of a wastewater is 350 mg/L, what are the maximum and minimum sample volumes that can be used for BOD measurement in an unseeded test? Mention any assumptions that you make.

4.12 The results of a BOD test conducted on a wastewater at 20°C are shown in Table 4.4. Calculate the ultimate BOD and the rate constant k.

Table 4.4 BOD test results (for Problem 4.12)

Time, d	0	1	2	3	4	5	6	7	8
BOD, mg/L	0	75	138	191	236	274	305	332	354

4.13 Explain the oxygen deficit of a stream. Write a mathematical formulation to describe the oxygen deficit.

4.14 What are the natural mechanisms by which oxygen can be replenished in surface waters? Describe them.

4.15 Why are algae not considered to be a reliable source of dissolved oxygen replenishment in surface waters?

4.16 An ice cream plant discharges its effluent wastewater into a stream. It is desired to determine the effects of the waste discharge on the dissolved oxygen concentration of the stream. k_1 at 20°C is 0.30 d^{-1}, and k_2 at 20°C is 0.45 d^{-1}. The characteristics of the stream and wastewater are given in Table 4.5. Calculate the critical values and the distance from the point of discharge at which the critical values will occur. The stream velocity is 0.2 m/s.

Table 4.5 Characteristics of the stream and wastewater (for Problem 4.16)

Characteristics	Stream	Industrial wastewater
Flow rate, m³/d	16,000	1200
Temperature, °C	12	40
Dissolved oxygen, mg/L	8.2	0
BOD$_5$ at 20°C	3.0	910.0

4.17 Using the data from problem 4.16, draw the dissolved oxygen sag curve for a 150 km reach of the stream downstream from the point of discharge.

4.18 Consider the ice cream plant described in problem 4.16. The BOD$_5$ of the wastewater is too high for discharge into the stream. It needs to be treated to reduce the BOD$_5$ to an acceptable level prior to discharge. Calculate the maximum BOD$_5$ that can be discharged from the ice cream plant, if a minimum of 4.0 mg/L dissolved oxygen has to be maintained in the stream at all times.

4.19 Use the data from problem 4.16. Calculate the maximum BOD$_5$ that can be discharged from the ice cream plant, if a minimum of 3.0 mg/L dissolved oxygen has to be maintained in the stream at all times.

4.20 Consider a one-mile stretch of a stream that has three separate sources of wastewater entering it at three different locations. Can you use the Streeter–Phelps model to calculate the BOD of the stream 1.5 miles downstream? Explain your process.

REFERENCES

AWWA, WEF, and APHA. 2017. *Standard Methods for the Examination of Water and Wastewater*. 23rd edn. Edited by Baird, R. B., Eaton, A. D., and Rice, E. W. American Water Works Association, USA.

Droste, R. L. and Gehr, R. L. 2018. *Theory and Practice of Water and Wastewater Treatment*. 2nd edn. John Wiley & Sons, Inc. Hoboken, NJ, USA.

Fujimoto, Y. 1961. Graphical Use of First-Stage BOD Equation. *Journal of Water Pollution Control Federation*, 36 (1):69–71.

Hammer, M. J. and Hammer, M. J. Jr. 2012. *Water and Wastewater Technology*. 7th edn. Pearson, Hoboken, NJ, USA.

Metcalf and Eddy, Tchobanoglous, G., Stensel, H., Tcuchihashi, R., and Burton, F. 2013. *Wastewater Engineering: Treatment and Resource Recovery*. 5th edn. McGraw-Hill, Inc., New York, NY, USA.

Mihelcic, J. R. and Zimmerman, J. B. 2010. *Environmental Engineering - Fundamentals, Sustainability, Design*. John Wiley & Sons, Inc., Hoboken, NJ, USA.

Moore, E. W., Thomas, H. A., and Snow, W. B. 1950. Simplified Method for Analysis of BOD Data. *Sewage and Industrial Wastes*, 22 (October):1343–1355.

Peavy, H. S., Rowe, D. R., and Tchobanoglous, G. 1985. *Environmental Engineering*. McGraw- Hill, Inc., New York, NY, USA.

Thomas, H. A. Jr. 1950. Graphical Determination of BOD Curve Constants. *Water & Sewage Works*, 97 (March):123–124.

Chapter 5

Wastewater treatment fundamentals

5.1 INTRODUCTION

The science and engineering of wastewater treatment has progressed tremendously over the last four to five decades. As the knowledge and understanding of the relationship between waterborne pathogens and public health have increased, so has the impetus for innovation of new technologies for the treatment of wastewater increased. In the last century, population growth and industrialization have resulted in significant degradation of the environment. Disposal of untreated wastes and wastewater on land, or in streams and rivers is no longer an option. Newer regulations are aimed at protecting the environment as well as public health.

Wastewater engineering has come a long way from the time when city residents had to place *night soil* (fecal waste) in buckets along the streets, and workers collected the waste and delivered them to rural areas for disposal on agricultural lands. With the invention of the flush toilet, *night soil* was transformed into *wastewater*. It was not feasible to transport these large liquid volumes for land disposal. So cities began to use natural drainage systems and storm sewers to transport the wastewater to streams and rivers, where they were discharged without any treatment. The common notion was "solution to pollution is dilution." However, with increasing urbanization the self-purification capacity of the receiving waters was exceeded, causing degradation of the water bodies and the environment. In the late 1800s and early 1900s, various treatment processes were applied to wastewater (Peavy et al., 1985). By the 1920s, treatment plants were designed and constructed for the proper treatment of wastewater prior to disposal. With newer and more stringent regulations, existing processes are modified and innovative technologies are introduced to achieve enhanced removal of pollutants.

The objectives of wastewater treatment are to reduce (i) the level of solids, (ii) the level of biodegradable organic matter, (iii) the level of pathogens, and (iv) the level of toxic compounds in the wastewater, to meet regulatory limits that are protective of public health and the environment.

DOI: 10.1201/9781003134374-5

5.2 SOURCES OF WASTEWATER

The following are common sources or types of wastewater.

- *Domestic or municipal wastewater*: This includes wastewater discharged from residences, institutions such as schools and hospitals, and commercial facilities such as restaurants, shopping malls, etc.
- *Industrial wastewater*: Wastewater discharged from industrial processes, e.g. pharmaceutical industry, poultry processing, etc.
- *Infiltration and Inflow*: This includes water that eventually enters the sewer from foundation drains, leaking pipes, submerged manholes, and groundwater infiltration, among others.
- *Storm water*: Rainfall-runoff and snowmelt.

Municipal wastewater is usually collected in *sanitary sewers* and transported to the wastewater treatment plant. Storm water may be collected in separate sewer lines called *storm sewers*. In some cities, especially older cities, storm water is collected in the same sewer line as the domestic wastewater. This type of system is called a *combined sewer* system. There are advantages and disadvantages of each system. Industrial wastewater may be treated onsite, or pretreated and then discharged to sanitary sewers, after appropriate removal of pollutants.

5.3 WASTEWATER FLOW RATE

The fundamental design parameter of wastewater treatment facilities is the determination of design capacity, which is directly related to the flow rate of wastewater. The wastewater flow rate varies, largely depending on the type of community and relative distribution of residential, commercial, and institutional facilities, and types of industries. It is a direct function of the water demand. The determination of a design flow rate depends on 1) design period, 2) population prediction, 3) estimation of wastewater flow, 4) estimation of infiltration and inflow (I&I), and 5) variability of wastewater flow.

5.3.1 Design period

The design period is the length of time the facility will be able to meet the design capacity. Wastewater treatment facilities are generally developed with federal, state, and local government funds; as a result, the selection of design periods is often governed by the regulatory constraints. The newly developed facilities are generally designed to meet the generated flow rate for the future; thus, it is important to estimate the future population, although other factors such as interest rate of bonds, life expectancy of the structure, and difficulties of expansion in the future are also considered. A typical design period of 20–25 years is considered for major facilities that are difficult or expensive to expand, whereas a 10–15-year design period is considered for equipment that are easy to replace. For major sewer lines (trunk and interceptors), a design period of 20–25 years is usually considered (Davis, 2011).

Generally for a long design period, special consideration must be given during the earlier years of the facility. The characteristics of the wastewater, efficiency of the unit operation, and energy efficiency of the equipment may be impacted due to low flow rates in the early stages. Modular units and equipment can be considered and installed as needed with an increase of flow rate over time.

5.3.2 Population projection

Population data and projection, although very complex, are beneficial for many policy decisions including the determination of wastewater flow rate for the design period of the facilities. Various population projection models are available for short-term (10–15 yr) and long-term (15–50 yr) forecasts using data extrapolation and other advanced techniques. Population data can be obtained from the Bureau of Census in the US, EUROSTAT in the European Union, United Nations, or World Bank Data for many developing countries around the world. Some of the models presented here can project population using data extrapolation with sufficient accuracy for short-term planning purposes, although there are more advanced models available (Roser, 2019; McJunkin, 1964) for long-term projections.

5.3.2.1 Constant growth method

This is also known as the algebraic projection or arithmetic progression method. The population is predicted using the census data with an equal incremental increase or constant growth rate. The prediction model is given by

$$P_t = P_0 + k\Delta t \tag{5.1}$$

where
P_0 = current population
P_t = projected population after Δt years
k = growth rate

5.3.2.2 Log growth method

This is also known as exponential growth or geometric growth method. A constant percentage of growth rate is assumed for equal intervals of time. The projection equation is

$$P_t = P_0 e^{k\Delta t} \qquad (5.2)$$

Or,

$$\ln P_t = \ln P_0 + k\Delta t \qquad (5.3)$$

All the terms are as defined previously.

5.3.2.3 Percent growth method

This model uses a percent growth rate to predict the population compounded in a specific time period, generally annually. The prediction equation is

$$P_t = P_0 (1+k)^n \qquad (5.4)$$

where
n = number of compounding periods

All other terms are as defined previously.

5.3.2.4 Ratio method

This method projects the population assuming the ratio of the population of the area under study to that of a larger region, will continue to change in the future in the same manner as in the past. The prediction equation is

$$\frac{P_2}{P_{2R}} = \frac{P_1}{P_{1R}} = k \qquad (5.5)$$

where
P_2 = predicted population
P_{2R} = predicted population of a larger region
P_1 = population at last census
P_{1R} = population of a larger region at last census
k = growth ratio constant

5.3.2.5 Declining growth method

This method assumes that a county or city has a limiting or saturation population, and the population growth is a function of the population deficit. The prediction equation is

$$P_t = P_0 + (P_{sat} - P_0)(1 - e^{-k\Delta t}) \qquad (5.6)$$

where

P_{sat} = saturation population
k = growth rate constant
$(P_{sat} - P_o)$ = population deficit

All other terms are as defined previously.

> Example 5.1: The population of the Kansas City Metro Area over a 40-year period is provided in Table 5.1. Estimate the population of Kansas City in 2030 using 1) constant growth method, 2) log growth method, 3) percent growth method, and 4) declining growth method with a saturation population of 2,000,000.

Table 5.1 Population of Kansas City from 1980 to 2020 (for Example 5.1)

Year	Population
1980	1,075,000
1990	1,233,000
2000	1,365,000
2010	1,524,000
2020	1,686,000

Source: United Nations (2021).

SOLUTION

1. Constant growth method:
   Assume the population will grow at the same rate as it did from 2010 to 2020.
   Use population data from 2010 and 2020,

$$P_t = P_o + k\Delta t$$

or $P_{2020} = P_{2010} + k(2020 - 2010)$

or $1,686,000 = 1,524,000 + k \times 10$

or $k = 16,000 / \text{year}$

Population in 2030,

$$P_{2030} = P_{2020} + k(2030 - 2020)$$

$$= 1,686,000 + 16,000 \times 10$$

$$= 1,846,000$$

2. Log growth method:
    Use population data from 2010 and 2020,

$$\ln P_t = \ln P_\mathrm{o} + k\Delta t$$

$$\text{or } \ln P_{2020} = \ln P_{2010} + k(2020 - 2010)$$

$$\text{or } 14.34 = 14.24 + k \times 10$$

$$\text{or } k = 0.01 \,/\, \text{year}$$

Population in 2030,

$$\ln P_{2030} = \ln P_{2020} + k(2030 - 2020)$$

$$= 14.34 + 0.01 \times 10$$

$$= 14.44$$

$$P_{2030} = e^{14.44} = 1,865,220$$

3. Percent growth method:
    Use population data from 2010 and 2020

$$P_t = P_\mathrm{o}(1 + k)^n$$

$$\text{or } P_{2020} = P_{2010}(1 + k)^{(2020 - 2010)}$$

$$\text{or } 1,686,000 = 1,524,000(1 + k)^{10}$$

$$\text{or } k = 0.0102$$

Population in 2030,

$$P_{2030} = P_{2020}\left(1+k\right)^{(2030-2020)}$$

$$= 1,686,000\left(1+0.0106\right)^{10}$$

$$= 1,865,160$$

4. Declining growth method:
   Use population data from 2010 and 2020

$$P_{2020} = P_{2010} + \left(P_{sat} - P_{2010}\right)\left(1 - e^{-k\Delta t}\right)$$

$$\text{or } 1,686,000 = 1,524,000 + \left(2,000,000 - 1,524,000\right) \times \left(1 - e^{-k(2020-2010)}\right)$$

$$\text{or } k = 0.0416$$

Population in 2030,

$$P_{2030} = P_{2020} + \left(S - P_{2020}\right)\left(1 - e^{-k(2030-2020)}\right)$$

$$= 1,686,000 + \left(2,000,000 - 1,686,000\right)\left(1 - e^{-(0.0416)(10)}\right)$$

$$= 1,792,860$$

5.3.3 Wastewater flow

5.3.3.1 Residential wastewater flow

Municipal or residential wastewater flow rates are generally determined from the total water demand for the projected population in the design period, and 60–90% of the water withdrawal rate can be used for the estimated wastewater production. Historic records of water use or wastewater production in the community or a nearby similar community can provide a good estimate in the absence of population data. Various factors such as geographic location, climate condition, presence of commercial and industrial facilities, water conservation practices and regulations, and population characteristics must be considered for calculating the wastewater flow for a community. For example, average household water use is 47 Lpcd (liters per capita per day) in Africa, 95 Lpcd in Asia, 337 Lpcd in the UK, and 578 Lpcd in the US (FAO, 2021).

Shammas and Wang (2011) recommended that 70% of the water brought into the community must be removed as spent water. The average flow in sanitary sewers is about 378 Lpcd, or 100 gpcd (gallons per capita per day), in North America. An average domestic daily flow of 380 Lpcd was recommended by GLUMRB (2004) for sizing of facilities receiving flows from new wastewater collection systems, plus wastewater flows from commercial, institutional, and industrial facilities.

5.3.3.2 Commercial and institutional wastewater flow

Wastewater flow from commercial and institutional facilities can range from 7.5 to 14 m³/ha.d (cubic meter per hectare per day) or from 800 to 1500 gal/ac.d (gallons per acre per day). Typical wastewater flow rates from various commercial, institutional, and recreational facilities are provided in Metcalf and Eddy et al. (2013), although considerations must be provided for the region, climate, and type of facilities.

5.3.3.3 Industrial wastewater flow

Wastewater from industrial sources vary largely with the type and size of facilities, degree of recycling practiced within the facilities, and the presence of onsite wastewater treatment facilities. Typical design values of 7.5–14 m³/ha.d, or 800–1500 gal/ac.d can be used for small industrial facilities, and 14–28 m³/ha.d, or 1500–3000 gal/ac.d for estimating large industrial wastewater flows. For a facility without internal water recirculation, 85–95% of the water used for various operations and processes can be considered as wastewater flow. Average domestic wastewater contribution from industrial facilities per employee may vary from 30 to 95 Lpcd (Metcalf and Eddy et al., 2013).

5.3.4 Infiltration and inflow

In addition to the domestic and industrial sources, a significant amount of wastewater flow can be contributed from I&I. Wastewater entering the sewer system through faulty service connection, defective pipes, pipe joints, and manhole walls are considered infiltration; whereas water discharged from cellar and foundation drains, roof downspouts, basement, yard and area drains, cooling water discharge, manhole covers, drains from springs and swampy areas, catch basins, storm water, surface runoff, or street wash water are defined as inflow. The presence of high groundwater levels and age of the sewer network can significantly impact the I&I contribution to the total wastewater flow. Infiltration values may range from 0.2 to 28 m³/ha.d (20–3000 gal/ac.d) based on the area served, and may exceed 500 m³/ha.d (50,000 gal/ac.d) during heavy rains (Metcalf and Eddy, 1981). New sewer lines have much less contribution from infiltration as compared to old sewer lines. A peak infiltration rate of 14 m³/ha.d can be considered for a new sewer line, where the contribution may increase up to 48.5 m³/ha.d for old sewer lines for a 10 ha service area. As many sources of inflow cannot be identified separately, they are generally included within the estimation of infiltration.

5.3.5 Variability of wastewater flow

Wastewater flow varies during different times of the day (diurnal flow), days of the week (weekday or weekends), seasons of the year (dry and wet seasons), and sources and rate of I&I. Flow from a smaller community and short-term variations have higher fluctuations, compared to large communities and long periods of time. The variations are normally reported as a factor of average day, and various flow rate terms are defined, e.g. average daily flow rate, minimum hourly flow rate, minimum daily flow rate, minimum monthly flow rate, peak hourly flow rate, maximum daily flow rate, and maximum monthly flow rate, each having a different design and operational applications. The average daily flow rate is used for estimating pumping, chemical costs, and sludge production in a wastewater treatment plant; maximum daily flow rate is used for equalization tank design, and peak hourly flow rate is used for sanitary sewer design as well as sizing of physical units such as grit chambers and filters. Figure 5.1 provides a ratio method for estimating different types of flow rates from the average daily flow rate.

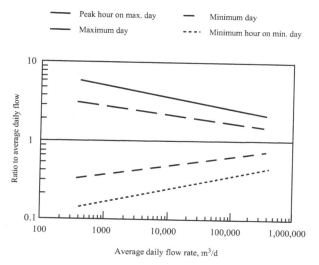

Figure 5.1 Ratio of different types of flow rates to average flow rate.

Source: Adapted from Metcalf and Eddy et al., 2013.

5.4 WASTEWATER CONSTITUENTS

The major constituents of municipal wastewater are suspended solids, organic matter, and pathogens. Nutrients such as nitrogen and phosphorus can cause problems when present in high concentrations. In recent years, the presence of EDCs (endocrine-disrupting compounds) has been recognized as an area of concern. Industrial wastewater can contain the above-mentioned contaminants, as well as heavy metals, toxic compounds, and refractory organics. Storm water may contain petroleum compounds, silt, and pesticides when it includes urban runoff and agricultural runoff. Table 5.2 provides the environmental impacts of the major constituents of wastewater.

Suspended solids consist of inert matter such as rags, silt, and paper, as well as food waste and human waste. Biodegradable organic matter is composed of 40–60% proteins, 25–50% carbohydrates, and about 10% lipids (Peavy et al., 1985). Proteins are mainly amino acids and contain nitrogen. Carbohydrates are sugars, starches, and cellulose. Lipids include fats, oils, and grease. All of these exert an oxygen demand. Table 5.3 presents the typical characteristics of untreated municipal wastewater.

The constituents of industrial wastewaters vary widely depending on the type of industry and the processes used in manufacturing the product. In the US, the Environmental Protection Agency (EPA) has grouped the pollutants into three categories. They are *conventional pollutants* such as pH, BOD_5, TSS, oil, and grease; *nonconventional pollutants* such as COD, ammonia, hexavalent chromium, phenols, etc.; and *priority pollutants* such as arsenic, cadmium, etc. The complete list can be found in the Code of Federal Regulations (e-CFR, 2021; Federal Register, 2010). Table 5.4 presents the selected characteristics of a number of industrial wastewaters.

Table 5.2 Environmental impacts of major wastewater pollutants

Pollutants	Source	Environmental impact on receiving waters
Suspended solids	Municipal wastewater, storm water	Scum layer on water surface, sludge deposits
Organic matter	Municipal wastewater, industrial wastewater	Dissolved oxygen depletion, possible anaerobic conditions, fish kills
Nutrients	Municipal wastewater, industrial wastewater	Eutrophication and impairment of water quality
Pathogens	Municipal wastewater	Transmission of diseases
Heavy metals	Industrial wastewater	Toxic to aquatic life
Refractory organics	Industrial wastewater	May be toxic or carcinogenic
Endocrine-disrupting compounds	Municipal wastewater	Feminization of fish, possible broader scope of impacts

Source: Adapted from Peavy et al. (1985).

Table 5.3 Typical characteristics of untreated municipal wastewater

Components	Concentration range
Biochemical oxygen demand, BOD_5 at 20°C	100–360 mg/L
Chemical oxygen demand, COD	250–1000 mg/L
Total organic carbon, TOC	80–300 mg/L
Total Kjeldahl nitrogen (TKN)	20–85 mg/L as N
Total phosphorus	5–15 mg/L as P
Oil and grease	50–120 mg/L
Total solids (TS)	400–1200 mg/L
Total dissolved solids (TDS)	250–850 mg/L
Total suspended solids (TSS)	110–400 mg/L
Volatile suspended solids (VSS)	90–320 mg/L
Fixed suspended solids (FSS)	20–80 mg/L
Settleable Solids	5–20 mL/L
Total Coliform	$10^6–10^{10}$ No./100 mL
Fecal Coliform	$10^3–10^8$ No./100 mL

Source: Adapted from Metcalf and Eddy et al. (2013).

Table 5.4 Typical characteristics of selected industrial wastewaters

Industry	BOD, mg/L	COD, mg/L	TSS, mg/L
Milk processing	1300	3100	300
Meat processing	1400	2500	950
Pulp and paper (kraft)	300	550	250
Tannery	4000	7500	15,000
Slaughterhouse (cattle)	2000	3600	800
Cheese production	3000	5500	950
Pharmaceuticals	280	390	160

Source: Adapted from Hammer and Hammer (2012), and Davis (2011).

5.5 WASTEWATER TREATMENT METHODS

Wastewater can be treated using any or a combination of the following types of treatment methods, depending on the nature of pollutants and the level of desired removal.

5.5.1 Physical treatment

This involves the removal of pollutants from the wastewater by simple physical forces, e.g. sedimentation, screening, filtration, etc. Physical treatment processes are used mainly for the removal of suspended solids.

5.5.2 Chemical treatment

This type of treatment involves the addition of chemicals to achieve conversion or destruction of contaminants through chemical reactions, e.g. coagulation–flocculation for solids removal, disinfection for pathogen destruction, and chemical precipitation for phosphorus removal.

5.5.3 Biological treatment

This involves the conversion or destruction of contaminants with the help of microorganisms. In municipal wastewater treatment plants, microorganisms indigenous to wastewaters are used in biological treatment operations. Examples of biological treatment include the activated sludge process, membrane bioreactor, trickling filter, among others. The primary purpose of biological treatment is to remove and reduce the biodegradable organic matter from wastewater to an acceptable level according to regulatory limits. Biological treatment is also used to remove nutrients such as nitrogen and phosphorus from wastewater.

5.6 LEVELS OF WASTEWATER TREATMENT

A wastewater treatment system is a combination of unit operations and unit processes that are designed to reduce the contaminants to an acceptable level. The term *unit operation* refers to processes that use physical treatment methods. The term *unit process* refers to processes that use biological and/or chemical treatment methods. Unit operations and processes may be grouped together to provide the following levels of treatment.

5.6.1 Preliminary treatment

This involves the physical removal of pollutant substances such as rags, twigs, etc. that can cause operational problems in pumps, treatment processes, and other appurtenances. Examples of preliminary treatment are screens for removal of large debris, comminutor for grinding large particles, grit chamber for removal of inert suspended solids, and flotation for removal of oils and grease.

5.6.2 Primary treatment

This involves the physical removal of a portion of the suspended solids from wastewater, usually by sedimentation. Primary clarifiers are used for this purpose. Primary clarifier effluent contains significant amounts of BOD and requires further treatment. The term primary treatment often includes preliminary as well as primary treatment operations.

5.6.3 Enhanced primary treatment

This involves the use of chemical treatment to obtain additional solids removal in a sedimentation process. Chemical coagulants are used to promote coagulation and flocculation of solids in a sedimentation tank, resulting in enhanced suspended solids removal. For example, Blue Plains Advanced Wastewater Treatment Plant in Washington, DC, uses an iron coagulant together with a polymer to achieve enhanced solids removal in their primary clarifiers (Neupane et al., 2008).

5.6.4 Conventional secondary treatment

This involves biological treatment for the degradation of organic matter and solids reduction. Efficiency is measured mainly in terms of BOD_5 and suspended solids removal. The treatment is carried out in a biological reactor followed by a sedimentation tank or secondary clarifier. Examples of secondary treatment are the activated sludge process, trickling filter, etc.

5.6.5 Secondary treatment with nutrient removal

When removal of nutrients, such as nitrogen and/or phosphorus, is required, it may be combined with the secondary treatment for BOD removal. Additional reactors may be required to achieve nitrogen removal through the nitrification–denitrification process. A combination of chemical and biological treatment can be used.

5.6.6 Tertiary treatment

This includes treatment processes used after the secondary, e.g. granular media filtration used for removal of residual suspended solids, and disinfection for pathogen reduction. Additional treatment for nutrient removal is also included in tertiary treatment.

5.6.7 Advanced treatment

Advanced treatment processes are used when additional removal of wastewater constituents is desired due to toxicity of certain compounds, or for potential water reuse applications. Examples include activated carbon adsorption for removal of volatile organic compounds, ion exchange for the removal of specific ions, etc.

5.7 RESIDUALS AND BIOSOLIDS MANAGEMENT

Each of the treatment processes described above generates a certain amount of waste solids. The waste generated is semi-solid in nature and is termed *sludge*. The waste generated from preliminary treatment includes grit and screenings. These waste residuals are low in organic content and are disposed of in landfills. The sludge generated from primary and secondary clarifiers has a significant amount of organic matter, and requires further treatment and processing prior to disposal. The term *biosolids* is used to denote treated sludge. The cost of treatment of sludge and disposal of biosolids can be equivalent to 40–50% of the total cost of wastewater treatment.

The main objectives of sludge treatment are (a) to reduce the organic content, (b) to reduce the liquid fraction, and (c) to reduce the pathogen content. If the sludge contains heavy metals or other toxic compounds, local or state regulations may require additional treatment depending on the final disposal of the biosolids produced.

The liquid fraction is reduced by a number of processes. These include gravity thickening, dissolved air flotation, centrifugation, belt filter press, etc. Organic content and pathogen reduction is achieved by processes that include anaerobic digestion, aerobic digestion, air drying, heat drying, thermophilic digestion, composting, lime stabilization, pasteurization, etc. A combination of these processes and some pretreatment may be used depending on the quality of biosolids desired.

Over the last four decades, most of the research has focused on the treatment of wastewater, while treatment of sludge has lagged behind. The traditional method of biosolids disposal in landfills is still used extensively. Land application of biosolids is practiced in some areas. In recent years, the concept of beneficial reuse of biosolids as a soil conditioner and fertilizer on agricultural lands has gained importance, both from the viewpoint of green engineering and necessity. As a result, we have seen increased research on biosolids for the purpose of further reducing pathogens for the safe reuse of the product. Detailed discussion on biosolids is provided in a subsequent chapter.

5.8 FLOW DIAGRAMS OF TREATMENT OPTIONS

Example 5.2: Draw a flow diagram for a process to treat a municipal wastewater that has a high concentration of suspended solids, organic matter, and pathogens. Also, illustrate a sludge treatment option.

SOLUTION

The wastewater can be treated with a conventional process consisting of primary and secondary treatment. The sludge can be treated using anaerobic digestion. The flow diagram is presented in Figure 5.2.

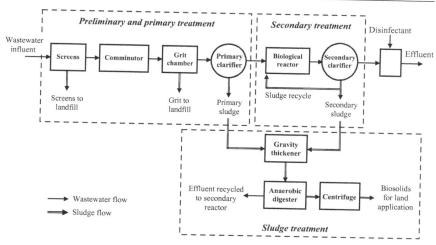

Figure 5.2 Flow diagram of a conventional wastewater treatment process with sludge digestion.

Example 5.3: Draw a flow diagram to treat a wastewater that has a high concentration of suspended solids, organic matter, pathogens, and a high concentration of ammonia-nitrogen.

SOLUTION

The wastewater can be treated using primary treatment followed by biological treatment to remove organic matter and ammonia-nitrogen. Ammonia is removed in a two-step biological process consisting of nitrification followed by denitrification. The nitrification step can be combined with BOD removal in an aerobic reactor. This is followed by denitrification in an anoxic reactor. The flow diagram is presented in Figure 5.3.

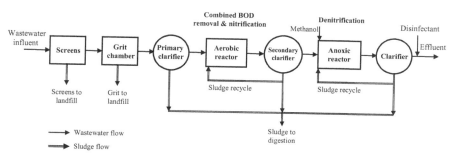

Figure 5.3 Flow diagram for treatment of wastewater with high nitrogen concentration.

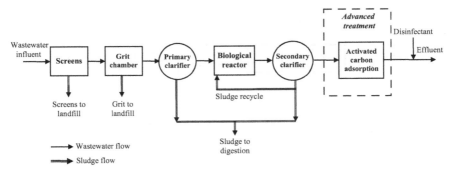

Figure 5.4 Flow diagram for an advanced wastewater treatment process.

Example 5.4: Draw a flow diagram for the treatment of a wastewater that has a high concentration of herbicides, as well as suspended solids and organic matter.

SOLUTION

The wastewater will be treated using primary, secondary, and tertiary advanced treatment. Activated carbon adsorption is used to remove the herbicide. The flow diagram is illustrated in Figure 5.4.

Example 5.5: Draw a flow diagram for the treatment of a wastewater that has a high concentration of suspended solids and organic matter. Effluent discharge regulations allow a very low concentration of suspended solids.

SOLUTION

The wastewater treatment will include tertiary treatment together with primary and secondary treatment. Tertiary treatment consists of dual media filtration to remove residual suspended solids. The flow diagram is presented in Figure 5.5.

5.9 TYPES OF BIOLOGICAL TREATMENT PROCESSES

There are two main types of wastewater treatment processes. They are:

1. *Suspended growth process* – The microorganisms are kept in suspension in a biological reactor by suitable mixing devices. The process

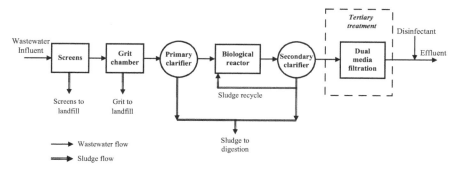

Figure 5.5 Flow diagram for a wastewater treatment system to achieve very low effluent solids concentration.

can be aerobic or anaerobic. Examples of suspended growth processes include activated sludge process, sequencing batch reactor, ponds and lagoons, digesters, etc.

2. *Attached growth process* – The microorganisms responsible for bio-conversion attach themselves onto an inert medium inside the reactor, where they grow and form a layer called *biofilm*. The wastewater flowing through the reactor comes in contact with the biofilm, where conversion and removal of organic matter take place. The inert medium is usually rock, gravel, slag, or synthetic media. The process can be operated aerobically or anaerobically. Examples are trickling filters, bio-towers, and rotating biological contactors (RBC).

Effective design and successful operation of the processes depend on a thorough understanding of the types of microorganisms involved, growth requirements, reaction kinetics, and environmental factors that affect their performance. The selection of a particular process should be based on bench-scale and pilot scale studies on the specific wastewater, investigating the effects of a variety of possible variables. A detailed discussion of each of these types of processes is provided in Chapters 8 and 9.

PROBLEMS

5.1   What are the common sources of wastewater? Name them.
5.2   Use the data provided in Table 5.1 and estimate the projected population of Kansas City in 2030 considering the 2000 and 2020 data using 1) constant growth method, 2) log growth method, 3) percent growth method, and 4) declining growth method with a saturation population of 2,000,000. Compare the results obtained with the four methods, as well as with the ones from worked-out Example 5.1.

5.3   What are the main objectives of wastewater treatment?

5.4   What are preliminary treatment and primary treatment? What do they remove?

5.5   What is biological treatment? What are the advantages of biological treatment?

5.6   What is chemical treatment? If you were given an option, would you prefer to use chemical treatment or biological treatment? Explain your reasons.

5.7   Define the terms *effluent, sludge*, and *biosolids*, as they pertain to wastewater treatment.

5.8   Define suspended growth and attached growth processes. Give an example of each.

5.9   What are the main objectives of the treatment of sludge?

5.10   Draw a flow diagram of a process to treat a wastewater that has a high concentration of suspended solids, BOD, pathogens, and phosphorus. Also, show the sludge treatment process.

5.11   Draw a flow diagram for a process to treat a wastewater that has a high concentration of synthetic organic compounds (SOCs) as well as suspended solids and organic matter.

REFERENCES

Davis, M. 2011. *Water and Wastewater Engineering: Design Principles and Practice.* McGraw-Hill, Inc., New York, NY, USA.

e-CFR. 2021. *Electronic Code of Federal Regulations.* Federal Government, USA, https://www.ecfr.gov/cgi-bin/text-idx?SID=15e352a79a295dd3e0f1699119f8 2c04&mc=true&node=se40.31.401_115&rgn=div8

FAO. 2021. http://www.fao.org/3/y4555e/y4555e00.htm

Federal Register. 2010. *Code of Federal Regulations*, Title 40. Federal Government, USA.

GLUMRB. 2004. *Recommended Standards for Wastewater Facilities*, Great Lakes – Upper Mississippi River Board of State and Provincial Public Health and Environmental Managers, Health Education Services, Albany, New York.

Hammer, M. J. and Hammer, M. J. Jr. 2012. *Water and Wastewater Technology.* 7th edn. Pearson-Prentice Hall, Inc., Hoboken, NJ, USA.

McJunkin, F.E. 1964. Population Forecast by Sanitary Engineers, *Journal of Sanitary Engineering Division*, American Society of Civil Engineers, 90, (4):31–58.

Metcalf and Eddy, Inc. 1981. *Wastewater Engineering: Collection and Pumping Wastewater*, McGraw-Hill, Inc., New York, NY, 61–95.

Metcalf and Eddy, Tchobanoglous, G., Stensel, H., Tcuchihashi, R., and Burton, F. 2013. *Wastewater Engineering: Treatment and Resource Recovery.* 5th edn. McGraw-Hill, Inc., New York, NY, USA.

Neupane, D., Riffat, R., Murthy, S., Peric, M., and Wilson, T. 2008. Influence of Source Characteristics, Chemicals and Flocculation on Chemically Enhanced Primary Treatment. *Water Environment Research*, 80 (4):331–338.

Peavy, H. S., Rowe, D. R., and Tchobanoglous, G. 1985. *Environmental Engineering.* McGraw- Hill, Inc., New York, NY, USA.

Roser, M. 2019. Future Population Growth. *Our World in Data.* https://ourworldin-data.org/future-population-growth

Shammas N.K., and Wang, L.K. 2011. *Fair, Geyer, and Okun's, Water and Wastewater Engineering: Water Supply and Wastewater Removal*, 3rd edn., John Wiley and Sons, Inc., Hoboken, NJ, USA.

United Nations. 2021. Kansas City Metro Area Population 1950-2021, https://www.macrotrends.net/cities/23028/kansas-city/population

Chapter 6

Preliminary treatment

6.1 INTRODUCTION

Preliminary treatment involves the removal of larger suspended solids and inert materials from the wastewater. Physical treatment processes are used to remove these particles and debris that may cause harm to pumps and other equipment, and for the removal of inert matter prior to secondary biological treatment. The unit operations used include screens, comminutors/grinders, grit chambers, and flow equalization. A typical layout is illustrated in Figure 6.1.

6.2 SCREENS

Raw wastewater contains a significant amount of suspended and floating materials. These include rags, weeds, twigs, organic matter, and a variety of solids. The solids can damage pumps and mechanical equipment, and interfere with the flow in pipes and channels. Screening devices are placed ahead of pumps, to remove the larger materials from the wastewater stream. The removed debris called *screenings* is usually disposed of in landfills, or by incineration. Different types of screens are available depending on wastewater characteristics and site requirements. The following sections describe the types of screens that are available, based on the size of the openings.

6.2.1 Trash racks

These are screens that have large openings to exclude larger debris and garbage. These consist of rectangular or circular steel bars arranged in a parallel manner, either vertically or at an incline to the horizontal channel, as illustrated in Figure 6.2. The size of the opening between bars ranges from 50 to 150 mm (2–6 in). Mechanical rakes are used to clear the solids collected on the trash racks. Rake machines are operated by hydraulic jacks. Trash racks are followed by coarse screens.

DOI: 10.1201/9781003134374-6

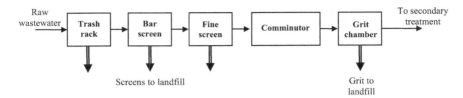

Figure 6.1 Typical layout of a preliminary treatment process.

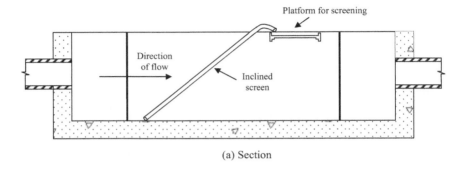

(a) Section

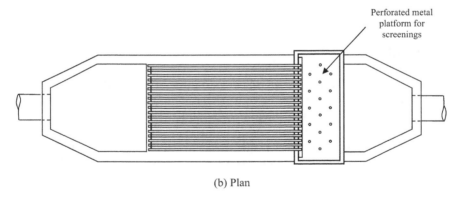

(b) Plan

Figure 6.2 Diagram of trash rack: (a) Section and (b) plan view.

6.2.2 Coarse screens or bar screens

These are similar to trash racks but have a smaller size opening ranging from 25 to 75 mm (1–3 in). Coarse screens can be manually cleaned or mechanically cleaned. Manually cleaned bar screens may be used in small-sized wastewater treatment plants. They are also used in bypass channels when other mechanically cleaned screens are being serviced, or in the event of a power failure. Manually cleaned screens should be placed on a slope of 30°–45° from the vertical. This increases the cleaning surface, makes cleaning easier, and prevents excessive headloss by clogging. Mechanically

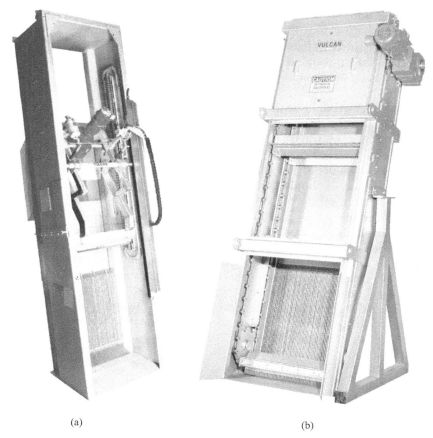

(a)   (b)

Figure 6.3 (a) Mensch Crawler™ bar screen and (b) VMR™ multi-rake bar screen.
Source: Courtesy of Vulcan Industries, Inc., Iowa.

cleaned screens can be placed at 0°–30° from the vertical. Lower maximum approach velocities are specified for manually cleaned screens (0.3–0.6 m/s, or 1.0–2.0 ft/s)), as compared to mechanically cleaned screens (0.6–1.0 m/s, or 2.0–3.25 ft/s)) (Metcalf and Eddy et al., 2013). There are four main types of mechanically cleaned bar screens. They are chain-driven, reciprocating rake, catenary, and continuous belt. Figure 6.3 presents two types of automatic, mechanically cleaned bar screens, manufactured by Vulcan Industries, Inc., of Iowa.

6.2.2.1 Design of coarse screens

The following parameters are important design considerations in the installation of coarse screens:

- Location
- Approach velocity
- Clear openings between bars or mesh size
- Headloss through the screen
- Disposal of screenings.

Coarse screens should be installed ahead of fine screens and grit chambers. For manually cleaned screens, the approach velocity should be limited to about 0.45 m/s (1.5 ft/s) at average flow. For mechanically cleaned screens, an approach velocity of at least 0.4 m/s (1.25 ft/s) is recommended to minimize solids deposition in the channel. At peak flow rates, the velocity through the screen should not exceed 0.9 m/s (3 ft/s), in order to prevent the pass-through of solids (Metcalf and Eddy et al., 2013; EPA, 1999). Velocity through the bar screen can be controlled by installing a downstream head control device, e.g. a Parshall flume. Two or more units should be installed, so that one unit may be taken out of service for maintenance. The headloss through mechanically cleaned screens is usually limited to about 150 mm (6 in) by operational controls. The headloss is measured as the difference in water level before and after the screen.

The headloss through a screen is a function of the approach flow velocity and the velocity through the bars. Bernoulli's equation is used to calculate the headloss (Droste and Gehr, 2018), which results in the following equation:

$$H_{\mathrm{L}} = \frac{1}{C_{\mathrm{d}}}\left(\frac{V_{\mathrm{s}}^2 - v^2}{2g}\right) \qquad (6.1)$$

where

H_{L} = Headloss through the screen

$h_1 - h_2$ = upstream depth of flow – downstream depth of flow

C_{d} = Coefficient of discharge, usually 0.70–0.84 for a clean screen, and 0.6 for clogged screen

V_{S} = Velocity of flow through the openings of the bar screen, m/s

v = Approach velocity in upstream channel, m/s

g = acceleration due to gravity, 9.81 m/s²

The velocity of flow through the bar screen openings can be calculated from the number of bars in the channel width, and the depth of the water level. The approximate number of bars is (Davis, 2020):

$$N_{\mathrm{bars}} = \frac{\text{channel width} - \text{bar spacing}}{\text{bar width} + \text{bar space}} \qquad (6.2)$$

$$\text{Number of bar spaces} = (N_{\text{bars}} + 1) \tag{6.3}$$

$$\text{Area of screen openings} = (\text{number of bar spaces}) \times (\text{bar spacing})$$
$$\times (\text{water depth}) \, \text{m}^2 \tag{6.4}$$

$$\text{Therefore,} \, V_s = (\text{flow rate} \, \text{m}^3/\text{s})/(\text{area of screen openings} \, \text{m}^2) \tag{6.5}$$

Example 6.1: A mechanically cleaned bar screen is used in preliminary treatment for the following conditions:

Wastewater flow rate = 100,000 m³/d
Approach velocity = 0.6 m/s
Open area for flow through the screen = 1.6 m²
Headloss coefficient for clean screen = 0.75
Headloss coefficient for clogged screen = 0.60
Incline from vertical = 0°
a. Calculate the clean water headloss through the bar screen.
b. Calculate the headloss after 40% of the flow area is clogged with solids.

SOLUTION
a. Calculate V_s for clean screen using Equation (6.5).
   Flow rate, Q = (100,000 m³/d)/(86,400 s/d) = 1.16 m³/s

$$V_s = (1.16 \, \text{m}^3/\text{s})/(1.6 \, \text{m}^2) = 0.725 \, \text{m}/\text{s}$$

Determine the clean water headloss using Equation (6.1).

$$H_L = \frac{1}{C_d}\left(\frac{V_s^2 - v^2}{2g}\right)$$

$$H_L = \frac{1}{0.75}\left(\frac{0.725^2 - 0.6^2}{2 \times 9.81}\right) = 0.01 \, \text{m}$$

b. Calculate V_s for clogged screen using Equation (6.5).

$$\text{Area available for flow,} \, A = 1.6 \times (1 - 0.4) = 0.96 \, \text{m}^2$$

$$V_s = \left(1.16\,\mathrm{m}^3\,/\,\mathrm{s}\right) / \left(0.96\,\mathrm{m}^2\right) = 1.21\,\mathrm{m}\,/\,\mathrm{s}$$

Calculate headloss for the clogged screen using Equation (6.1).

$$H_L = \frac{1}{0.6}\left(\frac{1.21^2 - 0.6^2}{2 \times 9.81}\right) = 0.09\,\mathrm{m}$$

Note: V_s for clogged screen exceeds the maximum suggested value of 0.9 m/s. This indicates that the screen should be cleaned. Screens are usually cleaned either at regular time intervals or when a specified maximum headloss value is reached.

6.2.3 Fine screens

These screens have openings less than 6 mm in size. Fine screens are used in preliminary treatment after coarse screens, in primary treatment prior to secondary trickling filters, and for treatment of combined sewer overflows. Different fabrication techniques are used to provide the small screen sizes. These include:

- Profile bars arranged in a parallel manner with openings from 0.5 mm (0.02 in).
- Slotted perforated plates with 0.8–2.4 mm (0.03–0.09 in) wide slots.
- Wedge-shaped bars welded together into flat panel sections.
- Looped wire construction with openings of 0.13 mm (0.005 in).
- Wire mesh with approximately 3.3 mm (0.013 in) openings.
- Woven wire cloth with openings of 2.5 mm (1.0 in).

A variety of fine screens are commercially available. Some of them are described below:

1. *Static Wedgewire Screen* – These screens have 0.2–1.2 mm openings, and are designed for flow rates of 400–1200 L/m².min of screen area (Metcalf and Eddy et al., 2013). Headloss ranges from 1.2 to 2 m. The screen consists of small stainless steel wedge-shaped bars, with the flat part of the wedge facing the flow.
2. *Stair Screen* – This type of screen has two step-shaped sets of thin vertical plates, one fixed and one movable. The fixed and movable plates alternate across the width of an open channel and together form a single screening surface. The movable plates rotate in a vertical motion, lifting the captured solids onto the next fixed-step landing, ultimately transporting them to the top of the screen from where they are discharged to a collection hopper. The range of openings between the screen plates is 3–6 mm (0.12–0.24 in). A stair screen manufactured by Vulcan Industries, Inc., of Iowa is presented in Figure 6.4(a).

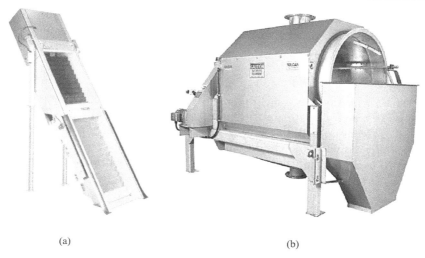

(a)   (b)

Figure 6.4 (a) ESR™ stair screen and (b) Liqui-Fuge™ rotary drum screen.

Source: Courtesy of Vulcan Industries, Inc., Iowa.

3. *Drum Screen* – The screening medium is mounted on a drum or cylinder that rotates in a flow channel. Depending on the direction of flow into the drum, the solids may be collected on the interior or exterior surface. Drum screens are available in various sizes ranging from 0.9 to 2 m (3–6.6 ft) in diameter, and from 1.2 to 4 m (4–13.3 ft) in length. A rotary drum screen is presented in Figure 6.4(b).

6.2.3.1 Design of fine screens

An installation should have a minimum of two screens; each should be capable of handling peak flow rates. Flushing water should be provided to remove the buildup of grease and other solids on the screen. The clear water headloss through a fine screen may be obtained from the manufacturer's rating tables. It can also be calculated from the following equation (Metcalf and Eddy et al., 2013):

$$H_L = \frac{1}{2g}\left(\frac{Q}{C_d A}\right)^2 \tag{6.6}$$

where
H_L = Headloss through the screen, m
C_d = Coefficient of discharge, usually 0.6 for a clean screen
Q = Wastewater flow rate, m³/d
A = Effective open area of the submerged screen, m²
g = acceleration due to gravity, 9.81 m/s²

The values of C_d and A depend on screen design factors and may be obtained from the screen manufacturer, or determined experimentally.

6.2.4 Microscreens

These screens have openings that are less than 50 µm. This type of screen is used in tertiary treatment to remove fine solids from treated effluents. It involves the use of variable low-speed rotating drum screens that are operated under gravity flow conditions. The fabric filter has openings ranging from 10 to 35 µm and is fitted on the periphery of the drum.

6.3 SHREDDER/GRINDER

Coarse solids, especially larger organic solids are reduced to smaller size solids by using shredding processes. These can be used in conjunction with mechanically cleaned screens to cut up the solids into smaller particles of uniform size, which are then returned to the flow stream for passage to secondary treatment units. There are three main types of shredding devices. They are:

1. *Comminutor* – These are used in small wastewater treatment plants, with flow rates less than 0.2 m³/s (5 MGD). A typical comminutor has a stationary horizontal screen to intercept the flow, and a rotating cutting arm to shred the solids to sizes ranging from 6 to 20 mm (0.25–0.77 in). A bypass channel with a medium screen is usually provided to maintain flow when the comminutor is taken offline for servicing. Figure 6.5 presents a diagrammatic layout of a comminutor. Headloss through a comminutor can range from 0.1 to 0.3 m (4–12 in) and can reach 0.9 m (3 ft) in large units at maximum flow rates (Metcalf and Eddy et al., 2013). Comminutors can create a string of rags and/or plastic, that can collect on downstream equipment and cause operational problems. Newer installations use macerators or grinders.
2. *Macerator* – These are slow-speed grinders that chop/grind solids to very small pieces. The macerator blade assembly is typically between 6 and 9 mm (Davis, 2020). Effective chopping action reduces the possibility of producing ropes of rags and plastics that can collect on downstream equipment.
3. *Grinder* – High-speed grinders are used to pulverize solids in the wastewater. They are also called Hammermills. The solids are pulverized as they pass through a high-speed rotating assembly. Wash water

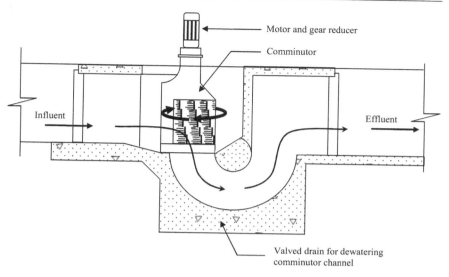

Figure 6.5 Diagrammatic layout of a comminutor.

is used to keep the unit clean and to transport solids back to the wastewater stream.

6.4 GRIT CHAMBER

Grit is defined as sand, gravel, or other mineral material that has a nominal diameter of 0.15–0.20 mm or larger (Droste and Gehr, 2018). Grit may also include ash, wood chips, coffee grounds, eggshells, and other non-putrescible organic matter. Some components such as coffee grounds are organic, but they are essentially non-biodegradable over the time span for grit collection and disposal. Grit chambers are sedimentation tanks that are placed after screens and before primary clarifiers. The purpose of a grit chamber is to remove materials that may form heavy deposits in pipelines, protect pumps, and other mechanical equipment from abrasion, reduce the frequency of digester cleaning caused by grit accumulation, and separate heavier inert solids from lighter biodegradable organic solids that are sent to secondary biological treatment. The amount of grit collected depends on the wastewater characteristics and the type of grit chamber used at the plant. Grit can be coated by grease or other organic matter. So, grit removed from the grit chambers are usually washed to remove organic matter and then transported to a sanitary landfill for disposal.

In general, grit chambers are designed to remove particles with a specific gravity of 2.65 (sand) and a nominal diameter of 0.20 mm or larger, at a wastewater temperature of 15.5°C (60°F). The settling velocity of these particles is about 2.3 cm/s (4.5 ft/min) based on curves of wastewater grit settling velocities developed by Camp (1942). Grit chambers are sometimes designed to remove 0.15 mm sand particles with a settling velocity of 1.3 cm/s (2.6 ft/min), based on Camp's curves. Subsequent research has revealed that the specific gravity of grit can range from 1.1 to 2.7 (Eutek, 2008; Metcalf and Eddy et al., 2013). Wilson et al. (2007) suggested a *sand equivalent size* (SES), where SES is the size of a clean sand particle that settles at the same rate as a grit particle.

There are mainly three types of grit chambers. They are:

1. *Horizontal Flow Grit Chamber* – It is a square or rectangular open channel with a sufficient detention time to allow sedimentation of grit particles, and a constant velocity to scour the organics. The velocity is controlled by channel dimensions, an influent distribution gate, and an effluent weir. It may be cleaned manually or by mechanical sludge scrapers. It is found in older installations.
2. *Aerated Grit Chamber* – This is used in newer installations. A spiral flow pattern is introduced in the wastewater as it flows through the tank, by supplying air from a diffuser located on one side of the tank. The air provides sufficient *roll* velocity to keep the lighter organic particles in suspension, while heavier grit particles settle at the bottom. The lighter organic particles are carried out of the tank with the wastewater. A hopper is provided along one site of the tank for grit collection. The advantages of the system include minimal headloss, and the fact that aeration helps reduce septic conditions in the wastewater. Disadvantages include high power consumption, labor-intense, and possible odor issues. Figure 6.6 illustrates the flow pattern in an aerated grit chamber. Aerated grit chambers are generally designed to remove 0.21 mm diameter or larger particles with a detention time of 2–5 minutes at peak hourly flow rates. The typical width-to-depth ratio is 1.5:1, and the length-to-width ratio is 4:1. Air supply ranges from 0.2 to 0.5 m³/min per m of length (3–8 ft³/ft·min) (Metcalf and Eddy et al., 2013). The design of an aerated grit chamber is shown in Example 6.2.
3. *Vortex Grit Chamber* – This type of device generates a vortex flow pattern, which tends to lift the lighter organic particles upward, while the grit settles in the hopper at the bottom. Settled grit is removed by a grit pump or air-lift pump. It has a small footprint with minimal headloss. The design is proprietary and the compaction of grit may be a problem. Vortex grit chambers are typically designed to handle peak flow rates of up to 0.3 m³/s (7 MGD) per unit. A vortex grit chamber is illustrated in Figure 6.7.

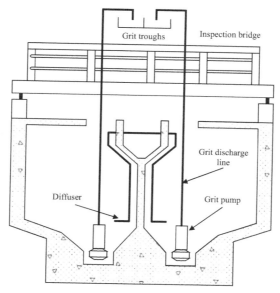

Figure 6.6 Aerated grit chamber.

Source: Adapted from Metcalf and Eddy et al., 2013.

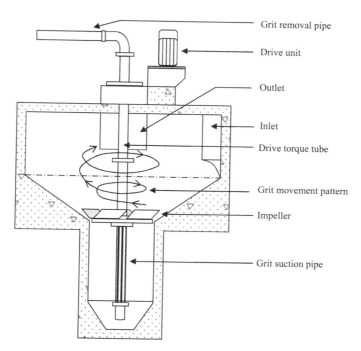

Figure 6.7 Vortex grit chamber.

Example 6.2: Design an aerated grit chamber for a municipal wastewater treatment plant. The average flow rate is 20,000 m³/d, with a peaking factor of 2.5. Use a depth of 3 m. Air is supplied at 0.35 m³/min per m of length. Assume grit collected is 0.10 m³/1000 m³ at peak flow. Determine the tank dimensions, total air supply required, and quantity of grit.

SOLUTION

a. Determine tank dimensions.
    Aerated grit chamber is designed for peak flow rates. Use two chambers in parallel.

$$\text{Peak flow rate} = 20,000 \times 2.5 = 50,000 \, \text{m}^3 / \text{d}$$

$$\text{Flow in each tank at peak flow}, Q = 50,000/2 = 25,000 \, \text{m}^3/\text{d}$$

$$\text{Assume detention time at peak flow}, t = 4 \, \text{min}$$

$$\text{Volume of each tank} = Q.t = \left(25,000 \, \text{m}^3/\text{d}\right) \times \left(4 \, \text{min}\right) / \left(1440 \, \text{min}/\text{d}\right)$$
$$= 70 \, \text{m}^3$$

$$\text{Assume width} - \text{to} - \text{depth ratio} = 1:1$$

$$\text{Depth} = 3 \, \text{m}$$

$$\text{Therefore, width} = 3 \, \text{m}$$

$$\text{Length} = 70 / \left(3 \times 3\right) = 7.7 \, \text{m} \approx 8 \, \text{m}$$

Tank dimensions are $8 \, \text{m} \times 3 \, \text{m} \times 3 \, \text{m}$.

b. Determine total air required.

$$\text{Air required} = \left(0.35 \, \text{m}^3/\text{min per m}\right) \times 8 \, \text{m} = 2.80 \, \text{m}^3/\text{min for each tank}$$

$$\textbf{Total air required} \text{ for 2 tanks} = 2.80 \times 2 = 5.60 \, \text{m}^3/\text{min}$$

c. Calculate volume of grit.

$$\text{Volume of grit} = \left(0.10 \, \text{m}^3/1000 \, \text{m}^3\right) \times 25,000 \, \text{m}^3/\text{d}$$
$$= 2.5 \, \text{m}^3/\text{d in each tank}$$

$$\textbf{Total grit volume} = 2.5 \times 2 = 5.0 \, \text{m}^3 / \text{d}$$

6.5 FLOW EQUALIZATION

The flow of wastewater entering a municipal wastewater treatment plant varies over the course of a 24-h day, with above-average flows occurring during mid-morning and late afternoon, and low flows occurring at night (11 pm to 5 am). The flow variation also occurs due to wet weather conditions, infiltration, and inflow, and may result in a significant increase in the concentration of suspended solids and organic loading (BOD_5). Flow equalization is a method of damping the variations in flow rates so that the unit processes receive nearly constant flow rates (Metcalf and Eddy et al., 2013). This is usually done to reduce peak flows and loads and to equalize combined storm sewer and sanitary sewer flows. The implementation of flow equalization can result in the following advantages: (i) Reduce the size requirements of the downstream primary and secondary treatment processes, (ii) increase the capacity of the plant, and (iii) significantly improve the overall plant performance.

Equalization tanks can be placed in-line before the primary clarifier, where it is used to eliminate diurnal flow variations, and to minimize shock loadings to the biological treatment process, as shown in Figure 6.8(a). Another possible arrangement is illustrated in Figure 6.8(b), where the equalization tank is kept off-line until the flow exceeds a pre-specified value, at which point the flow is passed through the equalization tank. Typically, the in-line equalization is used when the dampening of both flow and organic load is required.

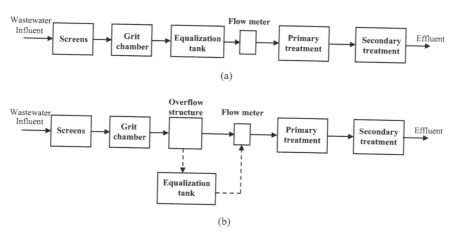

Figure 6.8 (a) In-line and (b) off-line equalization tank in a wastewater treatment plant.

6.5.1 Equalization tank design

The volume of the equalization tank is the principal factor considered for design, although the location, configuration, basin geometry, mixing, and pumping requirements are part of the overall design. The required volume is estimated from the diurnal flow data by performing a volume balance to determine the maximum volume required. A multiplier ranging from 1.1 to 1.25 is used to calculate the actual volume from the theoretical estimate to accommodate aeration and mixing equipment and other contingency for variations beyond the diurnal flow pattern. Example 6.3 illustrates a graphical method for estimation of tank volume required for flow equalization. However, a spreadsheet calculation is required to estimate the volume for in-line equalization using the mass balance technique and concurrent dampening of organic loading (BOD_5). Readers are referred to Davis (2020) for additional details of that method.

> **Example 6.3:** Determine the equalization tank volume (m^3) required for the wastewater entering a treatment plant, using the graphical method. The diurnal flow data is presented in Table 6.1.
>
> **SOLUTION**
>
> Design steps for calculation of equalization tank volume (graphical method).
>
> a. Calculate the volume converted from flow rate for every time period. This is the volume of influent wastewater that will enter the equalization tank.
>
> $$\text{For midnight to 1 am, Volume} = 0.047\,\frac{m^3}{s} \times 1h \times 3600\,\frac{s}{h} = 169.2\,m^3$$

Table 6.1 Diurnal flow data for a wastewater treatment plant (for Example 6.3)

Time period	Flow (m^3/s)	Time period	Flow (m^3/s)
Midnight–1 am	0.047	Noon–1 pm	0.08
1–2	0.036	1–2	0.075
2–3	0.026	2–3	0.071
3–4	0.017	3–4	0.073
4–5	0.016	4–5	0.081
5–6	0.019	5–6	0.087
6–7	0.038	6–7	0.088
7–8	0.058	7–8	0.085
8–9	0.064	8–9	0.075
9–10	0.072	9–10	0.064
10–11	0.076	10–11	0.058
11–noon	0.081	11–midnight	0.05

Table 6.2 Cumulative volume calculation (for Example 6.3)

Time (h)	Flow (m³/s)	Volume (m³)	Cumulative volume (m³)
1	0.047	169.2	169.2
2	0.036	129.6	298.8
3	0.026	93.6	392.4
4	0.017	61.2	453.6
5	0.016	57.6	511.2
6	0.019	68.4	579.6
7	0.038	136.8	716.4
8	0.058	208.8	925.2
9	0.064	230.4	1155.6
10	0.072	259.2	1414.8
11	0.076	273.6	1688.4
12	0.081	291.6	1980.0
13	0.08	288.0	2268.0
14	0.075	270.0	2538.0
15	0.071	255.6	2793.6
16	0.073	262.8	3056.4
17	0.081	291.6	3348.0
18	0.087	313.2	3661.2
19	0.088	316.8	3978.0
20	0.085	306.0	4284.0
21	0.075	270.0	4554.0
22	0.064	230.4	4784.4
23	0.058	208.8	4993.2
24	0.05	180.0	5173.2

$$\text{For } 1-2\,\text{am, Volume} = 0.036\,\frac{m^3}{s} \times 1\,h \times 3600\,\frac{s}{h} = 129.6\,m^3$$

b. Calculate the cumulative influent volume as a function of the time of day. This is the total volume of influent wastewater that enters the equalization tank from midnight up to the calculated time period. The calculated values for all time periods are shown in Table 6.2.

For midnight to 1 am   Cumulative volume $= 169.2\,m^3$

For $1-2$ am   Cumulative volume $= 169.2 + 1296 = 298.8\,m^3$

c. Plot the cumulative influent volume as a function of time of day. The line connecting the initial and final point on the plot indicates the average influent volume over the 24 h period, and the slope of the line indicates the average flow rate.

$$\text{Average flowrate} = \frac{\text{Cumulative volume}\left(m^3\right)}{24\,h} = \frac{5173.2}{24}$$

$$= 215.6\,\frac{m^3}{d}$$

Figure 6.9 illustrates (a) the diurnal flow pattern, (b) cumulative influent volume and average influent volume as a function of time.

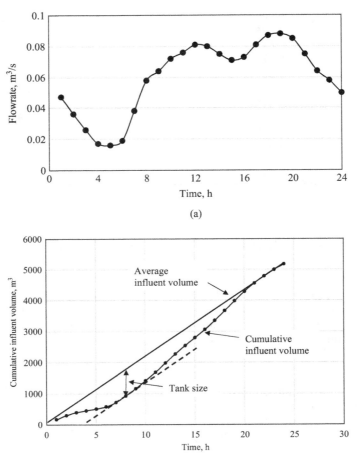

(a)

Figure 6.9 (a) Diurnal flow pattern and (b) cumulative influent volume and average influent volume as a function of time (for Example 6.3).

d. Draw a line parallel to the average influent volume line that is tangent to the cumulative influent volume curve. The tangent point on the cumulative influent volume curve is where the equalization tank is empty. The vertical distance between the average influent volume line and the parallel line is the estimated equalization tank volume.

Estimated equalization tank volume = 800 m³ from Figure 6.9(b)

e. Calculate the required volume of the equalization tank using a multiplier of 1.25 for consideration of various contingency factors.

$$\text{Required equalization tank volume} = 1.25 \times 800 = 1000\,m^3$$

PROBLEMS

6.1 Draw a flow diagram of a preliminary treatment process that consists of trash racks, two bar screens, two rotary drum screens, a macerator with a bypass channel, and an aerated grit chamber. What happens to the wastes removed from each unit?

6.2 What are the advantages of installing a bar screen at an incline?

6.3 Why should you provide a bypass channel with a comminutor?

6.4 How often does a screen have to be cleaned? What parameters and values are used for this decision?

6.5 The water elevations upstream and downstream of a bar screen are 0.89 m and 0.85 m. If the approach velocity is 0.45 m/s, what is the flow velocity through the screen? The discharge coefficient of the screen is 0.7.

6.6 A mechanically cleaned bar screen is used in preliminary treatment for the following conditions:

Incline from vertical = 30°
Wastewater flow rate = 150,000 m³/d
Approach velocity = 0.6 m/s
Open area for flow through the screen = 1.6 m²
Headloss coefficient for clean screen = 0.74
Headloss coefficient for clogged screen = 0.60

a. Calculate the clean water headloss through the bar screen.
b. Calculate the headloss after 50% of the flow area is clogged with solids.

6.7 Estimate the headloss for a bar screen set at a 30° incline from the vertical. The wastewater flow rate is 90,000 m³/d. The bars are 20 mm in diameter, with 25 mm clear spacing in between bars. The channel width is 1.5 m, the approach velocity is 0.65 m/s, and the headloss coefficient is 0.65.

6.8 What is grit? What are the objectives of grit removal from the wastewater? What is the disposal method for collected grit?

6.9 Find the name of the wastewater treatment plant that serves your locality. What unit operations are used as preliminary treatment at the plant? Draw a flow diagram of the preliminary treatment process.

6.10 Design a grit chamber for a wastewater treatment plant with an average flow rate of 25,000 m³/d and a peak flow rate of 55,000 m³/d. The detention time at peak flow is 3.0 min. The width-to-depth ratio is 2:1. Use a depth of 3 m. The aeration rate is 0.4 m³/min per m of tank length. Determine the total air required and dimensions of the grit chamber.

6.11 What is an equalization tank? What are the advantages of using an equalization tank? Draw a flow diagram to illustrate preliminary treatment with an on-line equalization tank.

REFERENCES

Camp, T. R. 1942. Grit Chamber Design. *Sewage Works Journal*, 14:368–381.

Davis, M. 2020. *Water and Wastewater Engineering: Design Principles and Practice.* 2nd edn. McGraw-Hill, Inc., New York, NY, USA.

Droste, R. L. and Gehr, R. 2018. *Theory and Practice of Water and Wastewater Treatment.* 2nd edn. John Wiley & Sons, Inc., Hoboken, NJ, USA.

EPA. 1999. Combined Sewer Overflow Technology Fact Sheet: Screens. EPA 832-F-99-040, United States Environmental Protection Agency, Office of Water, Washington, DC, USA.

Eutek. 2008. *Eutek Systems*, Hillsboro, Oregon, USA.

Metcalf and Eddy, Tchobanoglous, G., Stensel, H., Tcuchihashi, R., and Burton, F. 2013. *Wastewater Engineering: Treatment and Resource Recovery.* 5th edn. McGraw-Hill, Inc., New York, NY, USA.

Wilson, G., Tchobanoglous, G., and Griffiths, J. 2007. The Nitty Gritty: Grit Sampling and Analysis. *Water Environment & Technology*, (July) 64–68.

Chapter 7

Primary treatment

7.1 INTRODUCTION

The objective of primary treatment is to remove a significant fraction of the suspended solids and floating material from the wastewater by sedimentation. The suspended solids removed are organic in nature and thus contribute to the BOD (biochemical oxygen demand) of the sludge. The floating material can include oil, grease, rags, etc. that were not removed in upstream processes. These are removed as scum from the water surface in the tank. The removal of larger organic solids helps to reduce the load on the secondary biological reactors. The solids removed are further treated in digesters, or other processes, and stabilized before disposal.

Primary treatment mainly involves sedimentation or settling by gravity. The various types of settling that are observed in water and wastewater treatment operations are described in the following sections. In some cases, sedimentation is enhanced by the addition of coagulation and/or flocculation agents. The process is called Enhanced Clarification, or *Chemically Enhanced Primary Treatment* (CEPT). This is discussed in more detail at the end of the chapter.

7.2 TYPES OF SETTLING/SEDIMENTATION

There are mainly four types of gravitational settling observed in water and wastewater treatment operations. They are:

1. **Type I or discrete particle settling** – Particles whose size, shape, and specific gravity do not change with time are called *discrete particles* (Peavy et al., 1985). Type I sedimentation refers to the settling of discrete particles in a dilute suspension, where the particle concentration is low enough that the particles settle as individual entities. There is no interference of velocity fields with neighboring particles. This type of settling is usually observed in grit chambers.
2. **Type II or flocculent settling** – This refers to settling observed in a suspension with particles that coalesce, or flocculate as they come in

DOI: 10.1201/9781003134374-7

contact with other particles. This results in increasing the size, shape, and mass of the particles, thus increasing the settling rate. Type II sedimentation is observed in primary clarifiers, in the upper portion of secondary clarifiers in wastewater treatment, and also in clarifiers following coagulation–flocculation in water treatment operations.

3. **Type III or hindered settling** – This is also called *zone settling*. It refers to settling that occurs in a suspension of intermediate concentration, where interparticle forces are sufficient to hinder the settling of adjacent particles. The mass of particles settles as a unit, and a solid–liquid interface develops at the top of the mass (Metcalf and Eddy et al., 2013). Hindered or zone settling is observed in secondary clarifiers following biological treatment, such as activated sludge reactors.

4. **Type IV or compression settling** – This occurs in highly concentrated suspensions, where a structure is formed due to the high concentration, and settling can take place only by compression of the structure. As more particles are added to the structure from the liquid, the increasing mass causes compression settling. This type of settling is observed at the bottom of secondary clarifiers following activated sludge reactors, and also in solids thickeners.

7.3 TYPE I SEDIMENTATION

7.3.1 Theory of discrete particle settling

The settling of discrete particles in a fluid can be analyzed using Newton's law and Stokes' law. Consider a discrete particle falling in a viscous and quiescent body of fluid, as shown in Figure 7.1(a). The forces acting on the particle are (i) F_g due to gravity in a downward direction, (ii) F_b due to buoyancy in an upward direction, and (iii) F_D due to frictional drag in an upward

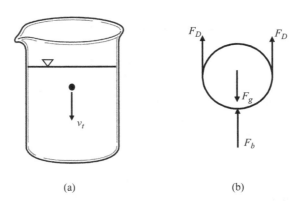

(a)   (b)

Figure 7.1 (a) Discrete spherical particle falling and (b) forces acting on the particle in a fluid.

direction (Droste and Gehr, 2018; Metcalf and Eddy et al., 2013), as shown in Figure 7.1(b). The effective gravitational force is given by

$$F_G = F_g - F_b = (\rho_p - \rho_w)gV_p \tag{7.1}$$

where

ρ_p = density of particle, kg/m^3
ρ_w = density of water, 1000 kg/m^3 at 4°C
g = acceleration due to gravity, 9.81 m/s^2
V_p = volume of particle, m^3

The drag force is given by

$$F_D = \frac{C_d A_p \rho_w v_p^2}{2} \tag{7.2}$$

where

C_d = drag coefficient
A_p = cross-sectional or projected area of particles in the direction of flow, m^2
v_p = particle settling velocity, m/s

The drag force F_D acts in a direction opposite to the driving force F_G, and increases as the square of the velocity. Acceleration occurs at a decreasing rate until a steady velocity is reached, where the driving force equals the drag force:

$$(\rho_p - \rho_w)gV_p = \frac{C_d A_p \rho_w v_p^2}{2} \tag{7.3}$$

For spherical particles with diameter, d_p,

$$\frac{V_p}{A_p} = \frac{2}{3}d_p \tag{7.4}$$

Substituting v_p for v_t, the terminal settling velocity of the particle, and using Equation (7.4), Equation (7.3) provides an expression for the terminal settling velocity:

$$v_t = \left\{ \frac{4gd_p}{3C_d} \left(\frac{\rho_p - \rho_w}{\rho_w} \right) \right\}^{\frac{1}{2}} \tag{7.5}$$

The expression for C_d depends on the flow regime (Peavy et al., 1985). The flow regime can be determined from Reynolds number,

$$Re = \frac{\Phi v_t \rho_w d_p}{\mu_w} \qquad (7.6)$$

where

μ_w = dynamic viscosity of water, $N.s/m^2$
Φ = shape factor depending on sphericity of particle. For perfect spheres, $\Phi = 1$.

For laminar flow, $Re < 1$, $C_d = \dfrac{24}{Re}$ $\qquad (7.7)$

For transitional flow, $1 < Re < 10^3$, $C_d = \dfrac{24}{Re} + \dfrac{3}{\sqrt{Re}} + 0.34$ $\qquad (7.8)$

For turbulent flow, $Re > 10^3$, $\qquad C_d = 0.4$ $\qquad (7.9)$

Determination of v_t involves the simultaneous solution of Equation (7.5) and an expression for C_d.

7.3.1.1 Stokes equation

For laminar flow and spherical particles, Equation (7.5) becomes

$$v_t = \frac{g(\rho_p - \rho_w)d_p^2}{18\mu_w} \qquad (7.10)$$

Equation (7.10) is known as Stokes equation, which can be used to calculate the terminal settling velocity of a discrete particle when the conditions of laminar flow and particle sphericity are satisfied.

Example problems are provided below to illustrate the use of the above equations.

Example 7.1: Calculate the terminal settling velocity of a spherical sand particle settling through water at 25°C. The diameter of the particles is 0.6 mm and the specific gravity is 2.65. At 25°C, ρ_w = 997 kg/m^3, and μ_w = 0.89 × 10^{-3} $N.s/m^2$.

SOLUTION

Assume laminar flow and use Equation (7.10) to calculate v_t
$\rho_p = 2.65 \times 1000$ kg/m³ = 2650 kg/m³, where density of water at 4°C = 1000 kg/m³

$$v_t = \frac{g(\rho_p - \rho_w)d_p^2}{18\,\mu_w}$$

$$= \frac{9.81\,\text{m/s}^2\,(2650-997)\text{kg/m}^3\,(0.6\times10^{-3})^2\,\text{m}^2}{18\times(0.89\times10^{-3}\,\text{N.s/m}^2)} = 0.36\,\text{m/s}$$

(Units of N are kg.m/s²)

Now we have to calculate Re and check whether our assumption of laminar flow was correct.

$$\text{Re} = \frac{\varPhi v_t \rho_w d_p}{\mu_w}$$

$$= \frac{1\times0.36\,\text{m/s}\times997\,\text{kg/m}^3\times0.6\times10^{-3}\,\text{m}}{0.89\times10^{-3}\,\text{N}\cdot\text{s/m}^2}$$

$$= 241.96 \rightarrow \text{transitional flow}$$

Our assumption was not correct. Therefore, we have to use Equation (7.5) to calculate v_t. Use a trial and error procedure, since both Re and v_t are unknown.

Trial #1: Assume Re = 241.96 (calculated from previous step)

$$\text{Calculate } C_d = \frac{24}{\text{Re}} + \frac{3}{\sqrt{\text{Re}}} + 0.34 = 0.63$$

$$\text{Calculate } v_t = \left\{ \frac{4gd_p}{3C_d}\left(\frac{\rho_p - \rho_w}{\rho_w}\right)\right\}^{\frac{1}{2}}$$

$$= \left\{ \frac{4\times9.81\,\text{m/s}^2\times0.6\times10^{-3}\,\text{m}}{3x0.63}\left(\frac{2650-997}{997}\right)\right\}^{\frac{1}{2}}$$

$$= 0.14\,\text{m/s}$$

Trial #2: With $v_t = 0.14$ m/s, calculate Re, C_d, and v_t.

$$\text{Re} = 94.1 \rightarrow \text{transitional flow}$$

$C_d = 0.90$

$v_t = 0.12\,\mathrm{m/s}$

Trial #3: With $v_t = 0.12\,\mathrm{m/s}$, calculate $Re, C_d,$ and v_t.

$Re = 80.66 \rightarrow$ transitional flow

$C_d = 0.97$

$v_t = 0.116\,\mathrm{m/s} = 0.12\,\mathrm{m/s}$

The terminal settling velocity of the particle is **0.12 m/s**.

7.3.2 Design of ideal sedimentation tank

An ideal sedimentation tank is designed to achieve complete removal of particles with a specified settling velocity v_0, such that all particles with a terminal settling velocity greater than v_o will be completely removed. Particles with a terminal settling velocity less than v_o will be fractionally removed. Earlier work by Hazen (1904) and Camp (1945) have provided the basis for the sedimentation theory and design of sedimentation tanks.

Let us consider an ideal horizontal flow, rectangular sedimentation tank as shown in Figure 7.2. The length, width, and height of the tank are L, W, and H, respectively. The wastewater flow rate is Q. The flow paths of two particles, P_1 and P_2 are illustrated, along with the horizontal and vertical components of velocity. Particle P_1 has a settling velocity of v_o and is

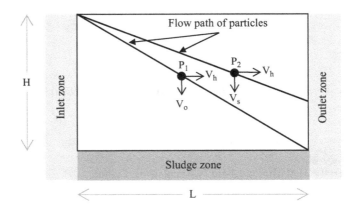

Figure 7.2 An ideal rectangular sedimentation tank.

Source: Adapted from Metcalf and Eddy et al., 2013.

completely removed in time t_d. t_d is the time taken by P_1 to travel the length of the tank, and be deposited in the sludge zone as the wastewater flows out through the outlet zone. The design *detention time* of the tank is thus t_d.

The following assumptions are made:

1. There is no settling of particles in the inlet and outlet zones.
2. Particles settle in the sludge zone and are not resuspended.
3. Plug flow conditions exist.

The horizontal component of velocity v_h is equal to the flow-through velocity, and is related to the flow rate (Q) and cross-sectional area (A_x) in the following manner:

$$v_h = \frac{Q}{A_x} = \frac{Q}{WH} \tag{7.11}$$

The detention time t_d is related to the flow rate and tank volume (V) as

$$t_d = \frac{V}{Q} = \frac{LWH}{Q} \tag{7.12}$$

$$\text{Also, } v_h = \frac{L}{t_d} \tag{7.13}$$

The design settling velocity is given by

$$v_0 = \frac{H}{t_d} \tag{7.14}$$

All particles with a settling velocity, $v_s > v_o$ will be removed 100%. Particles with a settling velocity, $v_s < v_o$ will be partially removed in the ratio

of $\frac{v_s}{v_o}$. The settling velocity of the particle, v_s can be calculated using Equation (7.5). Actual wastewater flows have a large gradation of particle sizes. To determine the removal efficiency for a given detention time, settling column tests can be performed to determine the range of settling velocities of the particles in the system. Settling velocity curves are constructed with corresponding removals, and integrated to determine the overall removal efficiency.

The rate v_o at which the particles settle in the tank is equal to the rate at which clarified water flows out from the tank. This rate is a design parameter and is called the *surface overflow rate* (SOR). It is defined as the flow rate per unit surface area (A_s) and is given by

$$v_o = \frac{Q}{A_s} = \frac{Q}{LW} \tag{7.15}$$

In actual practice, design factors have to be adjusted to account for the effects of inlet and outlet turbulence, short-circuiting, and velocity gradients caused by sludge scrapers.

> **Example 7.2:** A wastewater contains sand particles of three major sizes: 0.002 mm, 0.6 mm, and 6 mm. A settling basin is designed to achieve 100% removal of 0.6 mm diameter particles. Assume water temperature is 25°C. How much removal can be achieved for the other particle sizes?

SOLUTION

The settling velocity of 0.6 mm diameter particle, $v_t = 0.12$ m/s (calculated in Example 7.1)

Calculate the settling velocities of the other two particle sizes.

Assume laminar flow, spherical particles, and use Stokes Equation (7.10).

Calculate settling velocity of 0.002 mm particles, v_1

$$v_1 = \frac{g\left(\rho_p - \rho_w\right)d_p^2}{18\,\mu_w}$$

$$= \frac{9.81\,\text{m/s}^2\,(2650-997)\,\text{kg/m}^3\left(0.002\times10^{-3}\right)^2\,\text{m}^2}{18\times\left(0.89\times10^{-3}\,\text{N.s/m}^2\right)}$$

$$= 4\times10^{-6}\,\text{m/s} < v_t$$

Check Re

$$Re = \frac{\Phi v_1 \rho_w d_p}{\mu_w}$$

$$= \frac{1\times4\times10^{-6}\,\text{m/s}\times997\,\text{kg/m}^3\times0.002\times10^{-3}\,\text{m}}{0.89\times10^{-3}\,\text{N.s/m}^2}$$

$$= 8.96\times10^{-6} < 1, \text{laminar flow assumption is correct.}$$

There will be fractional removal, $\dfrac{v_1}{v_t} = \dfrac{4\times10^{-6}}{0.12}\times100\% = 0.003\%$

Calculate settling velocity of 6 mm particles, v_2, using Stokes equation

$$v_2 = \frac{9.81\,\text{m/s}^2\,(2650-997)\,\text{kg/m}^3\,(6\times10^{-3})^2\,\text{m}^2}{18\times(0.89\times10^{-3}\,\text{N}\cdot\text{s/m}^2)} = 56.28\,\text{m/s}$$

Check Re

$$\text{Re} = \frac{1\times56.28\,\text{m/s}\times997\,\text{kg/m}^3\times6\times10^{-3}\,\text{m}}{0.89\times10^{-3}\,\text{N}\cdot\text{s/m}^2}$$

$$= 3.78\times10^5 > 10^3, \text{turbulent flow}$$

Use $C_d = 0.4$

Calculate v_2 using Equation (7.5)

$$v_2 = \left\{\frac{4gd_p}{3C_d}\left(\frac{\rho_p-\rho_w}{\rho_w}\right)\right\}^{\frac{1}{2}} = \left\{\frac{4\times9.81\,\text{m/s}^2\times6\times10^{-3}\,\text{m}}{3\times0.4}\left(\frac{2650-997}{997}\right)\right\}^{\frac{1}{2}}$$

$$= 0.57\,\text{m/s} > v_t, \text{therefore}\,100\%\,\text{removal of 6 mm particles}$$
will be achieved.

7.4 TYPE II SEDIMENTATION

Type II sedimentation involves the settling of flocculent particles. As flocculation occurs, the shape and mass of the particles increase, resulting in an increase in settling velocities. A settling column test can be used to determine the settling characteristics and removal efficiency of flocculent particles. A settling column with sampling ports situated at regular intervals of depth is used, as illustrated in Figure 7.3 (a). The height of the column should be equal to the proposed tank depth. The test duration should be equal to the proposed detention time. Settling should take place under quiescent conditions. A suspension with solids concentration similar to the wastewater is introduced at the top of the column. At regular time intervals, samples are withdrawn from all the ports and analyzed for suspended solids concentration.

The percent removal at ith time interval for the jth port is given by

$$R_{ij} = \left(1 - \frac{C_{ij}}{C_0}\right)\times100\% \tag{7.16}$$

where
  C_o = initial concentration of suspension, mg/L
  C_{ij} = concentration at ith time interval for sample from the jth port, mg/L

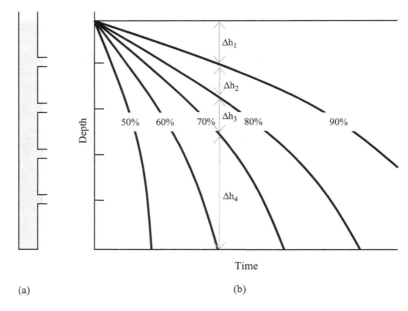

Figure 7.3 (a) Settling column and (b) isoremoval lines for settling column analysis.

The percent removals are plotted as points against time and depth. Then curves of equal percent removal or isoremoval lines are drawn, as illustrated in Figure 7.3 (b). The overflow rate for a particular curve is determined by noting the value where the curve intersects the x-axis (Metcalf and Eddy et al., 2013). The overflow rate or settling velocity, v_o is given by

$$v_o = \frac{H}{t_c} \tag{7.17}$$

where
H = height of settling column, m
t_c = time corresponding to point of intersection of an isoremoval line with the x-axis, min
The fraction of particles removed is given by

$$R = \sum_{n=1}^{n} \left(\frac{\Delta h_n}{H} \right) \left(\frac{R_n + R_{n+1}}{2} \right) \tag{7.18}$$

where
R = suspended solids removal, %
n = number of isoremoval lines

R_n = % removal of isoremoval line number n
R_{n+1} = % removal of isoremoval line number n+1
Δh_n = distance between two isoremoval lines, m
H = height of settling column, m

Conceptually, the shape of the isoremoval lines follows the trajectory of the particles. The slope at any point on any isoremoval line is the instantaneous velocity of the fraction of particles represented by that line (Peavy et al., 1985). It can be seen from Figure 7.3 (b) that the velocity increases with increasing depth, since the slope of the isoremoval line becomes steeper. This is due to the collision and flocculation of the particles, which results in increased mass and increased settling velocities. The settling column test enables us to obtain velocity and removal data at various depths of settling.

Two curves are plotted from the settling column analysis; suspended solids removal ($R\%$) with detention time (t_c), and suspended solids removal ($R\%$) with overflow rate (v_o). Eckenfelder (1980) has recommended the following scale-up factors for the design of a sedimentation tank:

Detention time: 1.75
Overflow rate: 0.65

Example 7.3: A batch settling test for flocculent particles using a 2.0-m column yields the following data:

Initial solids concentration = 430 mg/L
Concentrations in mg/L as a function of time and depth are shown in Table 7.1.
Calculate the detention time (t_o) and overflow rate (v_o) to remove 60% of the influent suspended solids.

SOLUTION

Calculate the percent removal at a time interval and depth, using Equation (7.16).

Table 7.1 Concentration (mg/L) as a function of time and depth (for Example 7.3)

Depth (m)	Time (min)					
	10	20	40	60	90	120
0.5	310	250	194	172	146	138
1.0	335	288	237	198	172	147
2.0	366	323	250	224	190	168

For t = 10 min, and depth of 0.5 m, $R_{10,0.5}$

$$= \left(1 - \frac{310}{430}\right) \times 100\% = 28\%$$

For t = 60 min, and depth of 2.0 m, $R_{60,2.0}$

$$= \left(1 - \frac{224}{430}\right) \times 100\% = 48\%$$

Calculate percent removal as a function of time and depth as shown in Table 7.2.

Plot the isoremoval lines as illustrated in Figure 7.4.

Calculate the overflow rates for each detention time (intersection point of isoremoval line and x-axis), using Equation (7.17).

Table 7.2 Percent removal as a function of time and depth (for Example 7.3)

Depth (m)	Time (min)					
	10	*20*	*40*	*60*	*90*	*120*
0.5	28	42	55	62	66	68
1.0	22	33	45	54	60	66
2.0	15	25	42	48	56	61

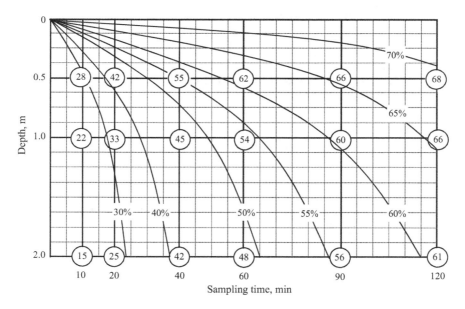

Figure 7.4 Isoremoval lines for settling column analysis (for Example 7.3).

For 50% isoremoval line, Detention time, t = 65 min

$$\text{Overflow rate}, v_o = \frac{2\,\text{m}}{65\,\text{min}} \times \frac{(24 \times 60)\,\text{min}}{\text{d}} = 44.3\,\text{m/d}$$

Calculate the fraction of particles removed using Equation (7.18) for each isoremoval line that intersect with the x-axis (30%, 40%, 50%, 55%, and 60%).

For 50% isoremoval line

$$R = \frac{1.0}{2.0} \times \left(\frac{50+55}{2}\right) + \frac{0.37}{2.0} \times \left(\frac{55+60}{2}\right) + \frac{0.30}{2.0}$$
$$\times \left(\frac{60+65}{2}\right) + \frac{0.20}{2.0} \times \left(\frac{65+70}{2}\right) + \frac{0.13}{2.0} \times \left(\frac{70+100}{2}\right)$$

$$= 58.5\%$$

Plot the percentage of suspended solids removal ($R\%$) versus detention time (t_c), and percentage of suspended solids removal ($R\%$) versus overflow rate (v_o), as shown in Figures 7.5 (a) and 7.5 (b).

Calculate the design values for detention time and overflow rate considering the scale-up factors.

$$\text{Detention time}, t_o = 70\,\text{min} \times 1.78 = 125\,\text{min}$$

$$\text{Overflow rate}, v_o = 40\,\text{m/day} \times 0.65 = 26\,\text{m/d}$$

7.5 PRIMARY SEDIMENTATION

Primary sedimentation tanks or clarifiers are designed to achieve 50–70% removal of suspended solids and 25–40% removal of BOD. The BOD removed is associated with the organic fraction of the suspended solids. Rectangular or circular tanks may be used as primary clarifiers. The type of clarifier selected depends on the site conditions, size of the plant, local regulations, and engineering judgment. Two or more tanks should be provided so that clarification remains in operation while one tank is taken offline for service or maintenance. At large plants, the number of tanks is dictated largely by size limitations. Typical design information for primary sedimentation tanks followed by secondary treatment is provided in Table 7.3.

As outlined in Table 7.3, the important design parameters for primary clarifiers are (i) detention time, (ii) overflow rate, and (iii) weir loading rate.

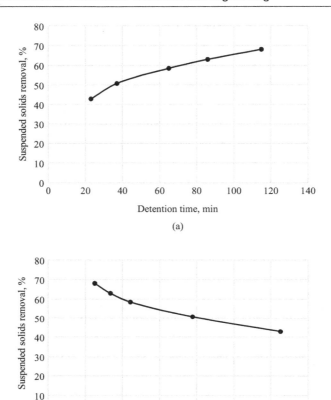

Figure 7.5 (a) Suspended solids removal (%) versus detention time (min) and (b) suspended solids removal (%) versus overflow rate (m/d) (for Example 7.3).

Historically, the tanks are designed for average flow rate conditions. Peak flow rates can be 2–3 times the average rates. For small communities or systems with combined sewers, peak rates can be 10–15 times the design average rates. If the objective is to maximize the primary clarifier efficiency and reduce the load on downstream biological processes, then the hydraulic design should address the peak flow (Davis, 2020). This may be done by sizing the clarifier for peak flows, and/or by using equalization tanks (discussed in Chapter 6).

7.5.1 Rectangular sedimentation tank

Rectangular sedimentation tanks have water flowing through in a horizontal manner. A rectangular sedimentation tank is illustrated in Figure 7.6. At a minimum, two tanks are placed longitudinally in parallel with a common

Table 7.3 Design criteria for primary sedimentation tanks

Parameter	Range	Typical value
Detention time, h	1.5–2.5	2.0
Overflow rate, m³/(m².d)		
At average flow	32–50	40
At peak hourly flow	78–120	100
Weir loading rate, m³/(m.d)	125–500	260
Rectangular tank		
Length, m	15–90	25–40
Width¹, m	3–24	5–10
Depth, m	3–5	4.5
Circular tank		
Diameter, m	3–60	12–40
Depth, m	3–5	4.5

Source: Adapted from Metcalf and Eddy et al. (2013), and Peavy et al. (1985).

¹ For widths greater than 6 m, multiple bays with individual sludge removal equipment may be used.

wall. The *inlet zone* or structure is designed to distribute the water over the entire cross section. The *settling zone* is usually designed based on overflow rates and detention times. In theory, the basin depth or side water depth (swd) is not a design parameter. However, clarifiers with mechanical sludge removal equipment are usually 3–5 m deep. This takes into account the minimum depth required for sludge removal equipment, control of flow through velocities, and prevention of scouring of settled particles. To provide plug flow and minimize short-circuiting, a minimum length-to-width ratio (*L:W*) of 4:1 is recommended. A preferred *L:W* is 6:1. In the *outlet zone*, collection channels called *launders* are placed parallel to the tank length. The clarified water flows into the launders/weirs and exits the tank through overflow weirs. The water level in the tank is controlled by overflow weirs, which may be V-notch weirs or broad crested weirs. The *weir loading rate* is the effluent flow rate over the weir divided by the weir length. The optimum weir loading rate depends on the design of prior and subsequent processes. It can range from 125 to 500 m³/m.d, with typical values around 250 m³/d per meter of weir length (Metcalf and Eddy et al., 2013; Peavy et al., 1985). In the *sludge zone*, the bottom of the tank is sloped toward a sludge hopper for solids collection. Solids collection is accomplished by chain and flight collectors, bridge collectors, or cross collectors. Scum is usually collected and removed from the water surface at the effluent end.

7.5.2 Circular sedimentation tank

A radial flow pattern occurs in circular sedimentation tanks. Circular clarifiers may be center-feed or peripheral feed. A center feed clarifier is illustrated in Figure 7.7. In the center feed tank, water enters a circular well at the center which is designed to distribute the water equally in all directions. It has

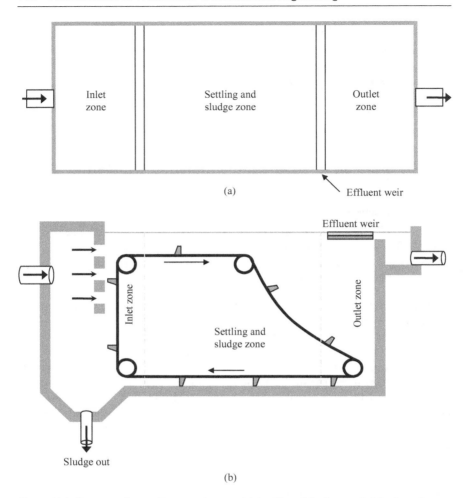

Figure 7.6 Rectangular sedimentation tank/clarifier: (a) plan and (b) elevation.

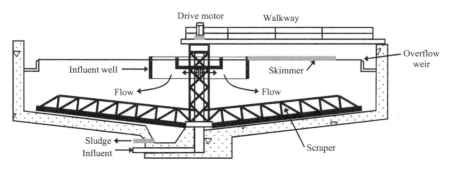

Figure 7.7 Circular sedimentation tank (center feed).

(a)

(b)

Figure 7.8 (a) Circular clarifier and (b) water flowing over V-notch weirs into the effluent launder. (Photos by Rumana Riffat)

an energy-dissipating inlet within the feedwell. In the peripheral feed tank, a suspended circular baffle forms an annular space in which the influent wastewater is discharged in a tangential direction. The water flows spirally around the tank, under the baffle, and clarified water is collected in a centrally located weir trough (Metcalf and Eddy et al., 2013). Circular tanks can range from 3.6 to 10.5 m (12–35 ft) in diameter or larger, depending on the flow rate and site specifications. Figure 7.8 (a) shows a circular clarifier, and (b) water flowing into the effluent launder over V-notch weirs.

Example 7.4: A wastewater treatment plant uses rectangular sedimentation tanks for primary clarification. The average design flow is 14,000 m³/d, with a peaking factor of 2.5. Two tanks are used. The length, width, and depth are 24 m, 7 m, and 4 m, respectively. Single effluent weirs are provided at the outlet zone. Calculate the SOR, detention time, and weir loading rates for the design flow. What happens at peak flow conditions? State regulations specify a minimum detention time of 1 h.

SOLUTION

(a) Determine parameters for average design flow conditions and compare them to values provided in Table 7.3.

$$\text{Flow in each tank}, Q = \frac{14,000}{2} = 7000 \, \text{m}^3 / \text{d}$$

$$\text{SOR}, v_0 = \frac{Q}{LW} = \frac{7000 \, \text{m}^3 / \text{d}}{24 \, \text{m} \times 7 \, \text{m}} = 41.67 \, \text{m}^3 / \text{m}^2.\text{d} \rightarrow \text{within range}$$

$$\text{Detention time}, t_d = \frac{LWH}{Q} = \frac{24 \, \text{m} \times 7 \, \text{m} \times 4 \, \text{m}}{7000 \, \text{m}^3/\text{d}} = 0.096 \, \text{d}$$

$$= 2.30 \, \text{h} \rightarrow \text{within range}$$

Weir length = W = 7 m

$$\text{Weir loading rate} = \frac{Q}{\text{weir length}} = \frac{7000 \, \text{m}^3/\text{d}}{7 \, \text{m}}$$

$$= 1000 \, \text{m}^3/\text{m.d} \rightarrow \text{very high}$$

A second set of weir may be added to reduce the weir loading rate to 500 m³/m.d.

(b) Determine parameters for peak flow conditions, and compare to values provided in Table 7.1.

$$\text{Peak flow} = 14,000 \times 2.5 = 35,000 \, \text{m}^3 / \text{d}$$

$$\text{Flow in each tank}, Q = \frac{35,000}{2} = 17,500 \, \text{m}^3 / \text{d}$$

$$\text{SOR}, v_o = \frac{Q}{LW} = \frac{17,500 \, \text{m}^3 / \text{d}}{24 \, \text{m} \times 7 \, \text{m}} = 104.17 \, \text{m}^3 / \text{m}^2.\text{d} \rightarrow \text{within range}$$

Detention time, $t_d = \dfrac{LWH}{Q} = \dfrac{24\,m \times 7\,m \times 4\,m}{17,500\,m^3/d} = 0.038\,d$

$= 0.92\,h \rightarrow$ slightly less than the specified 1 h minimum.

Total weir length from average flow conditions $= 2W = 14\,m$

Weir loading rate $= \dfrac{Q}{\text{weir length}} = \dfrac{17,500\,m^3/d}{14\,m}$

$= 1250\,m^3/m.d \rightarrow$ very high

One option is to add another effluent weir to reduce the weir loading rate. Another option is to add an equalization tank to store the additional flow during peak flow periods. This would increase the detention time in the primary clarifier, as well as reduce the weir loading rate.

Example 7.5: You have been assigned to design primary clarifiers for a wastewater treatment plant. The average flow rate of the wastewater is 32,000 m³/day with a BOD_5 of 220 mg/L and suspended solids concentration of 300 mg/L. The goal is to remove 30% BOD_5 and 60% suspended solids in primary treatment. Determine the following:

a. The diameter of the primary clarifier for a SOR of 40 m³/m²-day.
b. The detention time in the primary clarifier and the mass of solids removed in kg/day.

Assume a depth of 3.5 m.

SOLUTION

a. Use two circular clarifiers

Flow in each clarifier, $Q = \dfrac{32,000}{2} = 16,000\,m^3/d$

The surface area of each clarifier $= A_s$, with diameter D

$SOR, v_0 = \dfrac{Q}{A_s} = \dfrac{Q}{\dfrac{\pi}{4}D^2}$

Therefore, $A_s = \dfrac{16,000\,m^3/d}{40\,m/d} = 400\,m^2$

$$A_s = \frac{\pi}{4} D^2$$

$$\text{Therefore, } D = \sqrt{\left(\frac{4 \times 400 \, \text{m}^2}{\pi} \right)} = 22.56 \, \text{m} = \mathbf{23 \, m}$$

$$A_s = \frac{\pi}{4} 23^2 = 415.48 \, \text{m}^2$$

(b) Detention time, $t_d = \dfrac{A_s H}{Q} = \dfrac{415.48 \, \text{m}^2 \times 3.5 \, \text{m}}{16{,}000 \, \text{m}^3/\text{d}} = 0.091 \, \text{d} = \mathbf{2.18 \, h}$

$\text{Solids}_{in} = 300 \, \text{mg/L}$

Solids removed $= 60\%$

Mass of solids removed in primary $=$ flow $\times$ concentration

$$= 32{,}000 \, \text{m}^3/\text{d} \times 300 \, \text{mg/L} \times 0.60 \times 10^3 \, \text{L/m}^3 \times 10^{-6} \, \text{kg/mg}$$

$$= \mathbf{5760 \, kg/d}$$

Note : $\text{Solids}_{effluent} = 300 \, \text{mg/L} \times 0.40$

$= 120 \, \text{mg/L}$ going to secondary treatment

Mass of solids to secondary $= 32{,}000 \, \text{m}^3/\text{d} \times 120 \, \text{mg/L}$

$\times 10^3 \, \text{L/m}^3 \times 10^{-6} \, \text{kg/mg}$

$= 3840 \, \text{kg/d}$

$\text{BOD}_{5 \, \text{in effluent}} = 220 \, \text{mg/L} \times (1 - 0.30)$

$= 154 \, \text{mg/L}$ going to secondary treatment

Mass of BOD_5 going to secondary $= 32{,}000 \, \text{m}^3/\text{d} \times 154 \, \text{mg/L}$

$\times 10^3 \, \text{L/m}^3 \times 10^{-6} \, \text{kg/mg}$

$= 4928 \, \text{kg/d}$

7.6 CHEMICALLY ENHANCED PRIMARY TREATMENT (CEPT)

CEPT refers to the process which uses chemicals for coagulation, flocculation, and precipitation of particulate/dissolved solids in the wastewater as a primary step in clarification. Although CEPT was first used around 1840 in France, its use in the US started in the 1960s (Peric et al., 2008). A number of different chemicals were developed, tested, and used. A single chemical or a combination of chemicals can be used. In recent years, CEPT has been used at various wastewater treatment plants for phosphate removal, clarification of wastewater, reduction in sludge volume, and increase in SORs. Increasing the efficiency of primary treatment has dual benefits: (a) It reduces the load for downstream processes, and (b) it enhances the rate of secondary treatment because smaller, easily biodegradable particles are available after primary treatment (Odegaard, 1998). The selection of chemicals for CEPT depends on the primary objective of using them. The dose of chemical coagulant and the method of dosing have to be optimized for better clarification. Chemical coagulants such as ferric chloride are used, together with polymers as flocculating agents. Combined flocculator-clarifiers can be used for this process.

The performance of CEPT depends to a great extent, on the influent characteristics of wastewater. Influent characteristics include TSS (Total Suspended Solids), turbidity, BOD, COD (Chemical Oxygen Demand), particle size distribution, septicity, etc. Characterizing incoming wastewater can provide a vast array of benefits, e.g. feedback for chemical dosing, analysis, and prevention of operational inefficiencies, establishing the trends of seasonal variations, providing the benchmark on operational performance of the plant itself, and providing parameters for comparison of quality of wastewater with that of other plants in the region or country. It is necessary to study the source characteristics of the wastewater to design an optimized settling environment. Influent characteristics such as TSS, turbidity, and total COD were found to have a significant impact on CEPT, in an experimental study conducted by Neupane et al. (2008). Rapid mixing times did not impact performance, but increased flocculation time improved performance. A minimum flocculation time of 10 minutes was required for optimized CEPT performance, as observed by Parker et al. (2000) and Neupane et al. (2008).

PROBLEMS

7.1 Define Type I and Type II settling. What are the differences between these two types of settling? Where are they observed?

7.2 A one (1) cm diameter plastic sphere falls in a viscous liquid at a terminal velocity of 1 cm/sec. What would be the terminal velocity of a

10 cm diameter sphere (of the same plastic material) in the same fluid? Clearly state any assumptions that you make to arrive at your answer.

7.3　A water suspension contains coal particles with an average diameter of 300 μm and specific gravity of 1.47. Determine the settling velocity of the particles at 20°C. Clearly state any assumptions that you make.

7.4　The settling basin for a type-1 suspension is to operate at an overflow rate of 0.76 m³/m².h. The flow rate through the plant is 24,000 m³/day. Determine the dimensions for a long rectangular basin, using a length-to-width ratio of 4:1. Depth should not exceed 4 m. Use more than one tank. Determine the detention time in the tank and the horizontal velocity.

7.5　A rectangular sedimentation tank is designed with a depth of 3.5 m and a detention time of 1 h. The majority of particles in the water are between 0.01 and 0.10 mm in diameter. The specific gravity of the particles is 2.65. The flow rate and temperature of water are 10,000 m³/d and 20°C, respectively. At 20°C, the density of water is 998 kg/m³ and the dynamic viscosity of water is 10^{-3} kg/m.s. Is the design sufficient to achieve complete removal of all particles? State any assumptions you make to calculate your results.

7.6　What types of settling can be observed in a wastewater treatment plant? Draw a flow diagram of a conventional treatment plant treating municipal wastewater and label the unit processes together with the type of settling observed.

7.7　A settling column test was conducted with flocculent particles with a 3.0 m column. Table 7.4 presents the test data for percentage removal as a function of time and depth. Calculate the detention time and overflow rate to remove 65% of the influent suspended solids.

Table 7.4 Percent removal as a function of time and depth (for Problem 7.7)

Depth (m)	Time (min)					
	20	40	60	80	100	120
0.5	45	65	78	84	90	95
1.0	30	48	65	75	76	82
1.5	21	40	50	60	70	78
2.0	16	34	45	55	64	70
2.5	10	30	41	49	58	67
3.0	8	24	37	46	54	60

7.8　An industrial process wastewater contains a mixture of metal fragments and sand. The metal fragments range in diameter from 0.5 to 10 mm with a specific gravity of 1.65. The sand particles range in diameter from 0.04 to 2.0 mm with a specific gravity of 2.65. The wastewater discharge rate is 1,400 m³/d at 20°C. Design a settling tank to remove all the metal and sand particles. If the depth of the tank is 2 m, calculate the detention time.

7.9 A primary clarifier removes 35% of BOD_5 and 55% of suspended solids from the incoming wastewater. Calculate the mass of solids and mass of BOD_5 removed in kg/d for a plant processing 4500 m³/d of wastewater with 275 mg/L BOD_5 and 400 mg/L SS.

7.10 A secondary treatment plant treats 6000 m³/d of wastewater with 155 mg/L BOD_5 and 300 mg/L suspended solids. The primary clarifier removes 30% BOD_5 and 50% SS. Calculate the mass of BOD_5 that has to be removed in secondary treatment, in order to achieve an effluent BOD_5 of 30 mg/L.

7.11 Define the term *surface overflow rate*. How is it related to the settling velocity of the particles?

7.12 What is CEPT? What are the advantages of using CEPT?

7.13 A wastewater treatment plant has four primary clarifiers, each with a diameter of 15 m and a side water depth of 4 m. The average daily flow is 24,000 m³/d. The effluent weir is located on the periphery of each tank. Calculate the surface overflow rate, detention time, and weir loading rates. If one tank is taken out of service for maintenance, what happens to the overflow rate and detention time? Is the design adequate for the average flow with the 3 tanks in service?

7.14 A wastewater treatment plant is designed for a community with an average daily wastewater design flow of 2.0 MGD (million gal/d). The incoming wastewater has an average BOD_5 of 230 mg/L and 260 mg/L of suspended solids. After screening and grit removal, the wastewater is to be treated by primary sedimentation in two parallel treatment trains of circular clarifiers. Determine the diameter and depth of each tank for the following design criteria: maximum overflow rate of 700 gpd/ft² and detention time of 3 h at average flow.

7.15 A wastewater treatment plant has a design average flow of 15,000 m³/d. The engineer wishes to use rectangular clarifiers for primary treatment, with a design depth of 3 m. Design the rectangular tanks for a maximum overflow rate of 42 m³/m².d and a minimum detention time of 2 h. Which criterion governs the design?

7.16 A moderate amount of total BOD can be removed in a primary clarifier through settling. The relationship between percent removal of total BOD and the surface overflow rate in a circular clarifier can be described by the following straight-line equation:

$$y = 42 - 0.255x$$

where y is the percent removal of total BOD, and x is the overflow rate in m/d. The average wastewater flow rate is 5000 m³/d. Calculate the diameter of the clarifier that would result in the removal of one-third of the total BOD. Also, calculate the detention time in the primary clarifier, when the sidewall depth is 3.6 m.

7.17 You have been assigned to design a primary clarifier for a wastewater treatment plant. The average flow rate of the wastewater is 10,500 m^3/day with a BOD_5 of 200 mg/L and suspended solids concentration of 250 mg/L. Primary treatment must remove 30% BOD_5 and 45% suspended solids. Calculate the diameter of the primary clarifier for a surface overflow rate of 40 m^3/m^2-day. Calculate the detention time in the primary clarifier and the mass of solids removed in kg/day. Assume a depth of 2.5 m.

REFERENCES

Camp, T. R. 1945. Sedimentation and the Design of Settling Tanks. *Transactions of the American Society of Civil Engineers*, 111:895–936.

Davis, M. 2020. *Water and Wastewater Engineering: Design Principles and Practice.* 2nd edn. McGraw-Hill, Inc., New York, NY, USA.

Droste, R. L. and Gehr, R. 2018. *Theory and Practice of Water and Wastewater Treatment.* 2nd edn. John Wiley & Sons, Inc. Hoboken, NJ, USA.

Eckenfelder, W. W. 1980. *Industrial Water Pollution Control.* McGraw-Hill, Inc., NY, USA.

Hazen, A, 1904. On Sedimentation. *Transactions of the American Society of Civil Engineers*, 53:46–48.

Metcalf and Eddy, Tchobanoglous, G., Stensel, H., Tcuchihashi, R., and Burton, F. 2013. *Wastewater Engineering: Treatment and Resource Recovery.* 5th edn. McGraw-Hill, Inc., New York, NY, USA.

Neupane, D., Riffat, R., Murthy, S., Peric, M., and Wilson, T. 2008. Influence of Source Characteristics, Chemicals and Flocculation on Chemically Enhanced Primary Treatment. *Water Environment Research*, 80 (4):331–338.

Odegaard, H. 1998. Optimized Particle Separation in the Primary Step of Wastewater Treatment. *Water Science Technology* 37 (10):43–53.

Parker D.; Esquer M.; Hetherington M.; Malik A.; Robinson D.; Wahlberg E.; Wang J. 2000. Assessment and Optimization of a Chemically Enhanced Primary Treatment System. *Proc. 73rd Annual Conference and Exposition of Water Environment Federation*, WEFTEC, Anaheim, CA.

Peric, M., Riffat, R., Murthy, S., Neupane, D., and Cassel, A. 2008. Development of a Laboratory Clarifier Test to Predict Full-Scale Primary Clarifier Performance. *International Journal of Environmental Research*, 2 (2):103–110.

Peavy, H. S., Rowe, D. R., and Tchobanoglous, G. 1985. *Environmental Engineering.* McGraw-Hill, Inc., New York, NY, USA.

Chapter 8

Secondary treatment

Suspended growth process

8.1 INTRODUCTION

Secondary treatment usually consists of biological treatment of primary effluent wastewater. The objectives of secondary treatment are to reduce the BOD and suspended solids of the effluent to acceptable levels. In some cases, nutrient removal may also be an objective. Depending on discharge limits, the secondary effluent may be discharged to surface waters after disinfection, or proceed to tertiary treatment. Two major categories of biological treatment processes are (i) suspended growth and (ii) attached growth processes. In this chapter, aerobic suspended growth processes for BOD removal will be described in detail, with major emphasis on the activated sludge process and its variations. Land-based systems, such as ponds, lagoons, and septic tanks are presented in the latter part of this chapter. Attached growth processes are discussed in Chapter 9. Biological processes used for nitrogen and phosphorus removal are presented in detail in Chapter 13.

In a suspended growth process, the microorganisms are kept in suspension in a biological reactor by using a suitable mixing technique. The microorganisms use the organic matter as food and convert them to new biological cells, energy, and waste matter. Municipal wastewater contains a wide variety of organics, consisting of proteins, fats, and carbohydrates, among others. As a result, a variety of organisms or a mixed culture is required for complete treatment. Each type of organism in the mixed culture uses the food that is most suitable for its metabolism (Peavy et al., 1985). The larger species, in turn, feed on the smaller species. For example, the rotifers and crustaceans feed on the protozoa, while the protozoa feed on the bacteria, and so on. The microorganisms used in the biological treatment processes are essentially the same as those found in surface waters, performing degradation of organic matter during natural purification processes. Natural purification processes take place over an extended period of time, ranging from days to weeks, depending on the strength of the wastewater and the availability of a suitable microbial population, as described in Chapter 4. *The role of the engineer* is to design biological treatment processes using the same basic principles but providing suitable environmental conditions and

DOI: 10.1201/9781003134374-8

process parameters that enhance reaction rates, such that purification takes place within a short period of time, e.g. in a matter of hours. A thorough understanding of microbial growth kinetics, substrate utilization, principles of mass balance, reactor kinetics, and operational parameters is necessary for the design of biological treatment processes. These are discussed in more detail in the following sections.

8.2 MICROBIAL GROWTH KINETICS

The rate of microbial growth and rate of substrate utilization are among the fundamental kinetic parameters of biological treatment processes. A batch experiment can be conducted with a specific amount of food or substrate (S) in a laboratory reactor inoculated with a mixed culture of microorganisms (X). The rate of biomass growth dX/dt, and the corresponding rate of substrate utilization over time dS/dt, can be represented by the curves shown in Figure 8.1. The microbial growth curve has four distinct phases. These phases have been described previously in Chapter 3.

8.2.1 Biomass yield

From Figure 8.1, we can see that the rate of biomass growth increases with a corresponding decrease in the rate of substrate utilization. If all the substrate was converted to biomass, then the rate of biomass production would

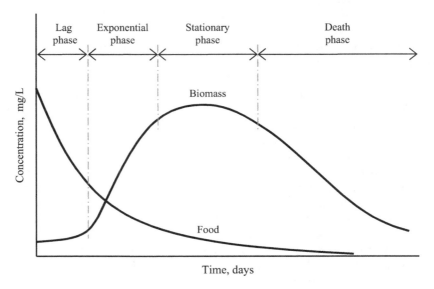

Figure 8.1 Relationship between microbial growth and substrate utilization.

equal the rate of substrate utilization. But a part of the food is converted to energy and waste products, as well as new cells. For this reason,

This can be expressed as

$$\frac{dX}{dt} \propto -\frac{dS}{dt} \tag{8.1}$$

or

$$\frac{dX}{dt} = Y\left(-\frac{dS}{dt}\right) \tag{8.2}$$

or

$$r_g = -Y\left(r_{su}\right) \tag{8.3}$$

where

$\frac{dX}{dt} = r_g$ = growth rate of biomass, mg/L.d

$\frac{dS}{dt} = r_{su}$ = rate of substrate utilization, mg/L.d

Y = biomass yield

or

$$Y = \frac{mg\, biomass\, produced}{mg\, substrate\, utilized} \tag{8.4}$$

The biomass in a reactor is usually measured in terms of the concentration of total suspended solids (TSS) or volatile suspended solids (VSS). The substrate concentration can be measured in terms of BOD or COD. Therefore,

$$Y = \frac{mg\, VSS\, produced}{mg\, BOD\, removed} \text{ or } Y = \frac{mg\, TSS\, produced}{mg\, COD\, removed}$$

The yield coefficient Y depends on the metabolic pathway used in the degradation process. Aerobic processes have a higher yield of biomass compared to anaerobic processes. Typical values of Y for aerobic processes range from 0.4 to 0.8 kg VSS/kg BOD_5, while they range from 0.08 to 0.2 kg VSS/kg BOD_5 for anaerobic processes.

8.2.2 Logarithmic growth phase

In the logarithmic growth phase, we can usually assume first order kinetics to obtain the following rate expression:

$$r_g = \frac{dX}{dt} = \mu X \tag{8.5}$$

where

$\frac{dX}{dt}$ = growth rate of biomass, mg/L.d

μ = specific growth rate, d^{-1}
X = concentration of biomass, mg/L

8.2.3 Monod model

A number of models have been developed to model the microbial growth in biological reactors. One of the earliest models was the *Monod model*, which has served as the basis for the development of numerous models that are in use today. The Monod model assumes that the rate of substrate utilization, and therefore the rate of biomass production, is limited by the rate of enzymatic reactions involving the limiting substrate. The Monod equation for microbial growth (Monod, 1949) is given by

$$\mu = \mu_{max} \cdot \frac{S}{K_s + S} \tag{8.6}$$

where

μ_{max} = maximum specific growth rate constant, d^{-1}
S = substrate concentration, mg/L
K_S = half-saturation coefficient, mg/L

K_S is the substrate concentration corresponding to growth rate $\mu = \frac{1}{2}\mu_{max}$. Figure 8.2 is a graphical representation of the Monod equation, which illustrates that the growth rate of biomass is a hyperbolic function of the substrate concentration.

Based on the Monod equation and Figure 8.2, at high substrate concentration, the system is considered to be enzyme-limited ($S \gg K_s$). In this case, the growth rate is approximately equal to the maximum growth rate, and Equation (8.6) becomes

$$\mu \approx \mu_{max} \tag{8.7}$$

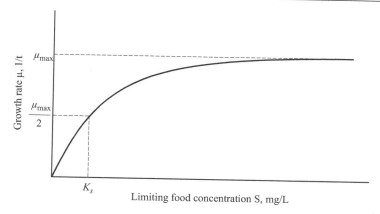

Figure 8.2 Graphical representation of the Monod model.

Another situation arises at low substrate concentrations, when the substrate is limiting ($S \ll K_s$). Equation (8.6) can then be written as

$$\mu = \frac{\mu_{max} S}{K_S} = K'S \qquad (8.8)$$

where

$$\frac{\mu_{max}}{K_S} = K'$$

The growth rate of biomass becomes independent of the concentration of biomass present. The specific growth rate becomes first order with respect to substrate concentration, as shown in Equation (8.8) and represented by the initial straight line segment of the curve in Figure 8.2.

8.2.4 Biomass growth and substrate utilization

Combining Equations (8.5) and (8.6), we can obtain an expression for the rate of biomass production as

$$r_g = \frac{dX}{dt} = \frac{\mu_{max} S X}{K_S + S} \qquad (8.9)$$

Combining Equations (8.3) and (8.9), we can write the following expression for the rate of substrate utilization

$$r_{su} = -\frac{r_g}{Y} = -\frac{\mu_{max} S X}{Y(K_S + S)} \qquad (8.10)$$

or

$$\frac{dS}{dt} = -\frac{\mu_{max} S X}{Y(K_S + S)} \tag{8.11}$$

or

$$r_{su} = -\frac{k S X}{K_S + S} \tag{8.12}$$

where $k = \frac{\mu_{max}}{Y}$, and is defined as the maximum rate of substrate utilization per unit mass of microorganisms.

8.2.5 Other rate expressions for r_{su}

Depending on the substrate and specific microorganisms involved, a number of rate expressions have been used to describe substrate utilization rates, in addition to the substrate-limited relationship presented in Equation (8.11). These are based on experimental results observed by various researchers. Some of the commonly used rate expressions are:

$$r_{su} = -k \tag{8.13}$$

$$r_{su} = -k S \tag{8.14}$$

$$r_{su} = -k S X \tag{8.15}$$

where k = substrate utilization rate coefficient, time^{-1}

8.2.6 Endogenous metabolism

In the death and decay phase of the microbial growth curve, some endogenous metabolism takes place. It is assumed that the decrease in biomass caused by death and predation is proportional to the concentration of microorganisms present. The endogenous decay is thus assumed to be first order and can be written as

$$\left(\frac{dX}{dt}\right)end = r_d = -k_d X \tag{8.16}$$

where
r_d = rate of decay, mg/L.d
k_d = endogenous decay coefficient, d^{-1}

8.2.7 Net rate of growth

Combining Equation (8.9) and (8.16), we can obtain an expression for the net rate of growth:

$$r_{g(net)} = r_g + r_d$$

$$\left(\frac{dX}{dt}\right)net = \left(\frac{dX}{dt}\right) + \left(\frac{dX}{dt}\right)end$$

$$\left(\frac{dX}{dt}\right)net = \frac{\mu_{max} S X}{K_S + S} - k_d X \tag{8.17}$$

The net biomass yield can be expressed as

$$Y_{net} = -\frac{r_{g(net)}}{r_{su}} \tag{8.18}$$

The net biomass yield is used as an estimate of the amount of active microorganisms in the system.

Table 8.1 presents typical values of the kinetic coefficients for the activated sludge process treating domestic wastewater.

8.2.8 Rate of oxygen uptake

The rate of oxygen uptake is stoichiometrically related to the rate of utilization of organic matter and the biomass growth rate. Based on the formula $C_5H_7O_2N$ for biomass, the oxygen equivalent of biomass (measured as VSS) is approximately 1.42 g COD utilized/g VSS produced (Metcalf and Eddy et al., 2013). Therefore, the oxygen uptake rate can be expressed as

$$r_o = r_{su} = -1.42 r_g \tag{8.19}$$

Table 8.1 Kinetic coefficients for activated sludge process treating municipal wastewater

Kinetic coefficient	Range	Typical value
μ_{max}, mg COD/mg VSS.d	1–10	5
K_s, mg/L COD	12–60	38
k_d, mg VSS/mg VSS.d	0.05–0.15	0.10
Y, mg VSS/mg COD	0.25–0.60	0.40

Source: Adapted from Metcalf and Eddy et al. (2013).

where
r_o = rate of oxygen uptake, g O_2/m^3.d
r_{su} = rate of substrate utilization, g COD/m^3.d
1.42 = COD of cell tissue, g COD/g VSS
r_g = rate of biomass growth, g VSS/m^3.d

8.2.9 Effect of temperature

Temperature has a significant effect on biological reactions. Temperature influences the metabolic activities of the microbial population, as well as gas transfer rates and settling characteristics of the biomass. The van't Hoff–Arrhenius model can be used to describe the effect of temperature on reaction rate coefficients as shown below:

$$k_T = k_{20}\theta^{(T-20)} \tag{8.20}$$

where
k_T = reaction rate coefficient at temperature T °C
k_{20} = reaction rate coefficient at 20°C
θ = temperature activity coefficient
T = temperature, °C

Values of θ range from 1.02 to 1.25 depending on the type of substrate and biological process (Metcalf and Eddy et al., 2013).

8.3 ACTIVATED SLUDGE PROCESS (FOR BOD REMOVAL)

The most widely used suspended growth process is the activated sludge process. It is used for the biological treatment of municipal and industrial wastewaters. The process concept dates back to the work of Dr. Angus Smith in the early 1880s, who investigated the aeration of wastewater tanks to accelerate biological oxidation. In 1912 and 1913, experiments were conducted by Clark and Gage with aerated wastewater to grow microorganisms in bottles and tanks, at Lawrence Experiment Station (Clark and Adams, 1914). These results were the motivation for additional research carried out at Manchester Sewage Works in England, by Ardern and Lockett (1914). They developed the process and named it *Activated Sludge* because it involved the production of an activated mass of microorganisms capable of aerobic stabilization of organic matter in wastewater (Metcalf and Eddy et al., 2013).

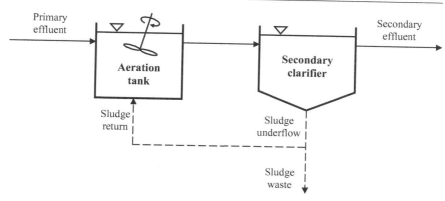

Figure 8.3 Activated sludge process.

   The basic activated sludge process consists of three components, as illustrated in Figure 8.3. They are, (i) a biological reactor where the microorganisms are kept in suspension and aerated, (ii) a sedimentation tank or clarifier, and (iii) a recycle system for returning settled solids from the clarifier to the reactor. Wastewater flows continuously into the aeration tank or biological reactor. Air is introduced to mix the wastewater with the microorganisms and to provide the oxygen necessary to maintain aerobic conditions. The microorganisms degrade the organic matter in wastewater and convert them to cell mass and waste products. The mixture then goes to the secondary clarifier, where clarification of effluent and thickening of settled solids takes place. The clarified effluent is discharged for further treatment or disposal. The thickened solids are removed as underflow. A portion of the underflow is wasted (called *waste activated sludge* WAS), while the remainder (20–50%) is returned to the aeration tank as *return activated sludge* (RAS). The return sludge helps to maintain a high concentration of *active biomass* in the aeration tank.
   A large number of variations of the activated sludge process have been developed and are currently in use. Descriptions of these processes are provided later in this chapter. The biological reactor may be operated as a continuous-flow stirred tank reactor (CSTR), or plug flow reactor (PFR). In recent times, activated sludge processes are used more frequently for BOD removal in conjunction with the removal of nitrogen and/or phosphorus. These are discussed in Chapter 13. A large body of knowledge exists based on past and present research on the microbial communities, operational parameters, process models, and removal capabilities of various pollutants in the activated sludge process (Jahan et al., 2011; Schmit et al., 2010; Plósz et al., 2010; Jones et al., 2009; Ma et al., 2009; Rieger et al., 2010; among others).

8.3.1 Design and operational parameters

The following are definitions of basic design parameters for biological treatment reactors.

MLSS – mixed liquor suspended solids concentration in the biological reactor. It is measured as the VSS or TSS concentration in the reactor, expressed as mg/L or kg/m^3. MLVSS represents the active biomass concentration in the reactor. The concentration of active biomass plus the inert solids is called MLSS. Usually, the term MLSS is used for both, with the units of measurement (VSS or TSS) indicating the difference.

SRT – solids retention time of the reactor. It is also called *sludge age*, or *mean cell residence time*. It is the amount of time spent by a unit mass of activated sludge in the reactor. It is defined as the ratio of the mass of solids in the reactor, to the mass of solids wasted per day. It is given by the equation below.

$$\theta_c = \frac{\text{Mass of solids in reactor}}{\text{Mass of solids wasted per day}} \tag{8.21}$$

For the activated sludge process illustrated in Figure 8.3, θ_c is given by

$$\theta_c = \frac{V\,X}{Q_w\,X_w + Q_e X_e} \tag{8.22}$$

where
  θ_c = solids retention time, d
  X = MLSS concentration in reactor, mg/L
  X_w, X_e = biomass concentration in waste sludge and effluent, respectively, mg/L
  Q_w = Rate of sludge wastage, m^3/d
  Q_e = Effluent flow rate, m^3/d

SRT is the most important design and operating parameter, as it affects process performance, aeration tank volume, sludge production, and oxygen requirements. For BOD removal, an SRT of 3 d may be used at temperatures ranging from 18°C to 25°C. At 10°C, SRT values of 5–6 d are required (Metcalf and Eddy et al., 2013).

F/M Ratio – the ratio of food to microorganisms in the reactor. It is calculated as the mass of BOD removed in the reactor, divided by the mass of microorganisms in the reactor. It is expressed as

$$\frac{F}{M} = \frac{Q(S_0 - S)}{V\,X} \tag{8.23}$$

where
  F/M = food to microorganism ratio, mg BOD/mg VSS.d
  S_0 = BOD$_5$ concentration of substrate entering the reactor, mg/L
  S = BOD$_5$ concentration of substrate leaving the reactor, mg/L

The F/M ratio is an important design variable that dictates the phase of operation on the microbial growth curve. A low F/M ratio of about 0.05 indicates operation in the decay phase, while a high F/M ratio around 1.0 and above indicates a log growth phase. Conventional activated sludge processes are operated at F/M ratios from 0.2 to 0.4. This indicates operation toward the end of the stationary phase and corresponds to a low substrate concentration. This is desired when the aeration tank is operated as a CSTR since the concentration in the reactor will be the same as the concentration in the effluent. These are illustrated in Figure 8.4.

The use of SRT and F/M ratio in design allows for a trade-off between reactor volume and MLSS concentration in the reactor.

Volumetric Loading Rate – the mass of substrate or food applied per unit volume of the reactor. It is given by

$$V_L = \frac{Q S_0}{V}$$

(8.24)

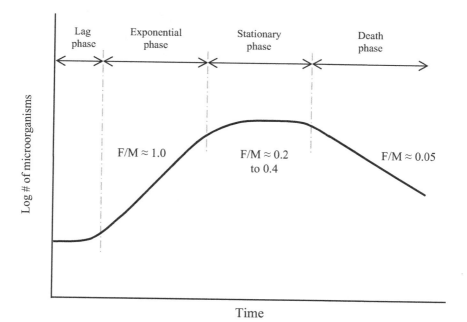

Figure 8.4 F/M ratios corresponding to various microbial growth phases.

where

V_L = volumetric loading rate, kg BOD_5/m^3
S_0 = substrate concentration entering the reactor, kg BOD_5/m^3
Q = flow rate entering the reactor, m^3/d
V = reactor volume, m^3

HRT – hydraulic retention time of the reactor. It is the time spent by a fluid particle in the reactor before it is discharged. The HRT is expressed as

$$\theta = \frac{V}{Q} \tag{8.25}$$

where

θ = hydraulic retention time, d
V = volume of the reactor, m^3
Q = volumetric flow rate, m^3/d

The HRT in conventional activated sludge reactors ranges from 3 to 8 h. It can be reduced in high rate processes.

Example 8.1: An activated sludge process is used to treat a wastewater with a flow rate of 1800 m^3/d and a BOD_5 concentration of 300 mg/L. The aeration tank is operated at an MLSS of 2500 mg/L, and HRT of 7 h. The sludge is wasted at 34 m^3/d with a solids concentration of 9000 mg/L. The effluent BOD_5 concentration is 25 mg/L. Calculate the volume of the aeration tank, the SRT, volumetric loading rate, and the F/M ratio of the process.

SOLUTION
Step 1. Calculate aeration tank volume.
HRT, θ = 7 h = 0.29 d
Using Equation 8.25,

$$\theta = \frac{V}{Q}$$

or

$$0.29\,d = \frac{V}{1800\,m^3/d}$$

or

$$V = 525\ m^3$$

Step 2. Calculate SRT.
Assume solids in the effluent is negligible, and using Equation 8.21,

$$\theta_c = \frac{\text{mass of solids in reactor}}{\text{mass of solids wasted per day}} = \frac{VX}{Q_w X_w}$$

or

$$\theta_c = \frac{525\,\text{m}^3 \times 2500\,\text{mg/L}}{34\,\dfrac{\text{m}^3}{\text{d}} \times 9000\,\text{mg/L}}$$

or

$$\theta_c = 4.29\,\text{d}$$

Step 3. Calculate F/M ratio using Equation (8.23).

$$\frac{F}{M} = \frac{Q(S_0 - S)}{VX}$$

or

$$\frac{F}{M} = \frac{1800\,\dfrac{\text{m}^3}{\text{d}}(300-25)\,\text{mg/L}}{525\,\text{m}^3 \times 2500\,\text{mg/L}}$$

or

$$\frac{F}{M} = 0.38\,\text{d}^{-1}$$

Step 4. Calculate volumetric loading rate using Equation 8.24.

$$V_L = \frac{Q S_0}{V}$$

or

$$V_L = \frac{1800\,\dfrac{\text{m}^3}{\text{d}} \times 0.3\,\text{kg/m}^3}{525\,\text{m}^3}$$

or

$$V_L = 1.03\,\text{kg BOD}_5\,/\,\text{m}^3.\text{d}$$

8.3.2 Factors affecting microbial growth

An in-depth knowledge of the factors that affect the growth of the mixed population of microorganisms is important for efficient reactor operation. Aerobic heterotrophic bacteria are predominant. Protozoa are also present, which consume bacteria and colloidal particles. The important factors include pH, temperature, alkalinity, type of substrate and concentration, presence of toxins, dissolved oxygen concentration, among others. Some of these factors have been discussed in detail in Chapter 3.

8.3.3 Stoichiometry of aerobic oxidation

The following general equations give a simplified description of the oxidation process (Davis, 2020). Assume CHONS represents organic matter and $C_5H_7O_2N$ represents new cells.

The synthesis reaction is given by

$$\text{CHONS} + O_2 + \text{nutrients} \xrightarrow{\text{bacteria}} C_5H_7O_2N + CO_2 + NH_3 \\ + \text{other products} \tag{8.26}$$

The endogenous respiration is given by

$$C_5H_7O_2N + 5O_2 \xrightarrow{\text{bacteria}} 5CO_2 + NH_3 + 2H_2O + \text{energy} \tag{8.27}$$

8.4 MODELING SUSPENDED GROWTH PROCESSES

In this section, the principles of mass balance will be used together with the kinetic relationships described previously to develop design equations for suspended growth processes. The examples are given for activated sludge reactors, but the principles are applicable to any suspended growth process. The mass balances for each specific constituent, e.g. substrate, biomass, etc., will be conducted across a defined volume of the system. The developed models will then be used for the prediction of effluent biomass and substrate concentrations, MLSS in the reactor, and oxygen requirements.

8.4.1 CSTR without recycle

Consider the completely mixed suspended growth reactor (CSTR) shown in Figure 8.5. Conduct a *mass balance for biomass X* around the system boundary represented by the dashed line.

$$\frac{\text{Rate of}}{\text{accumulation}} = \frac{\text{Rate of}}{\text{inflow}} - \frac{\text{Rate of}}{\text{outflow}} + \frac{\text{Rate of}}{\text{net growth}} \tag{8.28}$$

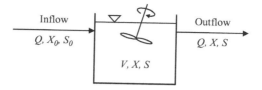

Figure 8.5 Completely mixed reactor without recycle.

or

$$\left(\frac{dX}{dt}\right)V = QX_0 - QX + V\left(\frac{dX}{dt}\right)\text{net}$$ (8.29)

where

Q = influent and effluent wastewater flow rate, m³/d

X_0, X = biomass concentrations in influent and effluent respectively, kg/m³

$\frac{dX}{dt}$ = growth rate of biomass, kg/m³.d

V = reactor volume, m³

The following *assumptions* are made to simplify Equation (8.29):

1. The reactor is at steady state condition. Therefore, accumulation = 0.
2. Complete mixing is achieved. Therefore, concentrations in reactor = concentrations in effluent.
3. The concentration of biomass in influent is negligible compared to the concentration of biomass in the reactor; i.e. $X_0 \approx$ negligible.

From Equation (8.17) we know that $\left(\frac{dX}{dt}\right)\text{net} = \frac{\mu_{max} S X}{K_S + S} - k_d X$
Equation (8.29) becomes

$$0 = -QX - V\left(\frac{\mu_{max} S X}{K_S + S} - k_d X\right)$$

or

$$\frac{Q}{V} = \frac{\mu_{max} S}{K_S + S} - k_d$$ (8.30)

Equation (8.30) can be rewritten using HRT, $\theta = \frac{V}{Q}$ or

$$\frac{\mu_{max} S}{K_S + S} = \frac{1}{\theta} + k_d$$ (8.31)

For a CSTR without recycle, SRT = HRT

Since, $\theta_c = \dfrac{V X}{Q X} = \dfrac{V}{Q} = \theta$

Conduct a *mass balance for substrate S* around the system.

$$\begin{matrix} \text{Rate of} \\ \text{accumulation} \end{matrix} = \begin{matrix} \text{Rate of} \\ \text{inflow} \end{matrix} - \begin{matrix} \text{Rate of} \\ \text{outflow} \end{matrix} + \begin{matrix} \text{Rate of mass} \\ \text{substrate utilization} \end{matrix} \qquad (8.32)$$

or

$$\left(\frac{dS}{dt}\right) V = Q S_0 - Q S - V \left(\frac{dS}{dt}\right) \text{su} \qquad (8.33)$$

where

$\dfrac{dS}{dt}$ = rate of substrate utilization, kg/m³.d

S_0 = substrate concentration in influent, kg/m³
S = substrate concentration in reactor and effluent, kg/m³

Note that the negative sign in Equations (8.32) and (8.33) indicates depletion of the substrate. Using the above-mentioned assumptions and Equation (8.11) for $\left(\dfrac{dS}{dt}\right)$su , Equation (8.33) becomes

$$0 = Q\left(S_0 - S\right) - \frac{V}{Y}\left(\frac{\mu_{\max} S X}{K_S + S}\right)$$

or

$$\frac{\mu_{\max} S}{K_S + S} = \frac{Q}{V} \frac{Y}{X}\left(S_0 - S\right)$$

or

$$\frac{\mu_{\max} S}{K_S + S} = \frac{Y}{\theta X}\left(S_0 - S\right) \qquad (8.34)$$

Equating Equations (8.31) and (8.34), we can write

$$\frac{1}{\theta} + k_d = \frac{Y}{\theta X}\left(S_0 - S\right)$$

Simplifying we get

$$X = \frac{Y(S_0 - S)}{1 + k_d \theta} \qquad (8.35)$$

Equation (8.35) gives us an expression for X in terms of the substrate, HRT (or SRT), and kinetic coefficients for a suspended growth CSTR without recycle.

To determine an expression for substrate S, substitute the value of X from Equation (8.35) into Equation (8.34).

$$\frac{\mu_{max}\, S}{K_S + S} = \frac{Y(S_0 - S)}{\theta\, \dfrac{Y(S_0 - S)}{1 + k_d \theta}} = \frac{1 + k_d \theta}{\theta}$$

$$\text{Or } \mu_{max}\, S = \frac{(K_s + S)(1 + k_d \theta)}{\theta}$$

Simplifying we obtain

$$S = \frac{K_s(1 + k_d \theta)}{\theta(\mu_{max} - k_d) - 1} \qquad (8.36)$$

Equation (8.36) is the design equation for substrate S in terms of the kinetic coefficients and HRT (or SRT), for a suspended growth CSTR without recycle.

8.4.2 Activated sludge reactor (CSTR with recycle)

Now we will develop the design equations for an activated sludge reactor (operated as CSTR) with recycle, using the same concepts of mass balance described above. Consider the activated sludge process presented in Figure 8.6. The flows, as well as biomass and substrate concentrations for the aeration tank and secondary clarifier, are shown.

Conduct a *mass balance for biomass X* around the system boundary represented by the dashed line.

$$\begin{array}{c} \text{Rate of} \\ \text{accumulation} \end{array} = \begin{array}{c} \text{Rate of} \\ \text{inflow} \end{array} - \begin{array}{c} \text{Rate of} \\ \text{outflow} \end{array} + \begin{array}{c} \text{Rate of} \\ \text{net growth} \end{array} \qquad (8.37)$$

or

$$\left(\frac{dX}{dt}\right)V = QX_0 - (Q_e X_e + Q_w X_u) + V\left(\frac{dX}{dt}\right)\text{net} \qquad (8.38)$$

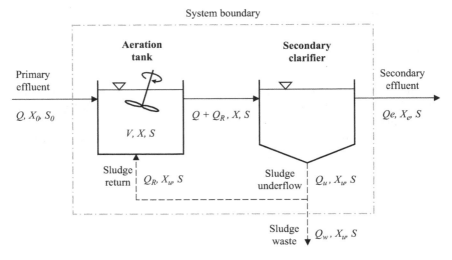

Figure 8.6 Activated sludge process operated as CSTR.

where
Q = influent wastewater flow rate, m^3/d
Q_e = effluent wastewater flow rate, m^3/d
Q_w = waste sludge flow rate, m^3/d
X_0, X_e = biomass concentration in influent and effluent, kg/m^3
X_u = biomass concentration in underflow, kg/m^3
$\dfrac{dX}{dt}$ = growth rate of biomass, $kg/m^3.d$
V = volume, m^3

Equation (8.38) can be simplified by making the following *assumptions*:

1. The reactor is at steady state condition. Therefore, accumulation = 0.
2. Complete mixing is achieved. Therefore, concentrations in reactor = concentrations in effluent.
3. Concentrations of biomass in influent and effluent are negligible compared to the concentrations at other points, i.e. X_0 and $X_e \approx$ negligible.
4. All reactions take place in the reactor or aeration tank. No further conversions of substrate or biomass occur in the clarifier.
5. The volume V represents the volume of the reactor only, based on the above assumption. It does not include the volume of the clarifier.

Using these assumptions and the expression for net growth rate from Equation (8.17), Equation (8.38) can be written as

$$0 = Q_w X_u - V\left(\frac{\mu_{max} S X}{K_S + S} - k_d X \right)$$

or

$$\frac{\mu_{max} S}{K_S + S} - k_d = \frac{Q_w X_u}{V X} \qquad (8.39)$$

Now, $\theta_c = \dfrac{V X}{Q_w X_u}$, using this in Equation (8.39) we get

$$\frac{\mu_{max} S}{K_S + S} = \frac{1}{\theta_c} + k_d \qquad (8.40)$$

Conduct a *mass balance for substrate S* around the system boundary represented by the dashed line.

$$\begin{array}{ccccc} \text{Rate of} & = & \text{Rate of} & \text{Rate of} & \text{Rate of mass} \\ \text{accumulation} & & \text{inflow} & \text{outflow} & \text{substrate utilization} \end{array} \qquad (8.41)$$

or

$$\left(\frac{dS}{dt}\right) V = Q S_0 - \left(Q_w S + Q_e S\right) - V\left(\frac{dS}{dt}\right) su \qquad (8.42)$$

where

$\dfrac{dS}{dt}$ = rate of substrate utilization, kg/m³.d

S_0 = substrate concentration in influent, kg/m³
S = substrate concentration in reactor and effluent, kg/m³

Note that the negative sign in Equations (8.41) and (8.42) indicates depletion of the substrate. From continuity, $Q_e = (Q - Q_w)$. Using the above-mentioned assumptions and Equation (8.11) for $\left(\dfrac{dS}{dt}\right) su$, Equation (8.42) becomes

$$0 = Q\left(S_0 - S\right) - \frac{V}{Y}\left(\frac{\mu_{max} S X}{K_S + S}\right)$$

or

$$\frac{\mu_{max} S}{K_S + S} = \frac{Q}{V} \frac{Y}{X}\left(S_0 - S\right)$$

or

$$\frac{\mu_{max} S}{K_S + S} = \frac{Y}{\theta X}\left(S_0 - S\right) \qquad (8.43)$$

where HRT $\theta = \dfrac{V}{Q}$

Combining Equations (8.40) and (8.43), we obtain the following

$$\frac{1}{\theta_c} + k_d = \frac{Y}{\theta X}(S_0 - S) \tag{8.44}$$

Simplifying Equation 8.44 we obtain the following expression for biomass X,

$$X = \frac{\theta_c Y (S_0 - S)}{\theta (1 + k_d \theta_c)} \tag{8.45}$$

Note: Equation (8.45) reduces to Equation (8.35) for a CSTR without recycle with $\theta = \theta_c$.

To determine an expression for substrate S, substitute the value of X from Equation (8.45) into Equation (8.44) and simplify to obtain

$$S = \frac{K_s (1 + k_d \theta_c)}{\theta_c (\mu_{max} - k_d) - 1} \tag{8.46}$$

Note: Equations (8.45) and (8.46) reduce to Equations (8.35) and (8.36) respectively, for a CSTR without recycle with $\theta = \theta_c$.

8.4.2.1 Other useful relationships

Equation (8.40) can be written as

$$\frac{1}{\theta_c} = \frac{\mu_{max} S}{K_S + S} - k_d$$

or

$$\frac{1}{\theta_c} = \mu \tag{8.47}$$

The *specific substrate utilization rate* U is defined as

$$U = -\frac{r_{su}}{X} = \frac{dS/dt}{X} = \frac{Q(S_0 - S)}{VX} = \frac{(S_0 - S)}{\theta X} \tag{8.48}$$

Using Equation (8.11), Equation (8.40) can be rewritten as

$$\frac{1}{\theta_c} = -Y\frac{r_{su}}{X} - k_d$$

or

$$\frac{1}{\theta_c} = YU - k_d \tag{8.49}$$

A plot of $1/\theta_c$ versus U will result in a straight line. The slope of the straight line will be the yield coefficient Y and the intercept will be k_d. The efficiency of substrate removal is given by

$$E\% = \frac{S_0 - S}{S_o} \times 100\% \tag{8.50}$$

U can also be expressed as

$$U = \frac{\left(\dfrac{F}{M}\right)E}{100} \tag{8.51}$$

8.4.3 Activated sludge reactor (Plug flow reactor with recycle)

An activated sludge process is illustrated in Figure 8.7 where the aeration tank is operated as a PFR. Plug flow may be achieved in long, narrow aeration tanks. In a true plug flow model, all the particles entering the reactor spend the same amount of time in the reactor. Some particles may spend more time in the reactor due to recycle. But while they are in the tank, all pass through in the same amount of time (Metcalf and Eddy et al., 2013). It is difficult to develop a kinetic model due to the varying concentrations of biomass and substrate in the reactor.

Lawrence and McCarty (1970) developed a model for the plug flow process by using two simplifying assumptions. They are:

1. The concentration of microorganisms in the influent to the aeration tank is approximately the same as that in the effluent from the aeration tank. This holds true when $\theta_c/\theta \geq 5$. The resulting average concentration of microorganisms in the reactor is denoted by X_{avg}.
2. The rate of soluble substrate utilization as the wastewater passes through the reactor is given by

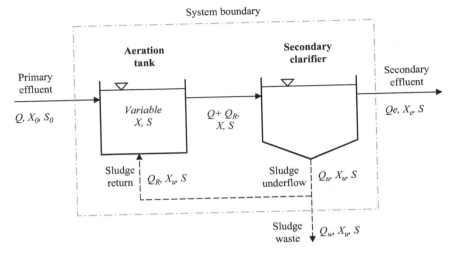

Figure 8.7 Activated sludge process operated as PFR.

$$r_{su} = -\frac{\mu_{max}\, S\, X_{avg}}{Y\left(K_S + S\right)} \qquad (8.52)$$

Integrating Equation (8.52) over the hydraulic retention time in the reactor, and substituting boundary conditions and recycle factor provides the following design equation:

$$\frac{1}{\theta_c} = \frac{\mu_{max}\left(S_0 - S\right)}{\left(S_0 - S\right) + K_s\left(1 + R\right)\ln\left(S_i / S\right)} - k_d \qquad (8.53)$$

where
  R = recycle ratio = $\dfrac{Q_R}{Q}$

  S_i = influent substrate concentration to reactor after dilution with recycle flow, mg/L

or

$$S_i = \frac{S_0 + R\, S}{1 + R} \qquad (8.54)$$

Other terms are the same as defined previously. One of the main differences between the design equations for an activated sludge CSTR (Equation 8.40) and activated sludge PFR (Equation 8.53) is that the SRT (θ_c) is a function of the influent substrate concentration S_0 for the PFR.

In practice, a true plug flow regime is almost impossible to maintain, because of longitudinal dispersion caused by aeration and mixing. The aeration tank may be divided into a series of reactors to approach plug flow kinetics. This also improves the treatment efficiency compared to a CSTR. The CSTR however, can handle shock loads better due to the higher dilution with influent wastewater, as compared to staged reactors in series (Metcalf and Eddy et al., 2013).

8.4.4 Limitations of the models

In practice, the assumptions of ideal CSTR or PFR are extremely difficult to achieve. Real-life reactors fall somewhere in between. The models described in the previous sections provide a useful starting point for the design and modeling of actual processes. The quantification of substrate concentration is also important. The substrate concentration S is the soluble COD concentration that is readily biodegradable, but it is not the total BOD. Some fraction of the suspended solids that remain in the secondary clarifier effluent also contributes to the BOD load of the receiving waters. The total BOD consists of a soluble fraction and an insoluble/particulate fraction. It is important to keep these in consideration when designing a treatment process. We can determine the effluent substrate concentration S in the following manner (Davis, 2020):

$$S = \text{Total allowable BOD} - \text{BOD in effluent suspended solids} \qquad (8.55)$$

Another assumption was that no biological reactions take place in the clarifier. Based on the concentration of biomass and the amount of time spent in the clarifier, this assumption may not be entirely correct. This may result in errors in the calculation of the volume V in the model. It is important to understand these limitations when using the models for the design of treatment processes.

Example 8.2: Develop an expression for the recycle flow Q_R for an activated sludge process using the concept of mass balances.

SOLUTION

Consider the activated sludge process illustrated in Figure 8.8. We will perform a mass balance around the secondary clarifier with the system boundary as shown. Conduct a mass balance for biomass around the secondary clarifier. Make all the assumptions that were stated previously in Section 8.4.2. Since all biological reactions take place in the aeration tank, there is no growth in the clarifier. Accumulation = 0

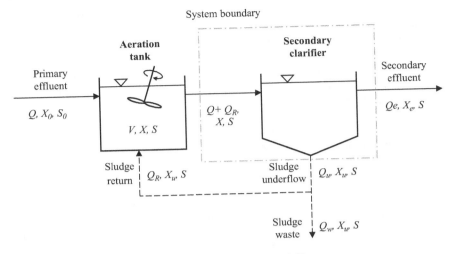

Figure 8.8 Activated sludge process (for Example 8.2).

Rate of accumulation = Rate of inflow – Rate of outflow

$$0 = \left(Q\, X + Q_R\, X \right) - Q_e\, X_e - Q_u\, X_u$$

or

$$Q\, X + Q_R\, X = 0 + Q_R\, X_u + Q_w\, X_u$$

or

$$Q_R \left(X_u - X \right) = Q\, X - Q_w\, X_u$$

or

$$Q_R = \frac{Q\, X - Q_w\, X_u}{X_u - X} \qquad (8.56)$$

Example 8.3: A completely mixed high rate activated sludge plant is to treat 15,000 m³/d of industrial wastewater. The primary effluent going to the activated sludge reactor has a BOD_5 of 1100 mg/L that must be reduced to 150 mg/L, prior to discharge to a municipal sewer. The flow diagram of the plant is given in Figure 8.6. Pilot plant analysis gave the following results: Mean cell residence time = 5 days, MLSS concentration in reactor = 6000 mg/L VSS, $Y = 0.7$ kg/kg, $k_d = 0.03$ d^{-1}. Determine the following:

a. The hydraulic retention time and volume of the activated sludge reactor.
b. The volumetric loading rate in kg BOD$_5$/m^3-d to the reactor.
c. The F/M ratio in the reactor.
d. The mass and volume of solids wasted each day, at an underflow solids concentration, $X_u = 12,000$ mg/L.
e. The sludge recirculation ratio.
f. The volume of solids that must be wasted each day, if the solids are wasted directly from the activated sludge reactor, instead of from the underflow.

SOLUTION

a. Use Equation 8.45 to calculate HRT (θ)

$$X = \frac{\theta_c Y (S_0 - S)}{\theta (1 + k_d \theta_c)}$$

Or $\theta = \dfrac{\theta_c Y (S_0 - S)}{X(1 + k_d \theta_c)} = \dfrac{(5d)(0.7)(1100 - 150)\,\text{mg}/\text{L}}{6000\,\text{mg}/\text{L}\left(1 + 0.03\,\text{d}^{-1} \times 5d\right)}$

Or $\theta = 0.48\,d = 11.56\,h$

Volume of aeration tank, $V = Q\theta = 15,000\,\text{m}^3/\text{d} \times 0.48\,d$
$$= 7200\,\text{m}^3$$

b. Use Equation 8.24 to calculate volumetric loading rate

$S_0 = 1100\,\text{mg/L} = 1.10\,\text{kg/m}^3$

$$V_L = \frac{Q S_o}{V} = \frac{15,000\,\text{m}^3/\text{d} \times 1.10\,\text{kg/m}^3}{7200\,\text{m}^3} = 2.29\ \text{kg BOD}_5/\text{m}^3 \cdot \text{d}$$

c. Use Equation 8.23 to calculate F/M ratio

$$\frac{F}{M} = \frac{Q(S_0 - S)}{V X} = \frac{15,000\,\text{m}^3/\text{d}\,(1100 - 150)\,\text{mg/L}}{7200\,\text{m}^3 \times 6000\,\text{mg/L}}$$
$$= 0.33\,\text{mg BOD}_5/\text{mg VSS.d}$$

d. At steady state, the SRT is given by

$$\theta_c = \frac{VX}{Q_w X_u}$$

Or $Q_w X_u = \dfrac{VX}{\theta_c} = \dfrac{7200 \, m^3/d \times 6 \, kg/m^3}{5 \, d} = 8640 \, kg/d$

Mass wasted each day $= 8640 \, kg \, VSS/d$

Volume wasted each day, $Q_w = \dfrac{Q_w X_u}{X_u} = \dfrac{8640 \, kg/d}{12 \, kg/m^3} = 720 \, m^3/d$

e. From mass balance around secondary clarifier and using Equation 8.56,

$$Q_R = \frac{QX - Q_w X_u}{X_u - X} = \frac{(15,000 \, m^3/d \times 6 \, kg/m^3) - (8640 \, kg/d)}{(12-6) \, kg/m^3}$$
$$= 13,560 \, m^3/d$$

Recirculation ratio, $R = \dfrac{Q_R}{Q} = \dfrac{13,560 \, m^3/d}{15,000 \, m^3/d} = 0.90$

f. From part (d) mass wasted each day $= 8640 \, kg/d$
    If solids are wasted directly from the aeration tank, then solids concentration in waste sludge = MLSS concentration

Therefore, volume wasted each day, $Q_w = \dfrac{\text{mass wasted each day}}{X}$
$$= \frac{8640 \, kg/d}{6 \, kg/m^3} = 1440 \, m^3/d$$

Example 8.4: It is desired to determine the kinetic coefficients Y and k_d for an activated sludge process treating a wastewater. Five bench-scale CSTRs were operated at different MLVSS concentrations and the data are shown in Table 8.2. Determine the kinetic coefficients.

SOLUTION
Step 1. Develop the equations for determination of Y and k_d.
    Using Equation 8.47 and 8.49 we can write

$$\mu = YU - k_d$$

Table 8.2 Data from five bench-scale CSTRs (for Example 8.4)

Reactor #	X, mg VSS/L	U, mg BOD$_5$/mg VSS.d	r$_g$, mg VSS/L.d
1	1000	0.39	194
2	1500	0.51	399
3	3000	0.60	960
4	5000	0.91	2530
5	6000	1.20	4080

Using Equation 8.5, $\dfrac{r_g}{X} = \mu$

Therefore, $\dfrac{r_g}{X} = YU - k_d$

Step 2. Prepare Table 8.3 for graphical plot of above equation.

Table 8.3 Data for graphical plot (for Example 8.4)

Reactor #	X, mg VSS/L	r$_g$/X, d^{-1}	U, mg BOD$_5$/mg VSS.d
1	1000	0.194	0.39
2	1500	0.266	0.51
3	3000	0.320	0.60
4	5000	0.506	0.91
5	6000	0.680	1.20

Step 3. Plot r_g/X versus U, as illustrated in Figure 8.9 with the best-fit line shown.

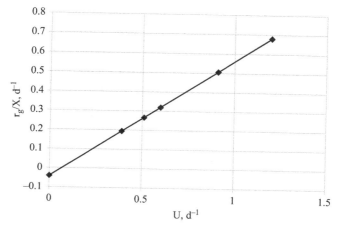

Figure 8.9 Plot of r_g/X versus U (for Example 8.4).

From the graph, (Figure 8.9)

Slope $= Y = 0.60\,\text{mg VSS / mg BOD}$

Intercept $= k_d = 0.04\,\text{d}^{-1}$

8.4.5 Aeration requirements

Air or oxygen is supplied to the aeration tank of the activated sludge process, to provide the oxygen required by the aerobic microorganisms for the degradation of organic matter. The amount of oxygen added should be sufficient to (i) match the oxygen utilization rate (OUR) of the microorganisms, and (ii) maintain a small excess in the tank, about 0.5–2 mg/L dissolved oxygen, to ensure aerobic metabolism at all times (Peavy et al., 1985). The OUR is a function of the characteristics of the wastewater and the type of reactor. In a conventional activated sludge process, the OUR is around 30 mg/L.h. For the extended aeration process, the OUR is about 10 mg/L.h, whereas for high rate processes, the OUR can be up to 100 mg/L.h.

The oxygen requirement may be estimated from the biodegradable COD (bCOD) of the wastewater, and the biomass wasted each day. The mass of oxygen required for BOD removal may be calculated from the following expression (Metcalf and Eddy et al., 2013):

$$M_o = Q(S_0 - S) - 1.42\,P_x \tag{8.57}$$

where

M_o = Mass of oxygen required for BOD removal, kg/d
Q = wastewater flow rate into aeration tank (without recycle flow), m³/d
S_0 = influent bCOD, kg/m³
S = effluent bCOD, kg/m³
P_x = biomass wasted, kg VSS/d
1.42 = COD of cell tissue, kg COD/g VSS

As an approximation, the oxygen requirement for BOD removal only will vary from 0.9 to 1.3 kg O_2/kg BOD removed for SRTs of 5–20 d, respectively (WEF, 1998). When nitrification is included in the process, the oxygen required for oxidation of ammonia and nitrite to nitrate is included as follows (Metcalf and Eddy et al., 2013):

$$M_{O+N} = Q(S_0 - S) - 1.42\,P_x + 4.33\,Q(NO_x) \tag{8.58}$$

where

NO_x = TKN concentration, kg/m³
M_{O+N} = oxygen required for BOD and nitrogen removal, kg/d

Other terms are the same as defined previously.

The aeration devices have to provide adequate oxygen to satisfy the demand for average and peak flows. A peaking factor of 2.0 is commonly used. However, a review of actual conditions should be performed. Each type of aeration device comes with a certain oxygen transfer efficiency and an oxygen transfer rate in pure water at standard temperature and pressure (standard oxygen transfer rate (SOTR)) specified by the manufacturer. The actual oxygen transfer rate (AOTR) varies from the SOTR due to wastewater characteristics, pressure variation, residual oxygen concentration, etc. The AOTR can be calculated from the following expression (Metcalf and Eddy et al., 2013):

$$AOTR = SOTR\left(\frac{\beta\, C_{s,T,H} - C_L}{C_{s,20}}\right)\left(\theta^{T-20}\right)(\alpha F) \qquad (8.59)$$

where

AOTR = actual oxygen transfer rate under field conditions, kg O_2/h or kg O_2/kWh (lb O_2/h or lb O_2/hp.h)

SOTR = standard oxygen transfer rate in clean water at 20°C and zero dissolved oxygen, kg O_2/h or kg O_2/kWh (lb O_2/h or lb O_2/hp.h)

β = salinity – surface tension correction factor, typically 0.95–0.98

$C_{s,T,H}$ = oxygen saturation concentration in clean water at wastewater temperature T, and diffuser depth H, mg/L

C_L = dissolved oxygen concentration in wastewater, mg/L

$C_{s,20}$ = dissolved oxygen saturation concentration in clean water at 20°C and 1 atm pressure, usually 9.17 mg/L

θ = correction factor for temperature = 1.024

T = wastewater temperature, °C

F = fouling factor, typically 0.65–1.0 for no fouling

α = oxygen transfer correction factor for wastewater, typically 0.3–0.4 for activated sludge reactors with BOD removal, and 0.45–0.75 for BOD removal and nitrification systems.

8.4.5.1 Types of aerators

Different types of aerators are used in aeration tanks. The selection of an aeration system depends on the site characteristics and the type of process used. The following are two of the commonly used aeration systems.

Air Diffusers – used to inject air into the aeration tank. The diffusers may be mounted along the side of the tank, or they may be placed in a manifold along the bottom of the aeration tank, as illustrated in Figure 8.10. Air diffusers may produce *coarse bubbles* or *fine bubbles*. Coarse bubbles may be up to 25 mm in diameter, while fine bubbles are 2–2.5 mm in diameter. There are advantages and disadvantages of both types of diffusers. Fine bubble diffusers have greater energy requirements and clog easily, even

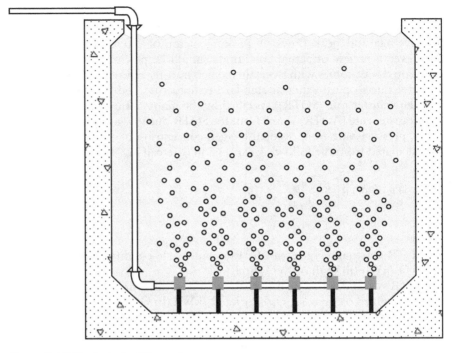

Figure 8.10 Typical air diffuser system.

though they have better oxygen transfer due to larger surface area per volume. Coarse bubble diffusers have lower oxygen transfer rates, but require less maintenance and have a lower headloss.

Mechanical Aerators – usually have impellers that produce turbulence at the air–water interface, which enhances the transfer of oxygen from air to water. High–speed impellers can add large quantities of air to relatively small quantities of water. The mixing of the aerated water with reactor contents takes place through velocity gradients. Brush-type aerators are used in oxidation ditches to promote air entrainment and also momentum to the wastewater.

Example 8.5: Consider the completely mixed high rate activated sludge plant from Example 8.3. Fine bubble membrane diffusers with total floor coverage are to be used for the aeration tank. The SOTR specified by the manufacturer is 3.5 kg O_2/kWh, with an αF of 0.5. The average wastewater temperature is 16°C. The residual DO in the aeration tank is 4 mg/L, β is 0.95, and saturation oxygen concentration at 16°C and tank depth elevation is 9.81 mg/L. Calculate the oxygen demand and the power required for aeration.

SOLUTION

Step 1. Calculate the oxygen demand for BOD removal.
Using Equation 8.57 and the data from Example 8.3,

$$M_o = Q(S_0 - S) - 1.42 P_x$$

$$= 15,000\,m^3/d\,(1.1 - 0.15)\,kg/m^3 - 1.42 \times 8640\,kg/d$$

$$= 1981.20\,kg/d$$

Use a peaking factor = 2.0

Oxygen demand = 1982

Oxygen demand = $1981.2 \times 2.0 = \textbf{3962.40\,kg/d}$

Step 2. Calculate the AOTR

$$SOTR = 3.5\,kg\,O_2/kWh$$

$$C_{s,T,H} = 9.81\,mg/L$$

$$C_{s,20} = 9.17\,mg/L$$

$$C_L = 4\,mg/L$$

Use Equation 8.59 to calculate AOTR

$$AOTR = SOTR \left(\frac{\beta\,C_{s,T,H} - C_L}{C_{s,20}} \right) (\theta^{T-20})(\alpha F)$$

Or

$$AOTR = 3.5 \left(\frac{0.95 \times 9.81 - 4}{9.17} \right) (1.024^{16-20})(0.5)$$

Or $AOTR = 0.92\,kg\,O_2/kWh$

Oxygen transfer efficiency = AOTR/SOTR × 100% = (0.95/3.5) × 100% = 27% which is typical for this type of aeration system.

Step 3. Calculate the power required for aeration.

$$Oxygen\ demand = 3962.40\,kg\,O_2\,/\,d = 165.10\,kg\,O_2\,/\,h$$

$$Power\ required = oxygen\ demand\,/\,AOTR$$

$$= \left(165.10\,kg\,O_2\,/\,h\right)/\left(0.92\,kg\,O_2\,/\,kWh\right)$$

$$= 179.46\,kW$$

$$\approx 180\,kW$$

8.5 TYPES OF SUSPENDED GROWTH PROCESSES

There are a wide variety of suspended growth processes in operation at wastewater treatment plants worldwide. Each type has its advantages and limitations. Wastewater characteristics, site characteristics, effluent limitations, regulatory requirements, and economic considerations, are some of the factors that influence the choice and selection of a particular process. Pilot studies should be conducted before making a final selection. Some of the more common types of suspended growth processes used for BOD removal are described below.

8.5.1 Conventional activated sludge

The conventional activated sludge process is the most widely used suspended growth process for the treatment of wastewater. The basic process has been described in detail in Section 8.3. Section 8.4 provides the development of design models for completely mixed and plug flow activated sludge processes. Typical design values for completely mixed systems are (Peavy et al., 1985) HRT 3–5 h, F/M 0.2–0.4, SRT 4–15 d, V_L 0.8–2.0 kg BOD_5/m^3.d, with BOD removal efficiency 85–95% and recycle ratio of 0.25–1.0. For plug flow systems the recycle ratio varies from 0.25–0.5, with an HRT of 4–8 h and V_L 0.3–0.6 kg BOD_5/m^3.d. The SRT and F/M ratios for the two systems are similar.

A number of variations of the conventional process have been developed and are in use. The following sections provide descriptions of a few of them.

8.5.2 Step aeration or step feed process

In this process, a long and narrow aeration tank is used for plug flow configuration. The influent wastewater enters the aeration tank at several locations

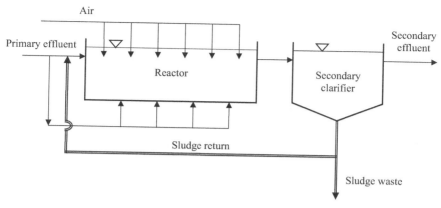

Figure 8.11 Step feed process.

along the length. This helps to reduce the oxygen demand at the head inlet point. At the same time, compressed air is injected into the tank at several locations along the length. This helps provide uniform aerobic conditions throughout the tank. This is illustrated in Figure 8.11. Typical design values are HRT 3–5 h, F/M 0.2–0.4, SRT 4–15 d, V_L 0.6–1.0 kg BOD_5/m^3.d, with BOD removal efficiency 85–95%.

8.5.3 Tapered aeration process

Plug flow configuration is used, with the wastewater entering the aeration tank at one end. The air flow is tapered with the higher flow toward the inlet, gradually tapering to low air flow toward the outlet. Air and maximum oxygen are provided at the inlet where the organic load is highest. As the wastewater flows through the tank, the substrate is degraded and the oxygen demand is lower toward the outlet. This results in efficient use of the air where it is needed most. The tapered aeration process is illustrated in Figure 8.12. Typical design values are HRT 4–8 h, F/M 0.2–0.4, SRT 5–15 d, V_L 0.3–0.6 kg BOD_5/m^3.d, with BOD removal efficiency 85–95%.

8.5.4 Contact stabilization process

This process uses two separate tanks for the treatment of wastewater and stabilization of activated sludge. The process consists of a *contact tank* with a short HRT of 30–60 min, followed by a clarifier. The readily biodegradable soluble COD is oxidized or stored, while particulate COD is adsorbed on the biomass at the same time. The treated wastewater is separated from the biomass in the clarifier. The settled biomass with the adsorbed organic matter is then transported to a *stabilization tank* (HRT 1–2 h), where the stored and adsorbed organics are degraded. The biomass is then returned to the contact tank as activated sludge. Overall BOD removal efficiency is 80–90%. Since the MLSS is very high in the stabilization tank, this results in

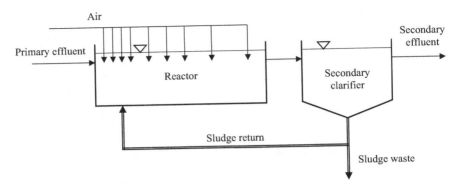

Figure 8.12 Tapered aeration process.

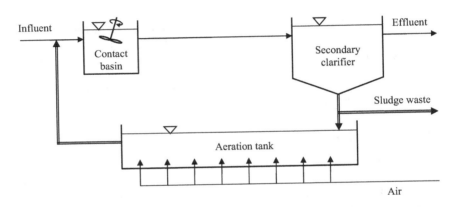

Figure 8.13 Contact stabilization process.

a lower tank volume. The advantage of this system is the reduction in overall tank volume. The contact stabilization process is presented in Figure 8.13.

8.5.5 Staged activated sludge process

A number of completely mixed reactors are placed in series followed by a final clarifier. The return activated sludge comes back to the first tank. This is illustrated in Figure 8.14. Three or more reactors in series approximate a plug flow system. The process is capable of handling high organic loads with high BOD removal efficiencies.

8.5.6 Extended aeration process

This process is used to treat wastewater from small communities that generate low volumes of fairly uniform characteristics. A completely mixed

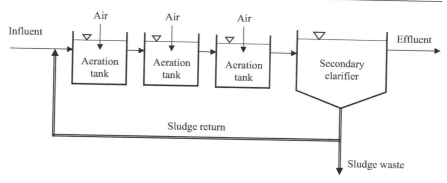

Figure 8.14 Staged activated sludge process.

activated sludge process configuration is used (Figure 8.3). Typical design values are HRT 18–24 h, F/M 0.05–0.15, SRT 20–30 d, V_L 0.16–0.4 kg BOD_5/m^3.d, with BOD removal efficiency 75–90%. The reactor is operated in the endogenous decay phase, as evidenced by the F/M ratio.

8.5.7 Oxidation ditch

This process consists of an oval-shaped aeration channel, where the wastewater flows in one direction, followed by a secondary clarifier. Brush-type mechanical aerators provide aeration and mixing and keep the water flowing in the desired direction. The influent enters the channel and is mixed with the return activated sludge. The flow in the channel dilutes the incoming wastewater by a factor of 20–30. Process kinetics approach that of a complete mix reactor, but with plug flow along the channels. The oxidation ditch is illustrated in Figure 8.15. This process is suitable for use in small rural communities where large land area is available. The oxidation ditch can be designed and operated to achieve both BOD and nitrogen removal.

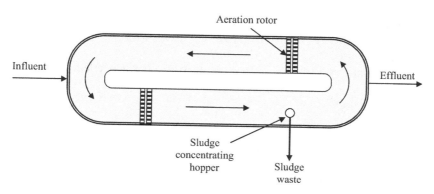

Figure 8.15 Oxidation ditch.

8.5.8 Sequencing batch reactor (SBR)

The SBR is a fill-and-draw-type of system where aeration, biodegradation, and settling all take place in a single reactor. The reactor sequences through a number of steps in one *cycle*. The reactor can go through 2–4 cycles per day. A typical cycle consists of the following steps: (i) Fill – where the substrate is added; (ii) react – mixing and aeration is provided; (iii) settle – for clarification of effluent; (iv) decant – for withdrawal of effluent. An idle step may also be included to provide flexibility at high flows. Aeration is accomplished by jet aerators or coarse bubble diffusers with submerged mixers.

8.5.9 Membrane biological reactor (MBR)

The MBR process uses a biological reactor with suspended biomass for BOD and/or nitrogen removal, and micro or ultrafiltration membranes for solids separation. The bioreactors may be aerobic or anaerobic. The effluent water quality is very high and makes this process attractive for water reuse applications. MBRs have been used for the treatment of municipal and industrial wastewater, as well as for water reuse. This type of reactor has been found suitable for the removal of a variety of contaminants from municipal wastewater, e.g. pharmaceutical products, and aromatic hydrocarbons, among others (Kimura et al., 2007; Francesco et al., 2011). Recent research has focused on the use of nanomaterials which are applied as a coating on the membranes to improve the hydrophilicity, selectivity, conductivity, fouling resistance, and anti-viral properties of membranes (Su et al., 2012; Kim and van der Bruggen, 2010; Lu et al., 2009; Zodrow et al., 2007; Bae and Tak, 2005).

The MBR process has the following advantages (Metcalf and Eddy et al., 2013):

- It can operate at high MLSS concentrations (15,000–25,000 mg/L).
- As a result of high MLSS, it can handle high volumetric loading rates at short HRTs.
- Longer SRT results in less sludge production.
- High-quality effluent in terms of low turbidity, TSS, BOD, and pathogens.
- Less space is required as no secondary clarifier is needed.

The disadvantages of the process include:

- High capital costs.
- Membrane fouling problems.
- Membrane replacement costs.
- Higher energy costs.

Membrane bioreactor systems have two basic configurations. They are (i) integrated bioreactor that has a membrane module immersed in the reactor and (ii) recirculated MBR where the membrane module is separately mounted outside the reactor. Immersed membranes use hollow fiber or flat sheet membranes mounted in modules. They operate at lower pressures and are used in activated sludge bioreactors. The separate systems use pressure-driven, in-pipe cartridge membranes. They are used more for industrial wastewaters (Davis, 2020). Additional discussion on membrane characteristics, membrane fouling, and flux calculations are provided in Chapter 13.

8.6 STABILIZATION PONDS AND LAGOONS

Ponds and lagoons are land-based suspended growth treatment systems. Usually, there are no primary or secondary clarifiers. All treatment and solids separation takes place in an earthen basin, where wastewater is retained and natural purification processes result in biological treatment. Mechanical mixing is provided in a lagoon, whereas there is no mechanical mixing in a pond. Ponds can be (i) aerobic – shallow pond, (ii) anaerobic – deep pond, and (iii) facultative. Lagoons can be (i) aerobic – with complete mixing and (ii) facultative, with mixing of the liquid portion. The majority of ponds and lagoons are facultative. A treatment system may consist of an aerobic lagoon followed by facultative ponds to achieve sufficient BOD removal. These types of systems are suitable for small communities, and onsite treatment of industrial wastewaters. Lagoons are used extensively for the treatment of livestock wastewaters at hog and poultry farms.

The advantages of these land-based treatment systems are:

- Low capital cost.
- Low operating cost.
- A large volume-to-inflow ratio provides enough dilution to minimize the effects of variable organic and hydraulic loadings.

The disadvantages include:

- A large land area is required.
- Odor problems are a concern.
- High suspended solids concentration in the effluent. In the US, the discharge limits for solids in the effluent are 75 mg/L as specified by EPA.
- At cold temperatures, biological activity is significantly reduced. In cold climates, it is often necessary to provide sufficient volume to store the entire winter flow.

8.6.1 Process microbiology

In this section, the microbiological processes taking place in a facultative pond will be discussed. This includes both aerobic and anaerobic processes. The facultative pond has a complex system of microbial processes that result in the degradation of organic matter. The facultative pond has an aerobic section in the top layers, an anaerobic section in the bottom layers, and some facultative reactions taking place in between. This is illustrated in Figure 8.16.

The wastewater enters the pond near the bottom. The biological and other solids settle at the bottom in a thin sludge blanket. Anaerobic bacteria

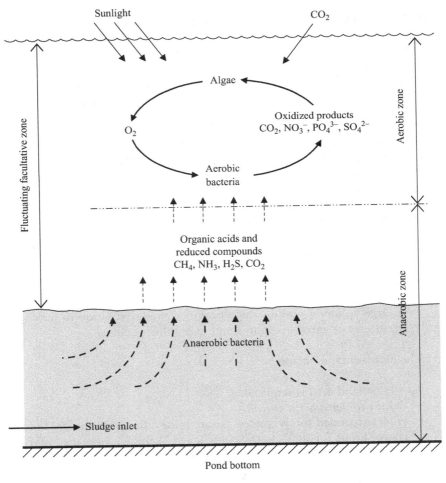

Figure 8.16 Microbiological processes in a facultative pond.

Source: Adapted from Peavy et al., 1985.

degrade the organic matter and release products of decomposition. These products are mainly organic acids and reduced compounds of carbon, nitrogen, sulfur, and phosphorus. There is a facultative region in the middle layers, where the bacteria can switch their metabolism from aerobic to anaerobic, or vice versa, depending on the loading conditions. The organic acids and reduced compounds are then used by the aerobic bacteria in the upper layers of the pond. The aerobic decomposition products are oxidized compounds of carbon, nitrogen, sulfur, and phosphorus, e.g. CO_2, NO_3^-, PO_4^{-3}, SO_4^{-2}, etc. Algae use these oxidized compounds as food in presence of sunlight and release O_2 as a by-product. The released oxygen helps to replenish the dissolved oxygen concentration of the pond and maintain aerobic conditions in the top layers. Thus a symbiotic relationship exists between the microbial communities in the pond.

The depth of the aerobic zone depends on the penetration of sunlight and wind action. Strong wind action and enhanced light penetration in clear waters can extend the depth of the aerobic zone downward. On the other hand, the absence of wind and cloudy skies can result in the anaerobic zone rising toward the surface. The facultative zone is the region where dissolved oxygen concentration fluctuates in the pond. Facultative microorganisms exist in this zone, which are capable of adjusting their metabolism in response to low or high dissolved oxygen concentrations.

8.6.2 Design of pond or lagoon system

A number of models are available for the design of ponds and lagoons. The most commonly used model assumes a completely mixed reactor without solids recycle. The rate of substrate utilization is assumed to be first order. A mass balance for the soluble portion of the substrate can be written, and the following design equation can be obtained (Metcalf and Eddy et al., 2013; Peavy et al., 1985):

$$\frac{S}{S_o} = \frac{1}{1+k\theta} \tag{8.60}$$

where
  S = effluent soluble BOD concentration, mg/L
  S_0 = influent soluble BOD concentration, mg/L
  k = first order rate coefficient, varies from 0.5 to 1.5 d^{-1}
  θ = HRT = V/Q, d

When a pond or lagoon system is used for municipal wastewater treatment, it is common practice to distribute the flow between two to three ponds in series. This is done to minimize short-circuiting that can occur in one large pond/lagoon. The first unit is usually designed as an aerated facultative lagoon, since it receives the wastewater with the highest BOD concentration. This is followed by two or more facultative ponds. The design equation for n number of equally sized ponds is given by

$$\frac{S_n}{S_o} = \frac{1}{\left(1 + k\theta / n \right)^n} \tag{8.61}$$

where
S_n = effluent soluble BOD concentration from nth pond, mg/L
θ = total HRT for the pond system, d
n = number of ponds/lagoons in series

Other terms are the same as described previously. The van't Hoff–Arrhenius model (Equation 8.20) is used to correct k values for temperature. Arrhenius coefficient can range from 1.03 to 1.12.

8.6.3 Design practice

The HRT of facultative ponds can vary from 7 to 30 d, with a BOD loading of 2.2–5.6 g/m².d (20 lb/acre.d to 50 lb/acre.d). The lower loading is for colder climates, where biological degradation is severely reduced in winter. Sufficient volume may have to be provided to store the entire winter flow. The HRT of facultative lagoons can vary between 7 and 20 d. The water depth ranges from 1 to 2 m, with 1 m of dike freeboard above the water level. A minimum water level of 0.6 m (2 ft) is required to prevent the growth of rooted aquatic weeds (Hammer and Hammer, 2012).

As mentioned previously, it is customary to use two or three ponds in series, and distribute the flow equally between the ponds. A three-cell system is illustrated in Figure 8.17. The first unit is called the *primary cell* which is operated as an aerated lagoon. The second cell may be operated as a primary or *secondary cell*, depending on the volume of flow. These may be operated in parallel or in series. The third cell provides additional treatment and storage volume. In the sizing of ponds, the secondary cell is not included in BOD loading calculations. However, the volume is included in the determination of hydraulic retention times.

Algae provide some dissolved oxygen replenishment to the facultative ponds. For lagoons, aeration is provided using mechanical aerators. Aeration requirements can be calculated as described in Section 8.4.5.

For ponds and lagoons treating municipal wastewater, the bottom and sides are sealed with bentonite clay to provide an impervious layer. For industrial wastewater treatment, the sides and bottom have to be covered with an impervious liner. In addition, state and local regulations for hazardous waste treatment would have to be followed.

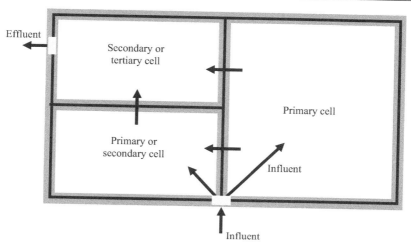

Figure 8.17 Three cell pond or lagoon system.

Example 8.6: A pond and lagoon system is to be designed for municipal and industrial wastewater treatment for a small community with
a population of 2500. The wastewater design flow is 400 L/capita.d
(Lpcd) with a BOD load of 70 g/capita.d. It is desired to use a three-cell
system similar to the one illustrated in Figure 8.18, with the first two
cells used as primary lagoons in parallel. The allowable BOD loading is
2.2 g/m².d. (a) Calculate the area of the pond system. (b) Calculate the
winter storage available if the water depth of the ponds is 2m. Assume
losses due to evaporation and seepage are 0.5 mm/d.

SOLUTION

Step 1. Calculate the required area of the pond system.

Design flow $= 400 \, \text{Lpcd}$

$Q = 400 \, \text{Lpcd} \times 2500 \, \text{people} = 1,000,000 \, \text{L/d} = 1000 \, \text{m}^3/\text{d}$

$\text{BOD produced} = 70 \, \text{g/capita.d} \times 2500 \, \text{people} = 175,000 \, \text{g/d}$

$= 175 \, \text{kg/d}$

Allowable BOD loading $= 2.2 \, \text{g/m}^2.\text{d}$

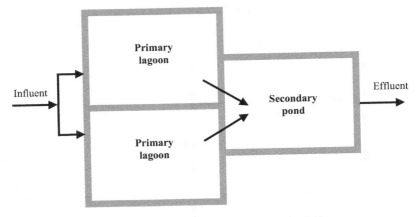

Figure 8.18 3-cell lagoon and pond system (for Example 8.6).

$$\text{Primary lagoon area required} = \frac{175,000\,\text{g/d}}{2.2\,\text{g/m}^2.\text{d}}$$

$$= 79,545.45\,\text{m}^2 \approx 80,000\,\text{m}^2$$

Note: BOD loading is calculated for primary cells only.
Select two primary lagoons of 40,000 m² area each. Add a third pond of equal area. The high water level is 2 m with 1 m of freeboard. So total depth is 3 m.

Step 2. Calculate winter storage available. The cross section of the pond is shown in Figure 8.19 with the low and high water levels.

Assume low water level = 0.6 m

Depth available between low and high water level = $2-0.6\,\text{m} = 1.4\,\text{m}$

Storage volume = depth × total area = $1.4\,\text{m} \times 3 \times 40,000\,\text{m}^2$

$$= 168,000\,\text{m}^3$$

Evaporation and seepage loss = $0.5\,\text{mm}\,/\,\text{d} = 0.0005\,\text{m}\,/\,\text{d}$

Evaporation and seepage loss volume = $0.0005\,\text{m/d} \times 3 \times 40,000\,\text{m}^2$

$$= 60\,\text{m}^3/\text{d}$$

Total storage time available $= \dfrac{168,000\,\text{m}^3}{(1000-60)\,\text{m}^3\,/\,\text{d}} = 179\,\text{d}$

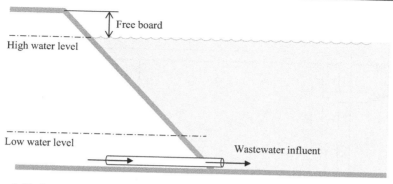

Figure 8.19 Pond cross section showing high and low water levels (for Example 8.6).

8.7 SEPTIC TANK SYSTEM

A decentralized wastewater treatment system used for many small and rural communities, as well as individual households is the septic tank system. Wastewater is treated at or near the source, rather than by collection and transport to a centralized facility. Septic systems are also known as onsite systems, cluster systems, package plants, or private sewage systems. US EPA has defined small communities as less than 10,000 people, or less than 1 MGD average wastewater flow rate. This provides a low-cost alternative for a home or small business. Nearly 25% of existing US households and 33% of new constructions of residential and commercial facilities use a septic system. In the UK, more than 75% of the households are not connected with a sewer network. Their wastewater is treated in septic tank systems. Individual septic systems are used in rural areas of France, serving more than 4 million people (Somogyi et al., 2009). The various types of decentralized wastewater treatment, when properly maintained and executed, can protect public health, preserve valuable water resources, and maintain economic vitality in a community. They are a cost-effective and long-term option for treating wastewater, particularly in less densely populated areas (EPA, 2021).

8.7.1 Process description

A septic system consists of a septic tank and a drainfield for absorption of the effluent. The septic tank can consist of one or two compartments. The underground tank is usually made of concrete, although fiberglass and polyethylene tanks are also used. The tank must be watertight to avoid leaking out and seepage into the groundwater. It works as a primary sedimentation tank with a long detention time. Both settling and anaerobic treatment reduce the solids and organic matter and moderately treat the wastewater. Digestion of organic matter and separation of fats, oil, and grease (FOG) takes place in the tank. The effluent is discharged into the drainfield through perforated pipes and chambers, or pumped to help trickle through sand, soil, or constructed

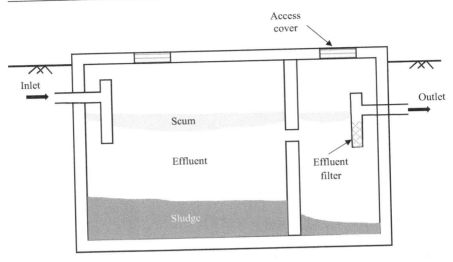

Figure 8.20 Septic tank with two compartments.

wetlands to remove the pathogens, nutrients, and other contaminants. The discharge of waste effluent in the soil may cause groundwater pollution; thus, some systems use evaporation or disinfection before discharge. A typical septic tank system with two compartments is illustrated in Figure 8.20.

First, the wastewater flows through a sewer pipe out of the house into the septic tank. Wastes like grease, hair, and soap float to the top and form a scum layer, and heavier solids settle at the bottom of the tank to create a sludge layer. Microorganisms degrade the organic matter of the sludge to simple nutrients, gas, and water. The solids are stored in the tank and the liquid waste or effluent is distributed in the drainfield.

The drainfield consists of a number of perforated pipes, laid in the ground for distribution of the effluent. The pipes are usually laid in gravel or sand-filled trenches for easy absorption of the effluent in the soil. The effluent flows through the pipe by gravity or by using a low-pressure pump. To avoid clogging of the perforated pipe by the solids carried out of the septic tank, newer systems have a removable filter at the tank outlet. The types of soil and depth available between the distribution pipes and groundwater are important considerations for the selection of the drainfield to avoid groundwater contamination. Coarse gravel and sandy soil allow the effluent to flow quickly and have a higher potential for groundwater contamination. Effluent flows very slowly in fine clay and compacted soil, and the absence of oxygen promotes anaerobic conditions which slow down the degradation of the solids and promote odorous compounds. Well-drained, medium-textured soils can provide sufficient oxygen for rapid degradation of the remaining organics and pathogens, and filter out the solids. Hazardous chemicals such as fuels, and solvents are not easily biodegradable and can kill the beneficial microorganisms in the tank, and eventually reach the environment through soil and groundwater contamination.

8.7.2 Types of septic systems

The design and size of the septic system vary widely, depending on the household size, soil type in the drainfield, the land area available, close proximity to the groundwater or surface water bodies, weather conditions, and local regulations. The following are some of the common types of septic systems (EPA, 2021).

1. Conventional system – consists of a septic tank and a drainfield for subsurface wastewater infiltration. They are generally installed for single families and small businesses. Gravel or stone drainfield are commonly used in the trenches. A large land area is required for these systems.
2. Chamber system – is similar to the conventional system but with a gravel-less drainfield. It can consist of open bottom chambers, fabric-wrapped pipe, or polystyrene materials that are used in place of gravel drainfield. The chamber system is used with a high groundwater table, and where the influent flow rate is variable. The construction is easy, and generally used where the plastic chambers are readily available.
3. Drip distribution system – can be used for different types of drainfield where the drip laterals are inserted into the top 6–12 in of soil, thus avoiding additional soil requirement. An additional dosing tank is used after the septic tank for wastewater distribution. A large land area and additional electricity and maintenance requirements can increase the total cost.
4. Aerobic treatment unit – injects oxygen into the septic tank to increase microbial activity for faster degradation of organic matter. Additional pathogen reduction may also be used. Such small-scale units are suitable for homes with small land areas, high water tables, and inadequate soil conditions. Regular maintenance of the system is required to achieve good effluent quality.
5. Recirculating sand filter system – includes a PVC or concrete sand filter box, above or below ground, for treatment of the wastewater effluent before discharge to the drainfield. They provide a high level of treatment for nutrients, making them suitable for high water table conditions. They are more expensive than conventional systems due to pumping and maintenance requirements.
6. Constructed wetland system – is similar to the natural wetland process. Wastewater flows through a wetland cell, where microbes, plants, and other media remove the pathogen and nutrients. The wetland has

an impermeable liner, gravel and sand fill, and appropriate plants. Wastewater flows through the wetland by gravity or pressure distribution and flows to the drainfield for further treatment.

8.7.3 Design considerations

The septic tank and drainfield must have adequate capacity to treat all the wastewater generated from the house or business, at times of peak use. Different states in the US have standard procedures for calculating the wastewater flow and sizing of the treatment system. A number of design guidelines of the New York State Department of Health (2010) are provided below.

1. A Minimum daily flow of 110 gpd per bedroom is considered for wastewater flow calculation. The tank capacity and minimum liquid surface area required are presented in Table 8.4.
2. The minimum and maximum liquid depth of 30 in and 60 in are allowed to calculate the volume of the tank.
3. The minimum distance between the inlet and outlet shall be 6 ft. The length-to-width ratio of the tank should be in the range of 2:1 to 4:1.
4. The tank must be able to support 300 psf pressure after installation.
5. Dual compartments are recommended for all tanks, and shall be required on all tanks with an interior length of 10 ft or more.
6. The first compartment or tank shall account for 60–75% of the total design volume.
7. Tanks in series should be connected by a single pipe with a minimum diameter of 4 in.
8. The maximum length of the absorption line in conjunction with gravity distribution shall be 60 ft.
9. The septic tank and distribution box should be at least 100 ft from the nearby well, stream, lake, and wetlands.

Table 8.4 Minimum septic tank capacity

Number of bedrooms	Minimum tank capacity, gal	Minimum liquid surface area, ft²
1, 2, 3	1000	27
4	1250	34
5	1500	40
6	1750	47

Source: New York State Department of Health, 2010

PROBLEMS

8.1 What is the Monod model? Graphically illustrate the Monod model. Show with the help of equations what happens at very low and very high substrate concentrations.

8.2 Define SRT and F/M ratio. Why are these considered to be important design parameters?

8.3 Name five factors affecting the microbial growth in an activated sludge reactor.

8.4 What are the basic components of an activated sludge reactor? Why is it called *activated sludge*?

8.5 Draw a diagram of a completely mixed suspended growth reactor without recycle and write down qualitative mass balance equations for biomass and substrate. Simplify the mass balance equations to obtain expressions for biomass (X) and substrate (S). Clearly state any assumptions that you make.

8.6 Develop the equation: $\dfrac{1}{U} = \left(\dfrac{K_s}{k}\right)\left(\dfrac{1}{S}\right) + \dfrac{1}{k}$

where $k = \dfrac{\mu_{max}}{Y}$, S = effluent substrate concentration, and all other terms are the same as defined previously. Describe how you can determine the values of the kinetic coefficients, K_s and k, by using this equation and experimental data.

8.7 A number of bench-scale reactors were operated in the laboratory as completely mixed reactors with recycle, to determine kinetic coefficients for a wastewater. The reactors were operated at the same HRT (θ) and initial soluble substrate concentrations, but at different SRT (θ_c) values. The initial soluble substrate concentration was 500 mg COD/L. The HRT was 6 h for all the reactors. The experimental data is provided in Table 8.5.

   a. Calculate the kinetic coefficients Y and k_d. Use a graphical procedure and use Equations 8.48 and 8.49.

   b. Calculate the kinetic coefficients k, K_s, and μ_{max}. Use a graphical procedure and the equation developed in Problem 8.6.

Table 8.5 Experimental data (for Problem 8.7)

Reactor #	S, mg/L	S₀, mg/L	X, mg VSS/L	SRT, d
1	500	300	295	0.45
2	500	200	472	0.50
3	500	150	584	0.55
4	500	100	746	0.60
5	500	60	990	0.75
6	500	30	1516	1.05

8.8 What is the difference between HRT and SRT? In what type of process is the HRT and SRT the same?

8.9 A completely mixed activated sludge plant is designed to treat 10,000 m³/d of an industrial wastewater. The wastewater has a BOD₅ of 1200 mg/L. Pilot-plant data indicates that a reactor volume of 6090 m³ with an MLSS concentration of 5000 mg/L should produce 83% BOD₅ removal. The value for Y is determined to be 0.7 kg/kg and the value of k_d is found to be 0.03 d⁻¹. Underflow solids concentration is 12,000 mg/L. The flow diagram is similar to Figure 8.6.

   a. Determine the mean cell residence time for the reactor.

   b. Calculate the mass of solids wasted per day.

   c. Calculate the volume of sludge wasted each day.

8.10 The city of Annandale has been directed to upgrade its primary wastewater treatment plant to a secondary treatment plant with sludge recycle, that can meet an effluent standard of 11 mg/l BOD₅. The following data are available:

Flow = 0.15 m³/s, MLSS = 2000 mg/L.
Kinetic parameters: K_s = 50 mg/L, μ_{max} = 3.0 d⁻¹, k_d = 0.06 d⁻¹, Y = 0.6
Existing plant effluent BOD₅ = 84 mg/L.

   a. Calculate the SRT (θ_c) and HRT (θ) for the aeration tank.

   b. Calculate the required volume of the aeration tank.

   c. Calculate the food to microorganism ratio in the aeration tank.

   d. Calculate the volumetric loading rate in kg BOD₅/m³-d for the aeration tank.

   e. Calculate the mass and volume of solids wasted each day, when the underflow solids concentration is 12,000 mg/L.

8.11 A completely mixed activated sludge plant is to treat 18,500 m³/d of industrial wastewater. The wastewater has a BOD₅ of 160 mg/L that must be reduced to 5 mg/L prior to discharge to a municipal sewer. Pilot-plant analysis indicates that a mean cell residence time (SRT) of 8 d with an MLSS concentration of 3000 mg/L produces the desired results. The value for Y is determined to be 0.5 kg/kg and the value of k_d is found to be 0.04 d⁻¹. Determine the following:

   a. The hydraulic retention time (HRT) and volume of the aeration tank.

   b. The F/M ratio in the reactor.

   c. Calculate the volumetric loading rate in kg BOD₅/m³-d for the aeration tank.

   d. The mass and volume of solids wasted each day, when the underflow solids concentration is 12,000 mg/L.

   e. It is desired to reduce the SRT to 5 d. Show with the help of calculations what changes you would make to the system to achieve that.

8.12 A completely mixed high rate activated sludge plant with recycle treats 17,500 m³/d of industrial wastewater. The influent to the activated sludge reactor has a BOD_5 of 1000 mg/L. It is desired to reduce the influent BOD_5 to 120 mg/L, prior to discharge to a municipal sewer. Pilot plant analysis gave the following results: Mean cell residence time = 6 d, MLSS concentration in reactor = 5500 mg/L, Y = 0.6 kg/kg, k_d = 0.03 d^{-1}. Determine the following:

a. The hydraulic retention time and volume of the activated sludge reactor.

b. The volumetric loading rate in kg BOD_5/m^3-d to the reactor.

c. The F/M ratio in the reactor.

d. The mass and volume of solids wasted each day, at an underflow solids concentration,
   X_u = 10,000 mg/L.

e. The sludge recirculation ratio.

8.13 Consider the completely mixed high rate activated sludge plant from Problem 8.12. Fine bubble membrane diffusers with total floor coverage are to be used for the aeration tank. The SOTR specified by the manufacturer is 3.2 kg O_2/kWh, with αF of 0.45. The average wastewater temperature is 18°C. The residual DO in the aeration tank is 4 mg/L, β is 0.90, and saturation oxygen concentration at 18°C and tank depth elevation is 9.54 mg/L. Calculate the oxygen demand and the power required for aeration.

8.14 The town of Orland Park uses stabilization ponds to treat its wastewater. The wastewater flow is 140,000 gpd with a BOD_5 of 320 mg/L. The total surface area of the pond is 14.8 acres. Water loss during the winter months is 0.11 in/day.

a. Calculate the BOD loading on the pond.

b. Calculate the days of winter storage available, when operating water depths range from 2 ft to 5 ft.

8.15 A small town in Virginia with a population of 2000 is considering a stabilization pond to treat its wastewater. The average wastewater flow is 140 gpcd with a BOD_5 of 200 mg/L. The maximum pond depth is limited to 8 ft including free board. The average annual temperature is 25°C. Design the stabilization pond for a BOD loading of 30 lb/acre · d, with a first order rate coefficient of 0.38 d^{-1} at 20°C. Calculate the following:

a. The area of the pond.

b. Available detention time.

c. Effluent BOD concentration.

REFERENCES

Bae, T., and Tak, T. 2005. Preparation of Tio2 Self-Assembled Polymeric Nanocomposite Membranes and Examination of their Fouling Mitigation Effects in a Membrane Bioreactor System, *Journal of Membrane Science*, 266 (1–2):1–5.

Davis, M. 2020. *Water and Wastewater Engineering: Design Principles and Practice.* 2nd edn. McGraw-Hill, Inc., New York, NY, USA.

EPA. 2021. https://www.epa.gov/septic/types-septic-systems

Francesco, F., Di Fabio, S., Bolzonella, D., and Cecchi, F. 2011. Fate of Aromatic Hydrocarbons in Italian Municipal Wastewater Systems: An Overview of Wastewater Treatment Using Conventional Activated-sludge Processes (CASP) and Membrane Bioreactors (MBRs), *Water Research*, 45 (1):93–104.

Hammer, M. J. and Hammer, M. J. Jr. 2012. *Water and Wastewater Technology.* 7th edn. Pearson-Prentice Hall, Inc., Hoboken, NJ, USA.

Jahan, K., Hoque, S., Ahmed, T., and Türkdoğan, P. 2011 Activated Sludge and Other Aerobic Suspended Culture Processes, *Water Environment Research*, 83 (10):1092–1149.

Jones, R., Parker, W., Zhu, H., Houweling, D., and Murthy, S. 2009. Predicting the Degradability of Waste Activated Sludge, *Water Environment Research*, 81 (8):765–771.

Kim, J., and van der Bruggen, B. 2010. The Use of Nanoparticles in Polymeric and Ceramic Membrane Structures: Review of Manufacturing Procedures and Performance Improvement for Water Treatment, *Environmental Pollution*, 158 (7):2335–2349.

Kimura, K., Hara, H., and Watanabe, Y. 2007. Elimination of Selected Acidic Pharmaceuticals from Municipal Wastewater by an Activated Sludge System and Membrane Bioreactors, *Environmental Science and Technology*, 41 (10):3708–3714.

Lawrence, A. W. and McCarty, P. L. 1970. A Unified Basis for Biological Treatment Design and Operation. *Journal of Sanitary Engineering Division*, American Society of Civil Engineers, 96 (3):757–778.

Lu, Y., Yu, S., and Meng, L. 2009. Preparation Of Poly(Vinylidene Fluoride) Ultrafiltration Membrane Modified by Nano-Sized Alumina and its Antifouling Performance, *Harbin Gongye Daxue Xuebao/Journal of Harbin Institute of Technology*, 41(10):64–69.

Ma, F., Guo, J., Zhao, L., Chang, C., and Cui, D. 2009. Application of Bioaugmentation to Improve the Activated Sludge System into the Contact Oxidation System Treating Petrochemical Wastewater, *Bioresource Technology*, 100 (2):597–602.

Metcalf and Eddy, Tchobanoglous, G., Stensel, H., Tcuchihashi, R., and Burton, F. 2013. *Wastewater Engineering: Treatment and Resource Recovery.* 5th edn. McGraw-Hill, Inc., New York, NY, USA.

Monod, J. 1949. The Growth of Bacterial Cultures. *Annual Review of Microbiology*, 3:371–394.

New York State Department of Health. 2010. *Wastewater Treatment Standards - Individual Household Systems*, Appendix 75-A, New York State Department of Health, USA.

Peavy, H. S., Rowe, D. R., and Tchobanoglous, G. 1985. *Environmental Engineering*, McGraw- Hill, Inc., New York, NY, USA.

Plósz, B., Leknes, H, Liltved, H., and Thomas, K. 2010. Diurnal Variations in the Occurrence and the Fate of Hormones and Antibiotics in Activated Sludge Wastewater Treatment in Oslo, Norway, *Science of the Total Environment*, 408 (8):1915–1924.

Rieger, L., Takács, I., Villez, K., Siegrist, H., Lessard, P., Vanrolleghem, P., Comeau, Y. 2010. Data Reconciliation for Wastewater Treatment Plant Simulation Studies—Planning for High-Quality Data and Typical Sources of Errors, *Water Environment Research*, 82 (5):426–433.

Schmit, C. G., Jahan, K., Schmit, K., Aydinol, F., Debik, E., Pattarkine, V. 2010. Activated Sludge and Other Aerobic Suspended Culture Processes, *Water Environment Research*, 82 (10):1073–1123.

Somogyi, V., Pitas, V., Domokos, E. and Fazekas, B. 2009. On-site Wastewater Treatment Systems and Legal Regulations in the European Union and Hungary. *Acta Universitatis Sapientiae Agriculture and Environment*. 1:57–64.

Su, Y., Huang, C., Pan, J. R., Hsieh, W., and Chu, M. 2012. Fouling Mitigation by TiO2 Composite Membrane in Membrane Bioreactors, *Journal of Environmental Engineering (ASCE)*, 138 (3):344–350.

WEF. 1998. *Design of Wastewater Treatment Plants*, 4th ed., Manual of Practice no.8, Water Environment Federation, Alexandria, VA, USA.

Zodrow, K., Brunet, L., Mahendra, S., Li, D., Zhang, A., Li, Q., and Alvarez, P. J. 2007. Polysulfone Ultrafiltration Membranes Impregnated with Silver Nanoparticles Show Improved Biofouling Resistance and Virus Removal, *Polymers for Advanced Technologies*, 18 (7):562–568.

Chapter 9

Secondary treatment
Attached growth and combined processes

9.1 INTRODUCTION

The two major categories of biological treatment are (i) suspended growth, and (ii) attached growth processes. The focus of this chapter is aerobic biological treatment using attached growth processes for BOD removal. A combination of suspended and attached growth processes may be used for the biological treatment of wastewater. Some of these hybrid processes are discussed at the end of the chapter. Suspended growth processes have been discussed in Chapter 8. Biological processes used for nitrogen and phosphorus removal will be described in detail in Chapter 13.

In attached growth systems, the microorganisms are attached to an inert medium, forming a *biofilm*. As the wastewater comes in contact with and flows over the biofilm, the organic matter is removed by the microorganisms and degraded to produce an acceptable effluent. A secondary clarifier is used, but sludge recirculation to the biological reactor is not necessary. The settled sludge consisting of sloughed biofilm is usually recirculated back to the wet well or primary clarifier. Attached growth systems are characterized by a high degree of liquid recirculation (100–300%) to the biological reactor. The media is usually an inert material with high porosity and surface area, e.g. rock, gravel, synthetic media, etc. The system can be operated as an aerobic or anaerobic process. The media can be wholly or partially submerged in the wastewater. The common types of attached growth processes include trickling filters, bio-towers, and rotating biological contactors (RBC). These will be described in the following sections.

The main advantages of the attached growth processes are (Metcalf and Eddy et al., 2013):

- Simplicity of operation.
- Low energy requirement.
- Low maintenance required.
- Ability to handle shock loads.
- Lower sludge production.

DOI: 10.1201/9781003134374-9

- No problems with sludge bulking in secondary clarifiers.
- Better sludge thickening properties.

The disadvantages include:

- Low efficiency at cold temperatures.
- Mass transfer and diffusion limitations can occur.
- Problems with biofilm maintenance due to excess sloughing.
- Higher BOD and solids concentration in the effluent.
- Odor problems can occur.

9.2 SYSTEM MICROBIOLOGY AND BIOFILMS

The microorganisms in the biofilm are similar to those found in activated sludge reactors. They are mostly heterotrophic, with facultative bacteria being predominant. Fungi and protozoa are present. If sunlight is available, algae growth is found near the surface. Larger organisms such as sludge worms, insect larvae, rotifers, etc., may also be present. When the carbon content of the wastewater is low, nitrifying bacteria may be present in large numbers.

In an attached culture reactor, microorganisms attach themselves to the inert media and grow into dense films. These are called *biofilms*. As the wastewater passes over the biofilm, suspended organic particles are adsorbed on the biofilm surface. The adsorbed particles are degraded to soluble products, which are then further degraded to simpler products and gases by the bacteria. Dissolved organics pass into the biofilm according to mass transfer principles, due to the presence of concentration gradients. Dissolved oxygen from the wastewater diffuses into the biofilm for the aerobic bacteria. Waste products and gases diffuse outward from the biofilm and are carried out of the reactor with the wastewater. This process is illustrated in Figure 9.1.

With the passage of time, the thickness of the biofilm increases. The biofilm grows in a direction outward from the media. As the thickness increases, the outer 0.1–0.2 mm of biofilm remains aerobic (Peavy et al., 1985). The inner layers of the biofilm become anaerobic, as oxygen cannot pass into the inner layers due to diffusion limitations. The total thickness may range from 100 μm to 10 mm. With increasing thickness, the biofilm attachment becomes weak, and the shearing action of the wastewater dislodges it from the media and transports it to the secondary clarifier. This process is known as *sloughing* of biofilm. Regrowth of biofilm occurs quickly in places cleared by sloughing. Sloughing is a function of the hydraulic and organic loading on the reactor. The hydraulic loading accounts for shear velocities, and the organic loading controls the rate of metabolism in the biofilm layer.

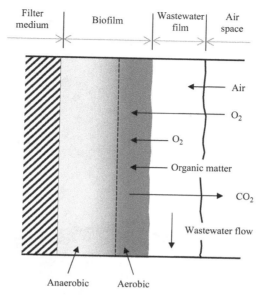

Figure 9.1 Mass transfer of organic matter and gases in a biofilm.

The rate of BOD removal depends on the following (Peavy et al., 1985):

- Wastewater flow rate.
- Organic loading rate.
- Temperature.
- Rates of diffusion of BOD and oxygen into the biofilm. The oxygen diffusion rate is usually a limiting factor.

9.3 IMPORTANT MEDIA CHARACTERISTICS

Selection of the media is an important aspect of the design of attached growth processes. In addition to the size and unit weight, the following are important characteristics of the media or packing material used in attached culture systems:

1. *Chemical and biological inertness* – The media material should not undergo any chemical or biological reactions with the constituents of the wastewater.
2. *Porosity* – It is defined as the ratio of the volume of voids to the total volume of a particle or material. It is given by

$$\text{Porosity} = \frac{V_v}{V_T}$$

(9.1)

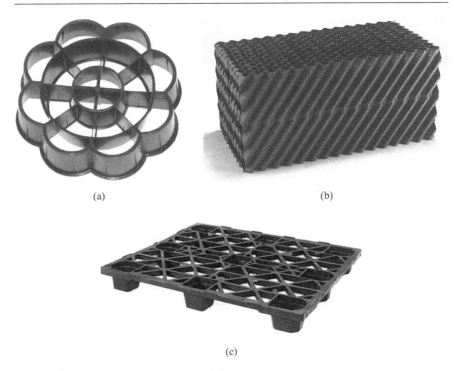

(a)   (b)

(c)

Figure 9.2 Trickling filter media: (a) Plastic media (Bio-Pac SF30™), (b) PVC sheet media (Dura-Pac XF31™), and (c) random media underdrain (ND 330™). (Photos courtesy of Jaeger Environmental of Virginia, and Jaeger Products, Inc. of Texas)

where
V_v = volume of voids
V_T = total volume

Porosity is often expressed as a percent. Stone media can have porosities ranging from 40% to 50%, while synthetic media can have porosities up to 95%. Higher porosity is desired, as that provides more passage for wastewater and gases. Figure 9.2 presents two types of media and an underdrain panel for trickling filters.

3. *Specific surface area* – It is defined as the amount of surface area of the media that is available for the growth of biofilm, per unit volume of media. Stone media can have specific surface areas ranging from 45 to 70 m²/m³. For synthetic media, this value can range from 100 to 200 m²/m³.

9.4 LOADING RATES

The organic loading or BOD loading is calculated using only the BOD load coming with the primary effluent. The BOD loading in the recycle flow is not added.

$$\text{BOD loading} = \frac{\text{Primary effluent BOD}}{\text{Volume of filter media}} = \frac{Q S_0}{V} \qquad (9.2)$$

where

Q = wastewater flow rate, m^3/d (mil gal/d)
S_o = BOD concentration of primary effluent, kg/m^3 (lb/mil gal)
V = volume of media, m^3 (1000s of ft^3)
BOD loading = kg BOD/m^3.d (lb BOD/1000 ft^3.d)

The organic loading for stone media trickling filters can vary from 0.08 to 1.8 kg BOD/m^3.d for low to high rate filters, respectively. The organic loading for plastic media filters ranges from 0.31 to above 1.0 kg BOD/m^3.d (Davis, 2020).

The hydraulic loading is the amount of wastewater applied to the filter surface including primary effluent and recycle flows.

$$\text{Hydraulic loading} = \frac{Q + Q_r}{A_s} \qquad (9.3)$$

where

Q = wastewater flow rate, m^3/d (mil gal/d)
Q_r = recirculation flow, m^3/d (mil gal/d)
A_s = surface area of the filter, m^2 (acres)
Hydraulic loading = m^3/m^2.d (mil gal/acre.d)

The hydraulic loading for stone media trickling filters can range from 4 to 40 m^3/m^2.d for low to high rate filters, respectively. For plastic media filters, the hydraulic loading can vary from 60 to 180 m^3/m^2.d.

9.5 STONE MEDIA TRICKLING FILTER

The trickling filter is one of the earliest types of attached growth processes that were used. It has been used for secondary treatment since the early 1900s (Metcalf and Eddy et al., 2013). The term trickling filter is misleading, as most of the physical processes involved in filtration are absent in this

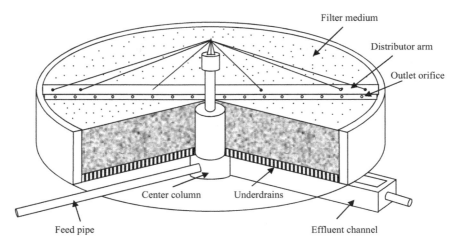

Figure 9.3 Section through a trickling filter.

process. Instead, sorption and subsequent degradation of organic matter are used for substrate removal (Peavy et al., 1985). A typical trickling filter is illustrated in Figure 9.3.

A trickling filter consists of a shallow tank filled with crushed stone, rocks, or slag as media. The tank depth ranges from 0.9 to 2.5 m for stone media trickling filters. These provide durable, chemically inert surfaces for the growth of biofilm. Media size ranges from 50 to 100 mm (2–4 in) with porosities of 40–50%. The wastewater is applied on the media by a rotary distributor arm from the top of the tank. As the water flows through the tank, organic matter is removed as it comes in contact with the biofilm. An underdrain system transports the treated wastewater and sloughed biofilm to the secondary clarifier.

Recirculation is an important aspect of trickling filters. Recirculation ratios ranging from 0.5 to 3 are used. Liquid recirculation is used to provide the desired wetting rate to keep the microorganisms alive and raise the dissolved oxygen of the influent. It helps to dilute the strength of shock loads. Typical diurnal variation of wastewater flow is illustrated in Figure 9.4. Recirculation is used to dampen the variation in loadings over a 24-h period (Davis, 2020).

When a portion of the effluent from the trickling filter is recycled back to the filter, while the remainder goes to the secondary clarifier, it is called *direct recirculation*. This is illustrated in Figure 9.5 (a). The recirculation of settled sludge from the secondary clarifier to the wet well or to the primary clarifier is termed *indirect recirculation*. Figure 9.5 (b) illustrates indirect recirculation of settled sludge and liquid effluent from the secondary clarifier.

Trickling filters may be used as a single stage system, or with two stages in series. An intermediate clarifier can be used in between the two filters. Using two filters in series aids in the improvement of efficiency.

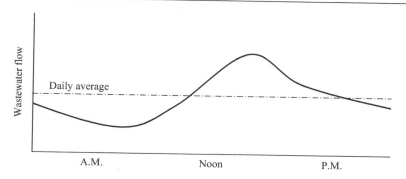

Figure 9.4 Variation of wastewater flow over a typical 24-h period.

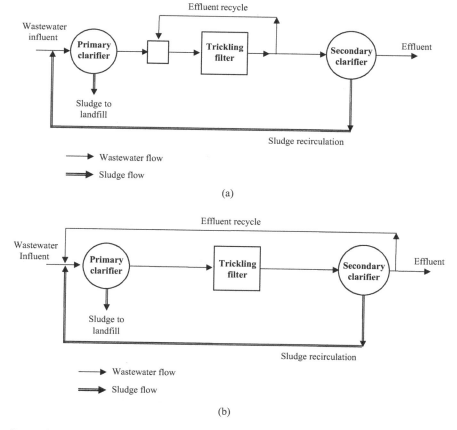

Figure 9.5 (a) Trickling filter with direct recirculation of effluent and (b) trickling filter with indirect recirculation.

9.5.1 Design equations for stone media

The first empirical design equations were developed for stone media trickling filters by the National Research Council, based on performance data at military installations treating domestic wastewater during World War II (Wolman et al., 1946). These are known as the National Research Council or NRC equations. These equations were used to predict the efficiency of trickling filters based on BOD load, volume of filter media, and recirculation ratio.

For a single-stage rock filter, or for the first stage of a two-stage filter, the efficiency is given by

$$E_1 = \frac{100}{1 + 0.4432 \sqrt{\dfrac{W_1}{V\,F}}} \tag{9.4}$$

where
   E_1 = BOD removal efficiency for the first-stage filter at 20°C including recirculation, %
   W_1 = BOD loading to filter, kg/d
   V = volume of filter media, m^3
   F = recirculation factor

The recirculation factor is calculated as

$$F = \frac{1 + R}{\left(1 + 0.1R\right)^2} \tag{9.5}$$

where R = recirculation ratio = $\dfrac{Q_r}{Q}$ = $\dfrac{\text{recycle flow rate}}{\text{wastewater flow rate}}$

For a two-stage trickling filter, the BOD removal efficiency of the second stage is given by

$$E_2 = \frac{100}{1 + \dfrac{0.4432}{1 - E_1} \sqrt{\dfrac{W_2}{V\,F}}} \tag{9.6}$$

where
   E_2 = BOD removal efficiency for the second-stage filter at 20°C including recirculation, %

W_2 = BOD loading to the second-stage filter, kg/d
E_1 = fraction of BOD removal in the first-stage filter

An intermediate clarifier is assumed to be situated between the first- and second-stage filters. The effect of wastewater temperature on BOD removal efficiency is calculated using a form of the van't Hoff–Arrhenius equation:

$$E_T = E_{20}(1.035)^{T-20} \tag{9.7}$$

where
  T = wastewater temperature, °C
  E_T = BOD removal efficiency at temperature T°C
  E_{20} = BOD removal efficiency at 20°C

When using the NRC equations it should be noted that the military installations which were the basis for the NRC study, had higher influent BOD concentrations than domestic wastewater today. The clarifiers were shallower and carried higher hydraulic loading than current practice today (Davis, 2020).

Example 9.1: A wastewater treatment plant uses a single-stage rock-media trickling filter for secondary treatment, as illustrated in Figure 9.5 (a). The wastewater flow rate is 2000 m³/d with a BOD_5 concentration of 400 mg/L. Primary clarification removes 30% of the BOD_5. The filter is 12 m in diameter, and 1.5 m in depth. A direct recirculation pump operates at 2.78 m³/min to the filter. Wastewater temperature is 20°C. Calculate the hydraulic loading rate, organic loading rate, effluent BOD_5 concentration, and overall plant efficiency.

SOLUTION
Step 1. Calculate the filter area and volume

Trickling filter area, $A_s = \dfrac{\pi}{4}(D)^2 = \dfrac{\pi}{4}(12)^2 = 113.09\,\text{m}^2$

Trickling filter volume, $V = A_s \times h = 113.09\,\text{m}^2 \times 1.5\,\text{m} = 169.65\,\text{m}^3$

Step 2. Calculate the hydraulic loading rate on filter

$$\text{BOD}_5 \text{ in to trickling filter}, S_o = 400(1-0.30) = 280\,\text{mg/L} = 0.28\,\text{kg/m}^3$$

$$\text{Recirculation flow}, Q_r = 2.78\,\text{m}^3/\text{min} = 2.78 \times 60 \times 24 = 4003\,\text{m}^3/\text{d}$$

$$\text{Recirculation ratio}, R = \frac{Q_r}{Q} = \frac{4003}{2000} = 2$$

$$\text{Hydraulic loading rate} = \frac{Q+Q_r}{A_s} = \frac{2000+4003\,\text{m}^3/\text{d}}{113.09\,\text{m}^2} = 53.08\,\text{m}^3/\text{m}^2.\text{d}$$

Step 3. Calculate the organic loading rate

$$\text{Organic / BOD loading rate} = \frac{Q\,S_0}{V} = \frac{2000\dfrac{\text{m}^3}{\text{d}} \times 0.28\,\text{kg/m}^3}{169.65\,\text{m}^3} = 3.30\,\text{kg/m}^3.\text{d}$$

Step 4. Calculate the filter efficiency

$$\text{Recirculation factor}, F = \frac{1+R}{(1+0.1R)^2} = \frac{1+2}{(1+0.1\times2)^2} = 2.08$$

$$W_1/V = 3.3\,\text{kg/m}^3.\text{d}$$

$$E_1 = \frac{100}{1+0.4432\sqrt{\dfrac{W_1}{V\,F}}} = \frac{100}{1+0.4432\sqrt{\dfrac{3.30}{2.08}}} = 64.17\%$$

$$\text{Plant effluent BOD concentration} = 280\,\text{mg/L}(1-0.6417) = 100.32\,\text{mg/L}$$

$$\text{Overall plant efficiency} = \frac{400-100.32}{400} \times 100\% = 74.92\%$$

Comment: The hydraulic and BOD loading rates are high. These can be reduced by adding more filters in parallel to the first one.

9.6 BIO-TOWER

Bio-towers are deep bed trickling filters with plastic or synthetic media. Depths up to 12 m can be utilized, since lightweight media is used. Various types and shapes of media are used for packing. Small plastic cylinders with perforated walls may be used as illustrated in Figure 9.2. The specific surface area ranges from 100 to 130 m²/m³ with a porosity of about 94%. Random packing allows the wastewater to be distributed throughout the media, allowing enough contact time between the substrate and biofilm. Modular media consisting of corrugated and flat polyvinyl chloride (PVC) sheets welded together in alternating patterns are also used. These are presented in Figure 9.2.

9.6.1 Design equations for plastic media

The design models for plastic media trickling filters or bio-towers were based on the early works of Velz (1948), Howland (1958), and Schulze (1960). Eckenfelder (1961) applied the Schulze equation to plastic media bio-towers as

$$\frac{S_e}{S_o} = e^{-\frac{kD}{q^n}}$$

(9.8)

where

S_e = soluble BOD concentration of settled filter effluent, mg/L
S_o = soluble BOD concentration of filter influent, mg/L
D = depth of media, m
q = hydraulic application rate excluding recirculation, m³/m².min
k = wastewater and filter media treatability coefficient, min⁻¹
n = coefficient related to packing media, taken as 0.5 for modular plastic media

The value of k ranges from 0.01 to 0.1 min⁻¹. The value of k at 20°C is around 0.06 for municipal wastewater on modular plastic media (Germain, 1966). k is primarily affected by temperature, and temperature corrections are performed using the van't Hoff–Arrhenius equation:

$$k_T = k_{20}(1.035)^{T-20}$$

(9.9)

where

k_T = treatability constant at temperature $T°C$
k_{20} = treatability constant at 20°C
T = wastewater temperature, °C

Equation (9.8) can be modified taking into account the effect of recirculation, as shown below:

$$\frac{S_e}{S_a} = \frac{e^{-\frac{kD}{q^n}}}{(1+R) - R e^{-\frac{kD}{q^n}}} \tag{9.10}$$

where
S_a = BOD concentration of the mixture of primary effluent and recycled wastewater, mg/L
S_e = effluent BOD concentration, mg/L
R = recirculation ratio
q = hydraulic loading rate with recirculation, m³/m².min

All other terms are as defined previously. From mass balance, S_a is calculated as

$$S_a = \frac{S_o + R S_e}{1 + R} \tag{9.11}$$

Other models have been proposed taking into account the specific surface area of the media. The modified Velz equation incorporates the specific surface area as shown below (Hammer and Hammer, 2012):

$$\frac{S_e}{S_o} = e^{-\frac{k_{20} A_s D}{q^n}} \tag{9.12}$$

where A_s = specific surface area of media, m²/m³
All other terms are as defined previously. Equation (9.12) can be modified to incorporate the effect of recirculation and obtain an equation similar to Equation (9.10).

> **Example 9.2:** An industry has decided to treat its process wastewater in a Bio-tower with a plastic modular medium (n = 0.6). The flow rate of the wastewater is 1000 m³/d with a BOD$_5$ of 500 mg/L and an average temperature of 18°C. The treatability constant k is 0.04 min⁻¹ for the system at 20°C. The depth of the medium is 5.5 m. The desired effluent BOD$_5$ is 15 mg/L. Calculate the following:
>
> a. The area of bio-tower required, without any recycle.
> b. The organic loading rate for the bio-tower without recycle.

c. The area of bio-tower required when direct recirculation ratio is 3:2.

d. The organic loading rate for the bio-tower with recycle.

e. Which of the above designs seem better to you and why?

SOLUTION

Step 1. Adjust k for temperature
Use the van't Hoff–Arrhenius Equation (9.9):

$$k_{18} = k_{20} (1.035)^{18-20}$$
$$= 0.04 (1.035)^{-2}$$
$$= 0.037 \text{ min}^{-1}$$

Step 2. Calculate the surface area for bio-tower without recycle.
Using Equation 9.8, calculate q

$$\frac{S_e}{S_o} = e^{-\frac{kD}{q^n}}$$

$$\text{Or,} \frac{15 \text{ mg/L}}{500 \text{ mg/L}} = e^{-\frac{0.037 \times 5.5}{q^{0.6}}}$$

$$\text{Or,} \log_e (0.03) = -\frac{0.037 \times 5.5}{q^{0.6}}$$

$$\text{Or,} q^{0.6} = 0.058$$

$$\text{Or,} q = 0.0087 \text{ m}^3/\text{m}^2.\text{min} = 12.528 \text{ m}^3/\text{m}^2.\text{d}$$

$$Q = 1000 \text{ m}^3/\text{d}$$

$$\text{Surface area reqd} = \frac{Q}{q} = \frac{1000 \text{ m}^3/\text{d}}{12.528 \frac{\text{m}^3}{\text{m}^2}.\text{d}} = 79.82 \text{ m}^2 \approx 80 \text{ m}^2$$

Step 3. Calculate the organic loading rate using Equation (9.2).

$$\text{BOD loading} = \frac{Q S_0}{V}$$

$$= \frac{1000 \frac{m^3}{d} \times 0.5 \, kg / m^3}{80 \, m^2 \times 5.5 \, m} = 1.14 \, kg \, BOD / m^3.d$$

Step 4. Calculate S_a using Equation 9.11, with $R = 3/2 = 1.5$

$$S_a = \frac{S_o + R \, S_e}{1 + R}$$

$$Or, S_a = \frac{500 \frac{mg}{L} + 1.5 \times 15 \, mg / L}{1 + 1.5} = 209 \, mg / L$$

Step 5. Calculate q for bio-tower with recycle
Use Equation (9.10) to calculate q

$$\frac{S_e}{S_a} = \frac{e^{-\frac{kD}{q^n}}}{(1 + R) - R e^{-\frac{kD}{q^n}}}$$

$$Or, \frac{15 \, mg / L}{209 \, mg / L} = \frac{e^{-\frac{0.037 \times 5.5}{q^{0.6}}}}{(1 + 1.5) - 1.5 e^{-\frac{0.037 \times 5.5}{q^{0.6}}}}$$

$$Or, 0.179 = e^{-\frac{0.2035}{q^{0.6}}} + 0.108 e^{-\frac{0.2035}{q^{0.6}}}$$

$$Or, \frac{0.179}{1.108} = e^{-\frac{0.2035}{q^{0.6}}}$$

Or,

$$\log_e (0.1616) = -\frac{0.2035}{q^{0.6}}$$

Or, $q^{0.6} = 0.1116$

Or,

$$q = 0.026 \, m^3 / m^2.min = 37.27 \, m^3 / m^2.d$$

Step 6. Calculate the surface area of bio-tower with recycle.

$$\text{Surface area reqd} = \frac{Q(1+R)}{q} = \frac{1000(1+1.5)\,\text{m}^3/\text{d}}{37.27\,\dfrac{\text{m}^3}{\text{m}^2}.\text{d}} = 67\,\text{m}^2$$

Step 7. Calculate the organic loading rate for bio-tower with recycle, using Equation (9.2).

$$\text{BOD loading} = \frac{Q\,S_0}{V}$$

$$= \frac{1000\,\dfrac{\text{m}^3}{\text{d}} \times 0.5\,\text{kg}/\text{m}^3}{67\,\text{m}^2 \times 5.5\,\text{m}} = 1.36\,\text{kg BOD}/\text{m}^3.\text{d}$$

Note: The bio-tower with recycle seems to be the better design. A smaller surface area is required, and at the same time, it can handle higher BOD loading due to recirculation. The construction cost would be lower, but additional pumping cost for recirculation will have to be considered.

9.7 ROTATING BIOLOGICAL CONTACTOR (RBC)

The rotating biological contactor or RBC was first installed in Germany in 1960 and later introduced in the US (Metcalf and Eddy et al., 2013). RBC is a type of attached growth process where the media is in motion as well as the wastewater. A series of closely spaced circular disks of polystyrene or PVC are mounted on a horizontal shaft, which is rotated in a tank through which the wastewater is flowing, as illustrated in Figure 9.6. The media is partially submerged in the wastewater. It comes in contact with air and wastewater in an alternating fashion thus maintaining aerobic conditions, as the shaft with the disks rotates in the tank. Rotational speed varies from 1 to 2 rpm. It must be sufficient to provide the hydraulic shear necessary for sloughing of biofilm, and to maintain enough turbulence to keep the solids in suspension in the wastewater (Peavy et al., 1985). The media disks have a diameter ranging from 2 to 4 m, and a thickness of about 10 mm. Spacing between the disks is about 30–40 mm. Each shaft with the medium, along with its tank and rotating device is called a *module*. Several modules are arranged in series or parallel to obtain the desired removal efficiencies. Figure 9.7 illustrates a flow diagram of an RBC process.

The RBC modules provide a large amount of surface area for biomass growth. One module of 3.7 m diameter and 7.6 m length, contains approximately 10,000 m² of surface area. The large amount of biomass is able to produce acceptable effluents within a short contact time. Recirculation of effluent through the reactor is not necessary. Advantages of the process include low power requirements, simple operation, and good sludge

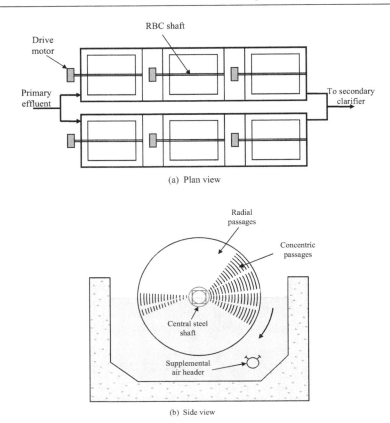

(a) Plan view

(b) Side view

Figure 9.6 (a) Plan and (b) side view of an RBC module.

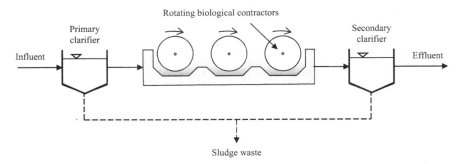

Figure 9.7 Flow diagram of an RBC process.

Source: Adapted from Peavy et al., 1985.

settleability. Disadvantages include high capital cost and susceptibility to cold temperatures. Covers have to be provided to maintain the biomass in winter. RBCs can be operated as aerobic or anaerobic processes.

Nitrification can be achieved in an RBC system by operating a number of modules in series. BOD removal takes place in the first stages. Then when the carbon content is low, nitrification takes place. Typically, five modules in series are required for complete nitrification.

Manufacturers of RBC systems often specify a soluble BOD loading rate for their equipment, since the soluble BOD is used more rapidly in the first stage of an RBC system. A soluble BOD loading in the range of 12–20 g sBOD/m².d (2.5–4.1 lb sBOD/1000 ft³.d) is commonly specified for the first stage (Metcalf and Eddy et al., 2013). The total BOD loading can range from 24 to 40 g sBOD/m².d, assuming a 50% soluble BOD fraction. To accommodate higher loading rates due to high-strength wastewaters, multiple modules are used in parallel for the first stage. A number of empirical design approaches have been used for RBC systems based on pilot plant and full-scale plant data. A review of these models was provided by WEF (2000).

9.8 HYBRID PROCESSES

Activated sludge systems that incorporate some form of media in the suspended growth reactor are termed *hybrid processes* (Davis, 2020). These include Moving Bed Biofilm Reactor (MBBR), Integrated Fixed-Film Activated Sludge (IFAS), Fluidized Bed Bio-Reactor (FBBR), among others. Hybrid activated sludge/MBBR processes have been investigated for the treatment of municipal wastewater in cold climates (Di Trapani et al., 2011).

9.8.1 Moving bed biofilm reactor (MBBR)

The MBBR process was developed in the late 1980s in Norway. Small cylinder-shaped polyethylene media elements are placed in the aeration tank to support biofilm growth. The tank may be mixed with aeration or mechanical mixers. A perforated plate or screen is placed at the outlet to prevent the loss of media with the effluent. Figure 9.8 illustrates a typical MBBR. The biofilm carrier elements are about 10 mm in diameter and 7 mm in height, with a density of about 0.96 g/cm³. The specific surface area ranges from 300 to 500 m²/m³. The biofilm in this process is relatively thinner and more evenly distributed over the carrier surface, as compared to other fixed-film processes. In order to obtain this type of biofilm, the degree of turbulence in the reactor is important (Ødegaard, 2006). The mixing or turbulence transports the substrate to the biofilm, and also maintains a low thickness of the biofilm by shearing forces.

Advantages of the MBBR process include (i) continuous operation of the reactor without the threat of clogging, (ii) no backwash requirement, (iii) no necessary sludge recirculation, (iv) low headloss, and (v) a high specific

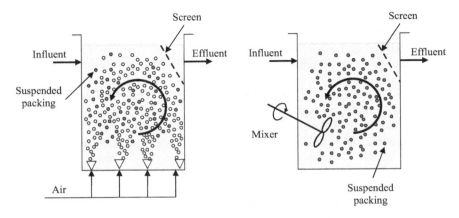

Figure 9.8 Moving bed biofilm reactor (MBBR): (a) Mixing with external air and (b) mechanical mixing.

Source: Adapted from Ødegaard, 2006.

biofilm surface (Ødegaard et al., 1994; Rusten and Neu, 1999; and Andrettola et al., 2003).

When a treatment plant needs to increase its capacity due to increased BOD loading, it can be achieved by adding more biofilm carrier elements to the reactor to increase the biofilm surface area (Aspegren et al., 1998, and Rusten et al., 1995). For the same reason, an existing activated sludge process can be upgraded to an MBBR process to handle increased loads without expansion of existing reactor volume. The cost of the synthetic media has to be considered with respect to other costs. The MBBR process may be used for aerobic, anoxic, or anaerobic processes for carbon and nitrogen removal.

9.8.2 Integrated fixed-film activated sludge (IFAS)

In the IFAS processes, a fixed packing material is placed in an activated sludge reactor. The packing can be in the form of frames, foam pads, etc., suspended in the aeration tank. A number of proprietary processes include BioMatrix® process, Bio-2-Sludge® process, Ringlace®, and BioWeb®. These processes differ from the MBBR in that they use a return sludge flow. The purpose of the fixed film media is to increase the biomass concentration in the reactor. This is advantageous in increasing the capacity of the activated sludge process without increasing tank volume.

9.8.3 Fluidized bed bioreactor (FBBR)

In an FBBR, wastewater enters at the bottom of the aeration tank and flows upward through a bed of sand or activated carbon. Activated carbon provides both adsorption properties and media surface for biofilm growth. The specific surface area is about 1000 m^2/m^3 of the reactor volume, with bed

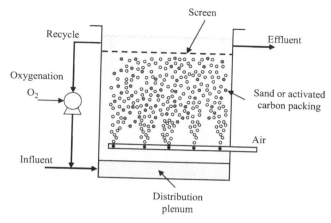

Figure 9.9 Fluidized bed bioreactor (FBBR).

depths of 3–4 m. Effluent recirculation is performed to provide the fluid velocity within the necessary detention times. A diagram of an FBBR is provided in Figure 9.9. As the biofilm increases in thickness, the media accumulates at the top of the bed from where it is removed and agitated to remove excess solids at regular intervals. In aerobic FBBRs, recirculated effluent is passed through an oxygen tank to saturate with dissolved oxygen. Air is not added directly to the reactor. The system has a number of advantages which include (i) long SRT for the microorganisms necessary to degrade toxic compounds, (ii) ability to handle shock loads, (iii) production of high-quality effluent that is low in TSS and COD concentration, (iv) system operation is simple and reliable (Metcalf and Eddy et al., 2013).

For municipal wastewater, FBBRs have been used for post-denitrification. The FBBR process is suitable for the removal of hazardous substances from groundwater.

9.9 COMBINED PROCESSES

A combination of trickling filter and activated sludge can be used for the treatment of wastewater. Combined processes have resulted as part of a plant upgrade where a trickling filter or activated sludge reactor is added to an existing system, or they have also been incorporated into new treatment plant designs (Parker et al., 1994). Combined processes have the advantages of each of the individual processes. These include the following:

- Volumetric efficiency and low energy requirement of the attached growth process for partial BOD removal.
- Stability and resistance to shock loads of the attached growth process.
- High quality of effluent with activated sludge treatment.
- Improved sludge settling characteristics.

Figure 9.10 presents flow diagrams of a number of combined processes. Figures 9.10 (a), (b), and (c) illustrates a trickling filter/activated sludge (TF/AS) process, a trickling filter/solids contact (TF/SC) process, and a series trickling filter-activated sludge process, respectively.

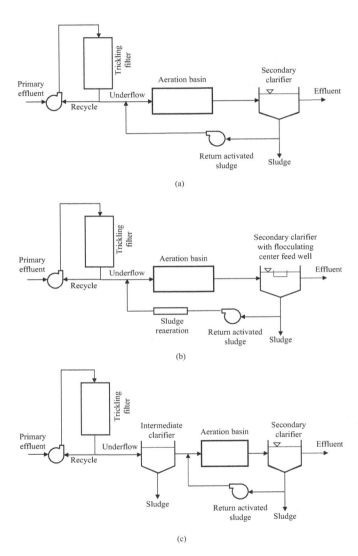

Figure 9.10 (a) Trickling filter/activated sludge (TF/AS) process, (b) trickling filter/solids contact (TF/SC) process, and (c) series trickling filter-activated sludge process.

Source: Adapted from Metcalf and Eddy et al., 2013.

PROBLEMS

9.1 What is the major difference between suspended growth and attached growth processes?

9.2 Explain with the help of flow diagrams, the meaning of the terms (i) direct recirculation and (ii) indirect recirculation.

9.3 A wastewater treatment plant uses a single-stage rock-media trickling filter for secondary treatment, as illustrated in Figure 9.5 (a). The wastewater flow rate is 2000 m³/d with a BOD$_5$ concentration of 400 mg/L. Primary clarification removes 30% of the BOD$_5$. The filter is 12 m in diameter, and 1.5 m in depth. A direct recirculation pump operates at 2.78 m³/min to the filter. It is observed, as shown in worked-out Example 9.1, that the use of a single filter produces a plant efficiency of about 75%, with very high hydraulic and BOD loading rates. In order to reduce the loading rates and increase plant efficiency, the engineer decides to add an identical filter in parallel to the first one. The direct recirculation ratio is maintained at 2.0 for each filter. Wastewater temperature is 20°C. Calculate the new hydraulic loading rate, organic loading rate, effluent BOD$_5$ concentration, and overall plant efficiency. Is the new design better than the old one?

9.4 Consider the single-stage trickling filter plant from worked-out Example 9.1. The use of a single-stage filter produced a plant efficiency of about 75%. In order to increase plant efficiency, the engineer decides to add an identical second-stage filter in series to the first one. The wastewater flow rate is 2000 m³/d with a BOD$_5$ concentration of 400 mg/L. Primary clarification removes 30% of the BOD$_5$. The filters are each 12 m in diameter and 1.5 m in depth. The direct recirculation ratio for each stage is 2.0. Wastewater temperature is 20°C. Calculate the second-stage hydraulic loading rate, and organic loading rates, effluent BOD$_5$ concentration, and overall plant efficiency. Comment on the advantages/disadvantages of using single-stage versus two-stage trickling filters.

9.5 Rework problem 9.3 for a wastewater temperature of 15°C.

9.6 Rework problem 9.4 for wastewater temperatures of 15°C and 22°C.

9.7 A single-stage rock trickling filter plant operates at the influent conditions of Q = 10,600 m³/d, BOD = 215 mg/L, and suspended solids = 240 mg/L. Total direct recirculation of 5300 m³/d is used. At the moment, there is no indirect recirculation. The primary clarifier removes 35% of the BOD and 50% of the suspended solids. There are two filters operated in parallel. Each filter is 18 m in diameter, and 1.5 m deep.

a. Calculate the overall plant efficiency for BOD removal.

b. It is desired to increase the overall BOD removal efficiency to 80% or above. One option being considered is the addition of a third filter in parallel, of the same dimensions as the other two filters.

Calculate the new BOD removal efficiency. Is this an effective option?

c. Another option to increase the overall BOD removal efficiency to 80% is to increase the recirculation of the existing filters. Show with the help of calculations what recirculation ratio would you select to achieve that?

9.8 A plastic media Bio-tower has vertical flow packing with $n = 0.5$ and k_{20} of 0.045 min^{-1}. The tower is cylindrical with a diameter of 10 m. The depth of the packing medium is 6 m. The primary effluent flow rate is 2500 m^3/d with a soluble BOD$_5$ of 250 mg/L. The average wastewater temperature is 16°C. Direct recirculation is practiced at a 1:1 ratio. Calculate the following:

a. Organic loading rate of the bio-tower.
b. Hydraulic loading rate of the bio-tower.
c. Recirculation ratio.
d. Effluent soluble BOD$_5$ concentration.

9.9 An industry has decided to treat its process wastewater in a Bio-tower with a plastic modular medium ($n = 0.55$). The flow rate of the wastewater is 1400 m^3/d with a BOD$_5$ of 630 mg/L. Average summer and winter temperatures are 22°C and 15°C, respectively. The treatability constant is 0.04 min^{-1} for the system at 20°C. The depth of the medium is 4 m. Calculate the area of the bio-tower required to produce an effluent with a BOD$_5$ of 20 mg/L, with a recycle ratio of 2:1.

9.10 Rework problem 9.9 using a winter recirculation ratio of 3:1. The summer recirculation ratio remains the same at 2:1. Comment on the pros and cons of using a higher recirculation ratio.

9.11 An industry treats its process wastewater in a Bio-tower with a plastic modular medium ($n = 0.50$). The flow rate of the wastewater is 2200 m^3/d with a soluble BOD$_5$ of 720 mg/L. The average winter temperature is 12°C. Treatability constant is 0.04 min^{-1} at 20°C. The depth of the medium is 4 m.

a. Calculate the area of the Bio-tower required to produce an effluent with soluble BOD$_5$ of 30 mg/L, with a recycle ratio of 3:1.
b. After operating the Bio-tower above for a few months, one of the recirculation pumps broke down. As a result, the system had to be operated at a reduced recirculation ratio of 3:2. How did this affect the effluent soluble BOD$_5$ concentration? Show with the help of calculations.

9.12 What is a hybrid process? Give an example.

REFERENCES

Andrettola, G., Foladori, P., Gatti, G., Nardelli, P., Pettena, M., and Ragazzi, M. 2003. Upgrading of a Small Overload Activated Sludge Plant Using a MBBR System. *Journal of Environmental Science and Health*, A38 (10):2317–2338.

Aspegren, H., Nyberg, U., Andersson, B., Gotthardsson, S., and Jansen, J. 1998. Post Denitrification in a Moving Bed Biofilm Reactor Process. *Water Science and Technology*, 38 (1):31–38.

Davis, M. 2020. *Water and Wastewater Engineering: Design Principles and Practice*. 2nd edn. McGraw-Hill, Inc., New York, NY, USA.

Di Trapani, D., Christensso, M, and Ødegaard, H. 2011. Hybrid Activated Sludge/ Biofilm Process for the Treatment of Municipal Wastewater in a Cold Climate Region: A Case Study. *Water Science and Technology*, 63 (6):1121–1129.

Eckenfelder, W. W. Jr. 1961. Trickling Filter Design and Performance. *Journal Sanitary Engineering Division*, 87 (4):33.

Germain, J. E. 1966. Economical Treatment of Domestic Waste by Plastic-Media Trickling Filters. *Journal Water Pollution Control Federation*, 38 (2):192.

Hammer, M. J. and Hammer, M. J. Jr. 2012. *Water and Wastewater Technology*. 7th edn. Pearson-Prentice Hall, Inc., Hoboken, NJ, USA.

Howland, W. E. 1958. Flow over Porous Media in a Trickling Filter. *Proceedings of the 12th Industrial Waste Conference*, Purdue University, Lafayette, Indiana, USA. p.435.

Metcalf and Eddy, Tchobanoglous, G., Stensel, H., Tcuchihashi, R., and Burton, F. 2013. *Wastewater Engineering: Treatment and Resource Recovery*. 5th edn. McGraw-Hill, Inc., New York, NY, USA.

Ødegaard, H. 2006. Innovations in Wastewater treatment: the Moving Bed Biofilm Process. *Water Science and Technology*, 53 (9):17–33.

Ødegaard, H., Rusten, B. and Westrum, T. 1994. A New Moving Bed Biofilm Reactor-Applications and Results, *Water Science Technology*, 29 (10–11):157–165.

Parker, D. S., Krugel, S., and McConnell, H. 1994. Critical Process Design Issues in the Selection of the TF/SC Process for a Large Secondary Treatment Plant. *Water, Science and Technology*, 29 (10–11):209–215.

Peavy, H. S., Rowe, D. R., and Tchobanoglous, G. 1985. *Environmental Engineering*, McGraw- Hill, Inc., New York, NY, USA.

Rusten, B., Hem, L.J. and Ødegaard, H. 1995. Nitrification of Municipal Wastewater in Moving-Bed Biofilm Reactors. *Water Environment Research*, 67 (1):75–86.

Rusten, B. and Neu, K. E. 1999. Down to Size. *Water Environment & Technology*, 11 (1):27–34.

Schulze, K. L. 1960. Load and Efficiency in Trickling Filters. *Journal Water pollution Control Federation*, 33 (3):245–260.

Velz, C. J. 1948. A Basic Law for the Performance of Biological Filters. *Sewage Works Journal*, 20 (4):607–617.

WEF. 2000. *Aerobic Fixed Growth Reactors: A Special Publication*, Water Environment Federation, Alexandria, VA, USA.

Wolman, A., Babbitt, H., Bishopp, F., et al. 1946. Sewage Treatment at Military Installations – Chapter V. Trickling Filters. National Research Council, Subcommittee Report, *Sewage Works Journal*, 18 (5):897–982.

Chapter 10

Secondary clarification and disinfection

10.1 INTRODUCTION

The term secondary clarification denotes clarification of effluent from secondary biological reactors in settling tanks. Secondary clarifiers are placed after the biological reactors and constitute secondary treatment together with the biological unit. The assumption is that all biochemical reactions take place in the bioreactor, and the function of the clarifier is separation of solids from the liquid fraction and thickening of settled solids in most cases.

The design of secondary clarifiers following suspended growth processes is different from clarifiers following attached growth processes. The characteristics of the biological solids in these two types of processes are significantly different. As a result, the design and operation of the secondary clarifiers for these systems are also different (Peavy et al., 1985).

According to the Urban Wastewater Treatment Directive (91/271/EEC) of the European Union (EU), the limits for secondary treatment BOD_5 and total suspended solids are 25 mg/L and 35 mg/L, respectively. The United States Environmental Protection Agency (USEPA) specifies an effluent BOD_5 less than or equal to 30 mg/L and an effluent suspended solids concentration less than or equal to 30 mg/L for secondary treatment. Multiple units capable of independent operation are required for all plants where design average flows exceed 380 m^3/d (GLUMRB, 2004). The hydraulic loading is usually based on peak flow rates.

10.2 SECONDARY CLARIFIER FOR SUSPENDED GROWTH PROCESS

Secondary clarifiers following suspended growth processes are designed to achieve two major functions: (i) Clarification of effluent and (ii) thickening of biological solids. The effluent from the activated sludge reactor or other suspended growth process has to be clarified to reduce the suspended solids concentration to meet discharge limits. At the same time, the sludge has to be thickened prior to recycling back to the activated sludge reactor, as

DOI: 10.1201/9781003134374-10

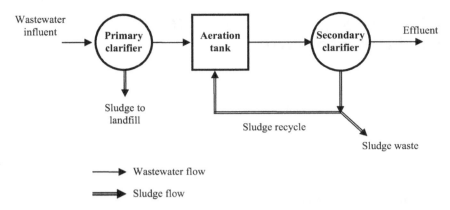

Figure 10.1 Secondary clarifier in activated sludge process with recycle.

shown in Figure 10.1, and before further treatment. The secondary clarifier has to achieve both of these criteria.

Clarification is due to the settling of the lighter flocculent particles, while thickening is due to the mass flux of solids in the hindered settling zone, where the solids concentration is much higher. It is not possible to select an overflow rate to represent the settling velocity of such a complex composition of biosolids. The surface areas for each of these functions have to be determined separately, and then the surface area that satisfies both criteria is selected. The surface area required for clarification is determined from the same principles as those used for a primary clarifier. The surface area required for thickening can be determined by one of the following methods: (i) *Solids flux analysis* and (ii) *state point analysis*. Both methods are based on the same principles. The solids flux method is discussed in more detail in the following sections. The state point analysis is used for the optimization of existing systems.

In a secondary clarifier, Type I, Type II, Type III, and Type IV settling may be observed at different depths, depending on the solids concentration. The secondary clarifier is designed to increase the incoming solids concentration X_i to a much higher underflow solids concentration X_u. As a result, the settling characteristics change, resulting in zones with different types of settling. This can be demonstrated with a batch settling column test. The settling column test can be used to determine the hindered settling velocity corresponding to the initial solids concentration (Metcalf and Eddy et al., 2013).

10.2.1 Settling column test

A clear plexiglass cylinder is used for the settling column test. The height of the cylinder should be equal to the height of the clarifier. The column is filled with a suspension with a solids concentration X_1, which is allowed to settle in an undisturbed manner. After a short time at t_1, four distinct

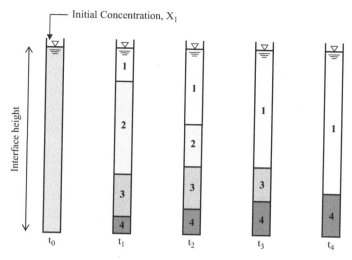

Figure 10.2 Settling column test from time t_0 to t_4: Zone 1 – clarified zone; Zone 2 – uniform settling zone at solids concentration X_1, Zone 3 – hindered settling with concentration gradient; and Zone 4 – compression settling.

zones will develop in the column, as illustrated in Figure 10.2. Zone 1 is the clarified effluent zone with a low concentration of particles, where discrete (type I) settling and some flocculent (type II) settling occurs. In zone 2, the initial concentration remains where the particles settle at a uniform velocity. Hindered (type III) settling is observed in zone 2. Due to the high concentration of particles, the water tends to move up through the interstices of the contacting particles. The contacting particles tend to settle as a blanket, maintaining the same relative position with respect to each other, resulting in hindered settling. Zone 3 represents a transition zone. The concentration increases from the interface of zones 2–3 to the interface of zones 3–4, creating a concentration gradient. Compression (type IV) settling is observed in zone 4. As solids settle to the bottom of the cylinder, particles immediately above fall on top of them, forming a zone where the solids are mechanically supported from below. Solids in the compression zone have an extremely low velocity that results mainly from consolidation.

After some time t_2, the height of the clarified zone 1 and compression zone 4 increases, while zone 2 decreases. With time, as more and more particles settle, zone 2 disappears (t_3). Eventually, zone 3 decreases until only zones 1 and 4 remain (t_4). The interface between zones 1 and 4 will travel downward at a very slow rate as the solids consolidate from their self-weight, and release water to the clarified zone. Only the interfaces involving zone 1 with the other zones will be visible upon observation. The interfaces between the other zones will not be readily visible as the concentration differences are not significant.

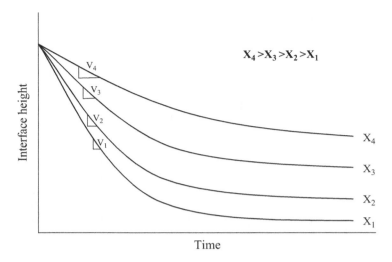

Figure 10.3 Settling curves corresponding to various initial solids concentration.

From the settling column test with X_1 initial solids concentration, the height of the interface is measured at regular time intervals and plotted versus time, to obtain a curve similar to that shown in Figure 10.3. The initial portion of the curve is a straight line. The slope of this portion is the hindered settling velocity corresponding to the initial concentration X_1. The straight-line portion represents hindered settling, while the horizontal flat end portion of the curve represents compression settling. A number of settling column tests can be run at different initial solids concentrations, to generate the settling curves illustrated in Figure 10.3. The corresponding hindered settling velocities can be calculated from the slopes of the straight-line portions. As the initial concentration increases, the curves become flatter with lower settling velocities. This is due to the presence of zone 2 for a very short period of time, with zones 3 and 4 becoming predominant at higher solids concentrations. Compression settling becomes more important at higher solids concentrations.

10.2.2 Solids flux analysis

The major process parameter for solids thickening is the solids loading rate. It is the rate of solids fed per unit cross-sectional area of clarifier (kg/m².d). The solids loading rate has to be determined based on sludge settling properties and clarifier return sludge flow rate. One of the methods for secondary clarifier analysis is the *Solids Flux* method.

Figure 10.4 presents a secondary clarifier at steady state conditions. Q_e represents the effluent or overflow from the clarifier, Q the plant flow rate, Q_r the recycle flow, and Q_u the underflow rate. The solids concentrations are

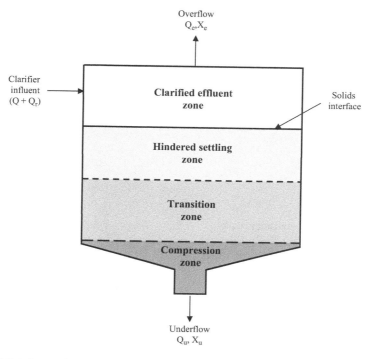

Figure 10.4 Secondary clarifier operating at steady state.

given by X the MLSS concentration, X_e the effluent solids concentration, and X_u the underflow solids concentration. The interface between zones 1 and 2 is stationary, as water in the clarified zone rises toward the overflow at a rate equal to the hindered settling velocity of the solids with concentration X (Peavy et al., 1985). This satisfies the clarification function.

The thickening function depends on the limiting solids flux that can be transported to the bottom of the clarifier. The solids flux is affected by the sludge characteristics, and settling column tests have to be conducted to determine the relationship between settling velocity and solids concentration. Data from the settling column test is then used to determine the area required for thickening using the *Solids Flux* method. The solids flux method was first proposed by Coe and Clevenger (1916) and later modified by a number of researchers, e.g. Yoshioka et al. (1957), Dick and Ewing (1967), and Dick and Young (1972), as reported by Peavy et al. (1985).

10.2.2.1 Theory

Solids flux is defined as the mass of solids passing per unit time through a unit area perpendicular to the direction of flow. It is calculated as the product of solids concentration (kg/m³) and the velocity (m/h), resulting in units of kg/m².h.

In a secondary clarifier at steady state, a constant flux of solids moves in a downward direction. The downward velocity of the solids has two components: (i) Transport velocity due to the withdrawal of the underflow sludge at a constant rate Q_u and (ii) gravity (hindered) settling of the solids.

The transport velocity v_u due to underflow is given by

$$v_u = \frac{Q_u}{A_s} \tag{10.1}$$

where
  v_u = underflow velocity, m/h
  A_s = surface area of clarifier, m²
  Q_u = underflow flow rate, m³/h

At any point in the clarifier, the resulting underflow solids flux G_u is

$$G_u = v_u X_i = \frac{Q_u}{A_s} X_i \tag{10.2}$$

where
  X_i = solids concentration at the point in question, kg/m³
  G_u = solids flux due to underflow, kg/m².h

At the same point in the clarifier, the mass flux of solids due to gravity settling is given by

$$G_g = v_g X_i \tag{10.3}$$

where
  G_g = solids flux due to gravity, kg/m².h
  v_g = settling velocity of solids at concentration X_i, m/h

The total mass flux is the sum of the underflow flux and the gravity flux, and is given by

$$G_T = G_u + G_g \tag{10.4}$$

$$G_T = v_u X_i + v_g X_i \tag{10.5}$$

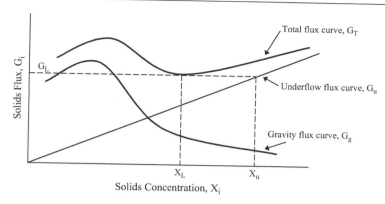

Figure 10.5 Total, underflow, and gravity flux curves for solids flux analysis.

Figure 10.5 illustrates the nature of the total, underflow, and gravity flux curves. The gravity flux depends on the solids concentration and the corresponding settling characteristics. At low solids concentration, the settling velocity is essentially independent of concentration. If the velocity remains the same as the solids concentration increases, then the gravity flux also increases. At very high solids concentration, as the solids approach the compression zone, the gravity (hindered) settling velocity becomes negligible, and the gravity flux approaches zero. So, the flux due to gravity must pass through a maximum value as the concentration is increased, as shown in Figure 10.5.

The solids flux due to underflow is a linear function of solids concentration. The slope of the underflow flux curve is equal to the underflow velocity v_u. The underflow velocity is used as a process control parameter. The total flux curve is drawn as the sum of the gravity and underflow flux curves. Increasing or decreasing the underflow flow rate can shift the total flux curve upward or downward. The lowest point on the total flux curve corresponds to a limiting solids flux for the clarifier. The *limiting solids flux* corresponds to the maximum solids loading that can be applied to the clarifier. This governs the thickening parameter, and is used to calculate the area required for thickening.

10.2.2.2 Determination of area required for thickening

The first step is to obtain data for the gravity flux curve. Settling column tests described in the previous section, are run with different initial solids concentrations. Graphs similar to Figure 10.3 are generated with the data. The slopes of the initial straight-line portions are calculated. These are the gravity (hindered) settling velocities (v_g) corresponding to the different initial solids concentrations (X_i). Equation (10.3) is used to calculate the gravity flux values (G_g) for different X_i values. The gravity flux curve is then plotted similarly to the one in Figure 10.5. As mentioned previously, the underflow velocity is used as a process control parameter. A value of v_u is selected, and a straight line is drawn through the origin to represent the underflow flux G_u. The total flux curve (G_T) is then drawn as the sum of the two flux curves. A horizontal line is drawn tangent to the lowest point on the total flux curve. Its intersection with the y-axis gives the *limiting solids flux* G_L that can be handled by the clarifier. The corresponding underflow solids concentration is obtained as the abscissa of the point where the horizontal line intersects the underflow flux curve. If the quantity of solids coming to the clarifier is greater than the limiting solids flux value, then solids will build up in the clarifier and may overflow from the top if adequate storage capacity is not available.

The surface area required for thickening is calculated as

$$A_T = \frac{\text{Total flow in to clarifier} \times \text{solids concentration}}{\text{Limiting solids flux}} \tag{10.6}$$

$$A_T = \frac{(Q + Q_r)X}{G_L} \tag{10.7}$$

where
   A_T = surface area required for thickening, m²
   Q_r = recycle flow, m³/h
   Q = plant flow rate, m³/h
   X = MLSS concentration, kg/m³
   G_L = limiting solids flux, kg/m².h

The depth of the thickening portion of the clarifier must be sufficient to (i) maintain an adequate sludge blanket depth so that thickened solids are not recycled, and (ii) temporarily store excess solids that may come into the clarifier (Metcalf and Eddy et al., 2013).

A slight modification to the above method was proposed by Yoshioka et al. (1957) to determine the limiting solids flux. This method is illustrated in Figure 10.6. The gravity flux curve is first plotted. A value of underflow solids concentration X_u is selected. Then a line is drawn from X_u on the x-axis and tangent to the gravity flux curve. The tangent is extended to the y-axis. The intersection point of the tangent with the y-axis provides the limiting flux value G_L. The absolute value of the slope of the tangent is the underflow velocity v_u. The ordinate value corresponding to the point of tangency is the gravity solids flux, while the intercept, $G_L - G_g$, is the underflow flux (Peavy et al., 1985). This method is useful to determine the effect of various underflow solids concentrations on the limiting solids flux. As illustrated in Figure 10.7, increasing the value of X_u results in decreasing the

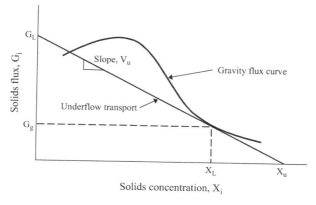

Figure 10.6 Alternative graphical method for determination of limiting solids flux.

Source: Adapted from Yoshioka et al., 1957

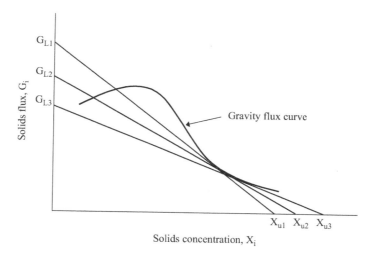

Figure 10.7 Effect of underflow solids concentration on limiting solids flux.

maximum solids loading that can be applied to the clarifier. Conversely, a higher solids loading can be applied to a clarifier with a lower desired under-flow solids concentration. These methods are illustrated in the following examples.

10.2.2.3 Secondary clarifier design based on solids flux analysis

Secondary clarifiers following suspended growth processes have to be designed to achieve two functions: (i) Clarification and (ii) thickening.

Table 10.1 Typical design data for secondary clarifiers for activated sludge systems

Type of system	Overflow rate, $m^3/m^2.d$		Solids loading, $kg/m^2.h$		Depth, m
	Average	Peak	Average	Peak	
Clarifier following air-activated sludge (excluding extended aeration)	16–32	40–64	4–6	8	3.5–6
Clarifier following oxygen activated sludge	16–32	40–64	5–7	9	3.5–6
Clarifier following extended aeration	8–16	24–32	1–5	7	3.5–6

Source: Adapted from Peavy et al. (1985) and Metcalf and Eddy et al. (2013).

The surface area required to achieve clarification of effluent is determined using the same principles as those used for primary clarifiers. The surface area required for thickening is determined using one of the methods mentioned previously. This requires data from settling column tests run with appropriate sludge samples. For a new plant, the activated sludge system that would produce the sludge may also be in the design phase. As a result, it would be very difficult to obtain representative sludge samples. The design procedure outlined so far is more applicable for the evaluation and optimization of an existing system, rather than the design of a new system. For the design of a new system, where analytical data are not available, design parameters from the literature may be used. Design values from prior installations that have worked successfully are presented in Table 10.1. However, careful consideration of wastewater characteristics and type of reactor should be made prior to the selection of design parameters from the literature.

The physical units used as secondary clarifiers are similar to those used as primary clarifiers. Circular or rectangular tanks may be used. The sludge removal mechanisms are somewhat different, due to the nature of the biological solids. The sludge should be removed as rapidly as possible, to ensure the return of active microorganisms to the activated sludge reactor. Effluent overflow rates based on peak flow conditions are commonly used, to prevent loss of solids with the effluent if design criteria are exceeded. An alternative method is to use the average dry weather flow rate with a corresponding surface loading rate, and also check for peak flow and loading conditions. Either condition may govern the design (Metcalf and Eddy et al., 2013).

Example 10.1: You have to design a secondary clarifier for an activated sludge process. The MLSS in the activated sludge reactor is 2500 mg/L. It is desired to thicken the solids to 10,000 mg/L in the secondary clarifier. The plant flow rate is 6500 m³/d. The sludge recirculation rate is 45%. Batch settling column tests were conducted at different initial solids concentrations, and corresponding settling velocities were calculated. The results are given in Table 10.2.

Table 10.2 Batch settling column test data (for Example 10.1)

Solids concentration, mg/L	Settling velocity, m/h
1000	5
2000	3.2
3000	2
4000	1.1
5000	0.5
6000	0.28
8000	0.11
10,000	0.075
12,000	0.06

SOLUTION

Step 1. Calculate the solids flux for the given data, as shown in Table 10.3.
Use equation 10.3, and note mg/L = g/m³

$$G_g\left(kg / m^2.h\right) = v_g\, X_i$$

$$= \text{settling vel}\left(m/h\right) \times \text{solids concn.}\left(g/m^3\right)/1000\,g/kg$$

Step 2. Use the alternative graphical method for *solids flux analysis*. Draw the gravity flux curve using the data from step 2. Since the desired underflow concentration is 10,000 mg/L or g/m³, use this value on the x-axis as the starting point and draw a tangent to the gravity flux curve. The tangent intersects the y-axis at 4.0 kg/ m².h, which is the *limiting solids flux* value or G_L for the clarifier. This is the maximum solids loading that can be applied to the clarifier and governs the thickening function. This is illustrated in Figure 10.8.

Step 3. Determine the area required for thickening.

Limiting solids flux = 4 kg/m².h

Table 10.3 Calculated solids flux (for Example 10.1)

Solids concentration, mg/L	Solids flux, kg/m².h
1000	5
2000	6.4
3000	6
4000	4.4
5000	2.5
6000	1.68
8000	0.88
10,000	0.75
12,000	0.72

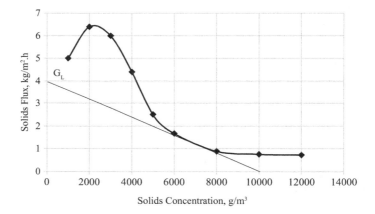

Figure 10.8 Solids flux method for calculation of G_L (for Example 10.1).

Plant flow rate $Q = 6500\,\text{m}^3\,/\,\text{d}$

Recycle flow rate $Q_r = 0.45\,Q$

Total flow to clarifier $= Q + Q_r = 1.45 \times 6500\,\text{m}^3\,/\,\text{d} = 9425\,\text{m}^3\,/\,\text{d}$

Solids loading to clarifier $= \text{MLSS} \times (Q + Q_r)$

$$= 2.5\,\text{kg/m}^3 \times 9425\,\text{m}^3/\text{d} \times \frac{\text{d}}{24\,\text{h}}$$

$$= 981.77\,\text{kg}\,/\,\text{h}$$

Surface area of clarifier required for thickening $= \dfrac{981.77\,\text{kg}\,/\,\text{h}}{4\dfrac{\text{kg}}{\text{m}^2}\,/\,\text{h}}$

$A_{sT} = 245.44\,\text{m}^2$

Step 4. Determine the area required for clarification.
The settling velocity of particles corresponding to the MLSS concentration can be used as the overflow rate for clarification. The MLSS concentration is 2500 mg/L. Draw settling velocity versus the solids concentration for the given data from settling column tests. This is illustrated in Figure 10.9.

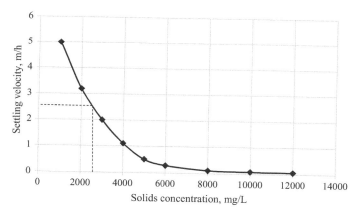

Figure 10.9 Settling velocity curve (for Example 10.1).

The settling velocity corresponding to MLSS of 2500 mg/L is determined as 2.6 m/h or 2.6 m³/m².h from Figure 10.9. This is the overflow rate for clarification of effluent.
Assume the rate of sludge wasting Q_w is negligible. Then effluent flow rate $Q_e = Q$

$$Q = 6500\,\text{m}^3\,/\,\text{d} \times \dfrac{\text{d}}{24\,\text{h}} = 270.83\,\text{m}^3\,/\,\text{h}$$

Surface area required for clarification $= \dfrac{270.83\,\text{m}^3/\text{h}}{2.6\,\text{m/h}} = 104.16\ \text{m}^2$

$A_{sC} = 104.16\ \text{m}^2$
$A_{sT} > A_{sC}$, therefore thickening function governs the design.
Design surface area of secondary clarifier = **245.44 m²**

Step 5. Calculate dimensions for clarifier.

Select depth = 4.5 m
Select a circular clarifier

$$\text{Diameter} = \sqrt{\frac{4 \times 245.44\,\text{m}^2}{\pi}} = 17.68\,\text{m} = 18\,\text{m}$$

$$\text{Therefore, design surface area} = \frac{\pi}{4}\left(18^2\right) = 255\,\text{m}^2$$

Example 10.2: Consider the activated sludge system described in Example 10.1. Calculate the underflow rate Q_u and the underflow velocity v_u, assuming that sludge wastage rate Q_w is negligible. Also estimate the maximum MLSS that can be maintained in the reactor.

SOLUTION
Step 1. Calculate the underflow rate Q_u considering Figure 10.10.

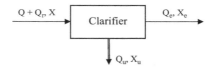

Figure 10.10 Underflow rate calculation for secondary clarifier (for Example 10.2).

$$Q_u = Q_w + Q_r \approx Q_r \left(\text{since } Q_w \text{ is negligible}\right)$$

$$\text{Underflow rate,} \, Q_u = Q_r = 0.45\,Q = 0.45 \times 6500\,\text{m}^3/\text{d} = \textbf{2925}\,\textbf{m}^3/\textbf{d}$$

Step 2. Calculate the underflow velocity v_u.
Underflow velocity is the slope of the limiting solids flux line in Example 10.2.

$$\text{Slope of line} = \left(4\,\text{kg}/\text{m}^2.\text{h}\right)/10\,\text{kg}/\text{m}^3 = 0.4\,\text{m}/\text{h}$$

$$v_u = 0.4\,\text{m}/\text{h}$$

Step 3. Estimate maximum MLSS "X" for reactor (X_m).
Consider the diagram of the secondary clarifier
From equation of continuity,
Inflow = Outflow

$$Q + Q_r = Q_e + Q_u = Q_e + Q_r$$

Therefore, $Q = Q_e$

Write a mass balance for solids around the secondary clarifier.

$$\left(\text{Mass rate of solids}\right)_{\text{in}} = \left(\text{Mass rate of solids}\right)_{\text{out}}$$

$$\left(Q + Q_r\right)X_m = Q_u\,X_u + Q_e\,X_e$$

Since $X_e \ll X_u$, X_e can be considered negligible. Also, $Q_u = Q_r$

$$\left(Q + Q_r\right)X_m = Q_r\,X_u$$

$$\left(1.45\,Q\right)X_m = \left(0.45\,Q\right)\left(10{,}000\,\text{mg}/\text{L}\right)$$

$$X_m = 3103.45\,\text{mg}/\text{L}$$

Example 10.3: Consider the activated sludge system described in Example 10.1. It is desired to operate the system at a higher MLSS of 3000 mg/L and increase the underflow solids concentration to 12,000 mg/L. Can this be done with the selected design surface area of 255 m²?

SOLUTION
Complete step 1 and step 2 similar to Example 10.1 to generate Figure 10.11.

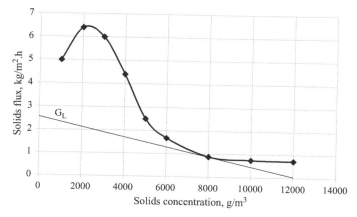

Figure 10.11 Solids flux method for calculation of G_L (for Example 10.3).

A tangent drawn through desired underflow solids concentration of 12,000 g/m³ gives a G_L of 2.6 kg/m².h. This is the maximum solids loading that can be applied to the clarifier.

Step 3. Check area required for thickening.

$$\text{Total flow to clarifier} = Q + Q_r = 1.45 \times 6500\,\text{m}^3/\text{d} = 9425\,\text{m}^3/\text{d}$$

$$\text{Solids loading to clarifier} = \text{MLSS} \times (Q + Q_r) = 3\,\text{kg/m}^3$$

$$\times 9425\,\text{m}^3/\text{d} \times \frac{\text{d}}{24\,\text{h}}$$

$$= 1178.13\,\text{kg/h}$$

$$\text{Surface area of clarifier required for thickening} = \frac{1178.13\,\text{kg/h}}{2.6\,\dfrac{\text{kg}}{\text{m}^2}/\text{h}}$$

$$A_{sT} = 453.13\,\text{m}^2$$

Step 4. Check area required for clarification.
Settling velocity corresponding to MLSS of 3000 mg/L is 2.0 m/h (given in settling column test data). This is equivalent to the overflow rate for clarification.

Assume the rate of sludge wasting Q_w is negligible. Then effluent flow rate $Q_e = Q$

$$Q = 6500\,\text{m}^3/\text{d} \times \frac{\text{d}}{24\,\text{h}} = 270.83\,\text{m}^3/\text{h}$$

$$\text{Surface area required for clarification} = \frac{270.83\,\text{m}^3/\text{h}}{2.0\,\text{m}/\text{h}} = 135.42\,\text{m}^2$$

$A_{sC} = 135.42$ m²
$A_{sT} > A_{sC}$, therefore thickening function governs the design.
Design surface area of secondary clarifier = **453.13 m²**

The surface area required is almost double that of Example 10.1. The surface area will have to be increased to 453.13 m² from 255 m², in order to increase the MLSS and underflow solids concentration to the new values.

10.2.3 State point analysis

The state point analysis is an extension to the solids flux analysis and is generally used for assessing the operating conditions of an existing clarifier. It allows the evaluation and optimization of an existing system for different mixed-liquor concentrations and operating conditions of the clarifier (e.g. hydraulic loading, recycle rate, etc.) relative to the solids-flux operating conditions (Metcalf and Eddy et al., 2013).

10.2.3.1 Theory

The *state point* is the intersection of the clarifier overflow solids flux, and the underflow solids flux, as illustrated in Figure 10.12. The aeration tank MLSS at any point along the overflow solids flux line is determined by constructing a vertical line to the x-axis.

The clarifier overflow solids flux is

$$G_{o} = \frac{QX}{A} \tag{10.8}$$

where
G_{o} = overflow solids flux rate, kg/m^2.d
Q = clarifier effluent flow rate, m^3/d
X = aeration tank MLSS, kg/m^3
A = clarifier surface area, m^2

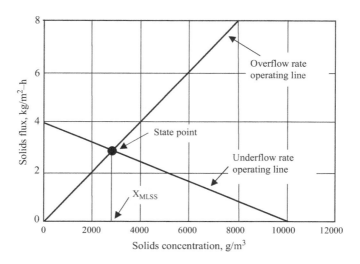

Figure 10.12 State point at the intersection of overflow and underflow lines.

The underflow operating line represents the negative slope of the clarifier underflow velocity (V_u) which can be controlled by controlling the flow rate of RAS. V_u can be calculated using the following equations:

$$V_u = \frac{Q_u}{A} = -\frac{Q_r}{A} \tag{10.9}$$

$$V_u = \frac{\dfrac{(Q+Q_r)X}{A} - \dfrac{QX}{A}}{-X} \tag{10.10}$$

where
 Q_u = underflow flow rate, m³/d
 Q_r = recirculation (RAS) flow rate, m³/d

In order to evaluate the clarifier loading condition, the overflow, underflow, and gravity flux curves are plotted on the same page. When the state point is within the flux curve, the clarifier can be underloaded, critically loaded, or overloaded, depending on whether the underflow line is below, tangential, or above the descending limb respectively, of the gravity flux curve. The RAS flow rate can be increased for critically loaded and overloaded clarifier conditions. If the state point is on or outside the gravity flux curve, the clarifier feed concentration must be reduced to prevent overloading conditions.

10.2.3.2 Clarifier evaluation based on state point analysis

The state point analysis can be used to evaluate and optimize clarifier overflow rates and mixed liquor concentration in the activated sludge reactor. The settling column test on activated sludge mixed liquor can be used with *state point* analysis to determine the optimal MLSS concentration and recycle ratio of RAS for a given influent flow condition. Example 10.4 illustrates the application of state point analysis.

Example 10.4: Consider the secondary system described in Example 10.1. Using state point analysis, determine whether the clarifier is overloaded, critically loaded, or underloaded for the underflow solids concentrations of 8000 mg/L, 10,000 mg/L, and 12,000 mg/L.

SOLUTION

Step 1: Develop the gravity solids flux curve, similar to Example 10.1.
Step 2: Draw the overflow rate and underflow rate lines. Determine the state point as shown in Figure 10.13.

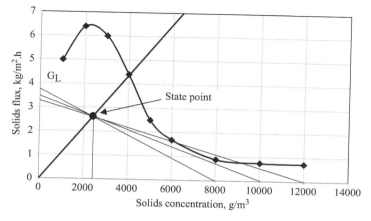

Figure 10.13 State point analysis for clarifier evaluation (for Example 10.4).

The area of the clarifier is calculated as 255 m² for the MLSS concentration of 2500 mg/L.

Overflow solids flux,

$$G_o = \frac{QX}{A} = \frac{6500\,\frac{m^3}{d} \times X\,\frac{kg}{m^3}}{255\,m^2}$$

For $X = 2500\,mg/L$,

$$G_o = \frac{6500\,\frac{m^3}{d} \times 2.5\,\frac{kg}{m^3}}{255\,m^2} \times \frac{1\,d}{24\,h} = 2.66\,\frac{kg}{m^2.h}$$

This is the state point, and the overflow rate line is drawn from the origin through 2.66 kg/m².h solids flux at 2500 mg/L.

Step 3: Evaluate the underflow conditions at 8,000, 10,000, and 12,000 mg/L.

For 8000 mg/L underflow concentration, draw a line from 8000 mg/L solids concentration through the state point. The line intercepts the y-axis at 3.8 kg/m².h. Similar lines are drawn for 10,000 mg/L and 12,000 mg/L underflow solids concentrations.

For 8000 and 10,000 mg/L, the state point is within the flux curve, and the underflow line is below the descending limb, thus the clarifier is underloaded, and the operating condition is considered feasible. For 12,000 mg/L, the state point is within the flux curve; however, the underflow line is above the descending limb of the flux curve, thus the clarifier is slightly overloaded. The RAS flow rate must be increased to reduce the clarifier loading.

10.3 SECONDARY CLARIFIER FOR ATTACHED GROWTH PROCESS

The primary objective of secondary clarifiers following attached growth processes is to achieve *clarification* of treated wastewater. Sludge thickening is not considered in the design. The goal is to settle the sloughed biofilm or humus, which exhibits Type I or Type II settling. As a result, this type of secondary clarifier is designed similar to primary clarifiers. Sidewater depths range from 2 to 5 m, with corresponding maximum overflow rates of 18–65 m³/m².d (Davis, 2020). GLUMRB (2004) specifies a peak hourly overflow rate of 2.0 m/h.

High-rate trickling filters and biotowers are usually designed with a high degree of liquid recirculation, which can range from 100% to 300%. If indirect recirculation is used, then the clarifier size is increased significantly. Direct recirculation may be an option to handle high recirculation rates, together with the use of modular synthetic media. There is no sludge recycle from the clarifier to the bioreactor. Sludge may be pumped to the primary clarifier where they are settled with the raw wastewater solids, then undergo further processing prior to disposal. Figure 10.14 illustrates a secondary clarifier for a biotower system with (a) direct recirculation, and (b) indirect recirculation.

Example 10.5: Design secondary clarifiers for a wastewater treatment plant using Biotowers for the treatment of municipal wastewater. The wastewater flow rate is 1500 m³/d with a BOD_5 of 180 mg/L and suspended solids of 200 mg/L. The bio-tower uses indirect recirculation and operates at a recycle ratio of 2:1. A design sidewater depth of 3 m is selected with a maximum overflow rate of 1.6 m³/m².h.

SOLUTION
Design two circular secondary clarifiers for the plant.

Total wastewater flow with recycle, $Q_T = 1500\,\text{m}^3/\text{d} \times (2+1)$
$$= 4500\,\text{m}^3/\text{d}$$

Use a peaking factor of 2.0

Total design flow, $Q = 4500\,\text{m}^3/\text{d} \times 2.0 = 9000\,\text{m}^3/\text{d}$

Flow in each clarifier $= \dfrac{9000}{2} = 4500\,\text{m}^3/\text{d} = 187.50\,\text{m}^3/\text{h}$

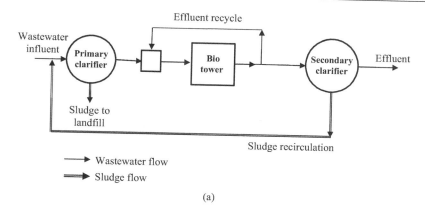

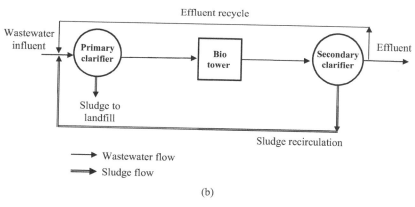

Figure 10.14 Secondary clarification for a bio-tower system with (a) direct recirculation and (b) indirect recirculation.

Surface area, $A_s = \dfrac{187.50\,\text{m}^3/\text{h}}{1.6\,\text{m}^3/\text{m}^2.\text{h}} = 117.19\,\text{m}^2$

Diameter of each clarifier, $D = \sqrt{\dfrac{4 \times 117.19\,\text{m}^2}{\pi}} = 12.21\,\text{m}$

$= \text{round off to } 12.5\,\text{m}$

Note: We should not round off to 12 m, as that will reduce the surface area and increase the overflow rate beyond the maximum value, so round off to the next 0.5 m increment.

Therefore, two circular clarifiers will be used with 12.5 m diameter.

10.4 DISINFECTION

Effluent from a wastewater treatment plant must be disinfected before discharging to the receiving water bodies. The requirements are less stringent as compared to drinking water treatment; however, the reduction of pathogens to acceptable levels is required according to NPDES (National Pollutant Discharge Elimination System) permits for most states. The European Community's (EC) environmental regulations are also aimed at reducing the pollution of surface water caused by municipal wastewater discharge (European Union, 1998).

A pathogen indicator is a substance that indicates the possible presence of pathogenic microorganisms. Most strains of *Enterococci* and *Escherichia coli* bacteria do not cause illness in humans; rather, they indicate the presence of fecal contamination. The US EPA Recreational Water Quality Criteria (RWQC) recommend using the bacteria *Enterococci* and *E. coli* as indicator organisms for freshwater. Many of the states use the 2012 RWQC (US EPA, 2020) as the basis for their wastewater treatment criteria. The EPA recommendation of a monthly average concentration of 126 cfu/100 mL with a maximum daily limit of 410 cfu/100 mL for *E. coli* is used by several states in the US and many European countries as treatment standards (Collivignarelli et al., 2018). Canadian guidelines for Recreational Water Quality recommend a geometric mean concentration less than 200/100 mL, with a single sample concentration less than 400/100 mL for *E. coli* (Droste and Gehr, 2018).

Chlorine is the most commonly used disinfectant in water and wastewater treatment plants. Various factors such as contact time, concentration of disinfectant, types of microorganisms present in wastewater, temperature, and upstream treatment processes must be considered prior to selecting the chlorine dose. The dose also depends on the treatment objective and the degree of purification that must be achieved for the receiving water bodies. The demand for chlorination is significantly reduced if the receiving water body is mostly used for recreational purposes, as compared to drinking water sources. Chlorination is typically applied toward the end of the treatment system, before discharging the treated water to the receiving waters. However, chlorination can also be applied to raw sewage to control odor, slime growth, and corrosion. For biological treatment, chlorination can improve sludge settling and control foaming (Droste and Gehr, 2018). The typical chlorine dose required for wastewater disinfection is based on the type of wastewater, microorganism concentration, and treatment objective. These are presented in Table 10.4.

Excess or residual chlorine level in the wastewater effluent may exhibit environmental toxicity by reacting with organic constituents in water bodies. Chlorine can also damage membranes and ion exchange resins if used as a pretreatment for ultrafiltration, reverse osmosis, or ion exchange processes. For these reasons, dechlorination of the treated effluent may be necessary. Sulfur dioxide (SO_2) is most commonly used for dechlorination,

Table 10.4 Typical chlorine dose for different types of wastewater

Types of wastewater	Initial coliform count, MPN/100 mL	Chlorine dose, mg/L			
		Effluent Standard, MPN/100 mL			
		1000	200	23	≤ 2.2
Raw wastewater	10^7–10^9	16–30			
Primary effluent	10^7–10^9	8–12	18–24		
Trickling filter effluent	10^5–10^6	6–7.5	12–15	18–22	
Activated sludge effluent	10^5–10^6	5.5–7.5	10–13	13–17	
Filtered activated sludge effluent	10^4–10^6	2.5–3.5	5.5–7.5	10–13	13–17
Nitrified effluent	10^4–10^6		0.02–0.03	0.03–0.04	0.04–0.05
Microfiltration effluent	10^1–10^3			0.02–0.03	0.03–0.04
Septic tank effluent	10^7–10^9	16–30	30–60		

Source: Adapted from Metcalf and Eddy et al. (2013)

although sodium sulfite (Na_2SO_3), sodium bisulfite ($NaHSO_3$), and sodium metabisulfite (NaS_2O_5) are also used. In practice, about 1.0–2.0 mg/L of dechlorination compounds are required for 1.0 mg/L of chlorine residual (Metcalf and Eddy et al., 2013).

Besides chlorine, other disinfectants such as chlorine dioxide, chloramine, ozone, and UV (ultraviolet) radiation are also used. For wastewater treatment, UV disinfection has become more popular in recent years. In 1979, almost 95% of wastewater treatment plants in the US, used chlorination for disinfection. By 2003, the use of chlorine dropped to 72%, of which 56% used chlorine gas, and the remainder used hypochlorite (Leong, 2008). Most new treatment plants and a majority of the existing plants are installing UV irradiation systems. Among the advantages, UV disinfection significantly reduces the hazard potential, operation, maintenance, and other difficulties associated with chemical disinfection.

The UV system may be enclosed for a small plant, or placed in an open channel for medium-to-large plants. The dose to achieve regulatory standards is usually in the range of 50–140 mJ/cm² (Davis, 2020). The number, type, and rating of the UV lamps depend on wastewater characteristics, hydraulic loading, and fouling or formation of opaque films on the lamps. The presence of hardness, iron, manganese, and total dissolved solids facilitate fouling and reduce the efficiency of the lamps. Pretreatment of the effluent, combined with continuous cleaning, is practiced. Mechanical wiping is used for maintenance due to low labor costs, as well as periodic acid cleaning to reduce fouling.

10.5 POST-AERATION

The treated effluent from a wastewater treatment plant may contain low dissolved oxygen, and cause an immediate depression or oxygen sag in the receiving water body. Several states in the US have included dissolved oxygen (DO) monitoring in the NPDES permit in recent years, and a high DO value in the range of 5–8 mg/L is required before final discharge. The commonly used post-aeration methods include diffused or mechanical aerators and cascade aeration.

Diffused or mechanical aerators are appropriate for large treatment plants. The mechanism is similar to that used in aeration tanks with high transfer efficiency achieved with fine pore diffusers. Air diffusers are typically used when sufficient elevation is not present at the discharge point to ensure air incorporation.

Cascade aeration is the simplest and least costly alternative when sufficient elevation is available at the discharge location. Wastewater flows through a series of concrete steps in thin films, and the drops create turbulence to facilitate aeration. The height of the cascade can be calculated using the following equation (Barrett et al., 1960).

$$H = \frac{R-1}{0.361ab(1+0.046T)} \tag{10.11}$$

where

H = height of the cascade, m

R = oxygen deficit ratio $= \dfrac{C_s - C_0}{C_s - C}$

C_s = saturation dissolved oxygen at wastewater temperature T°C, mg/L
C_o = influent dissolved oxygen to the cascade, mg/L
C = effluent dissolved oxygen, mg/L
a = 0.8 for wastewater treatment
b = weir geometry parameter, (b = 1 for broad-crested weir, b = 1.1 for steps, b = 1.3 for step weir)

Metcalf and Eddy et al. (2013) recommend a cascade height of 2–5 m (6–16 ft) with a step height of 200 mm (8 in) and a step width of 450 mm (18 in) as typical design considerations. Rectangular or V-notch weirs can also be used with cascade steps.

> Example 10.6: Design a step cascade aeration system to increase the DO value from 2 to 6 mg/L for a wastewater effluent, with a temperature of 24°C.

SOLUTION

From Table A-2, saturation dissolved oxygen at 24°C, C_s = 8.53 mg/L

Influent DO to the cascade, C_o = 2 mg / L

Effluent DO, C = 6 mg / L

$$\text{Oxygen deficit ratio}, R = \frac{8.53 - 2}{8.53 - 6} = 2.58$$

$$\text{Height of cascade}, H = \frac{2.58 - 1}{0.361 \times 0.8 \times 1.1 \times (1 + 0.046 \times 24)} = 2.37\,\text{m}$$

Use steps with 200 mm height and 450 mm length:

$$\text{Number of steps} = \frac{2.37\,\text{m}}{\dfrac{200}{1000}\,\text{m}} = 11.8 \approx 12$$

PROBLEMS

10.1 What is the main design objective of secondary clarifiers for attached growth processes? Why is it different from the design of secondary clarifiers following suspended growth processes?

10.2 Design secondary clarifiers for a trickling filter plant treating wastewater from a municipality. The average flow rate is 500 m³/d, with a recirculation ratio of 1.5 to 1. The maximum overflow rate is 2.1 m³/m².h.

10.3 What are the major design considerations for secondary clarifiers following activated sludge processes?

10.4 What is a settling column test? How can you use it to determine the settling velocity?

10.5 How can you design a settling column test to locate the various settling zones within the column? Illustrate your design.

10.6 Write an expression for the total solids flux in a secondary clarifier. Define each of the component fluxes.

10.7 Illustrate with the help of a graph what happens to the limiting solids flux and underflow velocity, when the underflow solids concentration is increased or decreased from its design value.

10.8 An activated sludge plant with recycle is evaluating its secondary clarification system. Settling column tests are conducted and analyzed for settling velocities at different initial solids concentrations. The data

Table 10.5 Batch settling column test
data (for Problem 10.8)

Solids concentration, mg/L	Settling velocity, m/h
1500	6.5
3000	5.1
4500	4
6000	2.95
7500	2
9000	1.4
12,000	0.5
15,000	0.27
17,000	0.2

are provided in Table 10.5. The plant flow rate is 10,000 m³/d, with a recirculation ratio of 0.5. The MLSS in the reactor is 4000 mg/L. The desired underflow concentration is 15,000 mg/L.

a. Calculate the required surface area if one clarifier is used.
b. Calculate the underflow rate, underflow velocity, and overflow rate for single clarifier design. Clearly state your assumptions.
c. Calculate the required surface area if two clarifiers are used.

10.9   Design secondary clarifiers for the activated sludge plant of Problem 10.8, with a recycle ratio of 0.6. Calculate the required surface area if two clarifiers are used. Also calculate underflow rate, underflow velocity, and overflow rates. Clearly state your assumptions.

10.10 A wastewater treatment plant consists of a primary clarifier, and an activated sludge reactor (with recycle) followed by a secondary clarifier. The average inflow of wastewater to the primary clarifier is 14,500 m³/day with a BOD$_5$ of 250 mg/L and suspended solids concentration of 300 mg/L. The recirculation ratio in the secondary system is 0.75, with an underflow solids concentration of 12,000 mg/L. Calculate the area required for secondary clarification, when the slope of the limiting solids flux line is 0.5 m/h from a plot of solids flux (kg/m²-h) versus solids concentration (kg/m³). The settling velocity of the solids at the MLSS concentration of 6500 mg/L is 0.7 m/h.

10.11 Dissolved oxygen value of a secondary clarifier effluent is measured at 1.2 mg/L. It is desired to increase the DO concentration to 6.5 mg/L before discharging to a recreational lake. The average wastewater temperatures during summer and winter are 28°C and 12°C, respectively. Design a cascade aeration system (cascade height, step width, step height, and number of steps) with a broad-crested weir.

REFERENCES

Barrett, M. J., Gameson, A. L., and Ogden, C. G. 1960. Aeration Studies at Four Weir Systems. *Water and Water Engineering*, 775 (64):507.

Coe, H. S., and Clevenger, G. H. 1916. Determining Thickener Unit Areas. *Trans AIME*, 55 (3):356.

Collivignarelli, M. C., Abbà, A., Benigna, I., Sorlini, S. and Torretta, V. 2018. Overview of the Main Disinfection Processes for Wastewater and Drinking Water Treatment Plants. *Sustainability*, 10 (1):86.

Davis, M. 2020. *Water and Wastewater Engineering: Design Principles and Practice*. McGraw-Hill, Inc., New York, New York, USA.

Dick, R. I. and Ewing, B. B. 1967. Evaluation of Activated Sludge Thickening Theories. *Journal of Sanitary Engineering Division, Proc. ASCE*, 93, SA4:9.

Dick, R. I. and Young, K. W. 1972. Analysis of Thickening Performance of Final Settling Tanks. *Proceedings of 27th Industrial Waste Conference*, Purdue University, Indiana, p. 33.

Droste, R. L. and Gehr, R. L. 2018. *Theory and Practice of Water and Wastewater Treatment*. 2nd edn. John Wiley & Sons Inc., Hoboken, NJ, USA.

European Union. 1998. Directive 98/15/EC. Official Journal L 067: 29–30. European Union, Brussels, Belgium. https://eur-lex.europa.eu/legal-content/EN/TXT/?uri=celex%3A31998L0015

GLUMRB. 2004. *Recommended Standards for Wastewater Facilities*. Great Lakes-Upper Mississippi River Board of State and Provincial Public Health and Environmental Managers, Health Education Services, New York, NY, USA, 70–73.

Leong, L. 2008. Disinfection of Wastewater Effluent – Comparison of Alternative Technologies. *Water Intelligence Online*, 7. doi:10.2166/9781780403670.

Metcalf and Eddy, Tchobanoglous, G., Stensel, H., Tcuchihashi, R., and Burton, F. 2013. *Wastewater Engineering: Treatment and Resource Recovery*. 5th edn. McGraw-Hill, Inc., New York, NY, USA.

Peavy, H. S., Rowe, D. R., and Tchobanoglous, G. 1985. *Environmental Engineering*, McGraw- Hill, Inc., New York, NY, USA.

US EPA, 2020. 2012 Recreational Water Quality Criteria Documents. https://www.epa.gov/wqc/2012-recreational-water-quality-criteria-documents

Yoshioka, N., Hotta, S., Tanaka, S., Naito, S. and Tsugami, S. 1957. Continuous Thickening of Homogeneous Flocculated Slurries. *Chemical Engineering*, Tokyo, Japan, 21:66.

Chapter 11

Anaerobic wastewater treatment

11.1 INTRODUCTION

The biological treatment of wastewater and sludge in absence of oxygen is termed *anaerobic treatment*. Louis Pasteur was the first scientist to discover anaerobic life during his research on fermentation processes in 1861 (Madigan et al., 2021). He observed that the *Clostridium* bacteria which caused butyric fermentation was strictly anaerobic. Exposure to oxygen was toxic to the bacteria. Pasteur introduced the terms *aerobic* and *anaerobic* to designate biological life in the presence and absence of oxygen, respectively. Pasteur observed that there was a difference in yield between aerobic and anaerobic processes. Anaerobic fermentation resulted in lower microbial mass in yeast production than in aerobic conditions.

Historically, anaerobic treatment has been used more for the treatment of sludge or biosolids rather than for wastewater. The *septic tank* was one of the first forms of anaerobic treatment used for sewage sludge or biosolids. The development and use of the first septic tank date back to 1896 at Exeter, England, as reported by Fuller (1912). Wastewater clarification and digestion took place in the same tank. This was widely used for waste treatment in Europe and the US. In 1904, William Travis developed a two-story septic tank in Germany, where suspended material was separated from the wastewater by settling in the first stage. The second stage was a hydrolyzing chamber through which the supernatant was allowed to flow. The *Travis hydrolytic tank* was modified by Karl Imhoff in 1907 to provide a treatment system, which later became known as the *Imhoff tank*. The Imhoff tank did not allow the wastewater to flow through the hydrolyzing tank. Instead, the sludge was kept in the hydrolyzing tank for a long period of time to allow for digestion and stabilization. The Imhoff tank reduced the cost of sludge disposal and rapidly became popular both in Europe and the US (Dague et al., 1966; McCarty, 1981).

The importance of temperature on anaerobic treatment was observed and investigated by a number of researchers, as early as the 1920s. Rudolfs (1927) observed that the total amount of gas produced from a gram of organic matter under anaerobic conditions was not dependent on temperature, but the rate of gas production was temperature-dependent. Eventually

DOI: 10.1201/9781003134374-11

the mesophilic (35°C) and thermophilic (55°C) temperature ranges were identified for anaerobic treatment (Heukelekian, 1933). Extensive studies were conducted by researchers to gain a better understanding of the microbiology of anaerobic treatment, as well as the biochemical and environmental factors that affect the process (Babbitt and Schlenz, 1929; Heukelekian, 1958; Fair and Moore, 1932 and 1937; Sawyer et al., 1954; McCarty et al., 1963; Dague et al., 1966). In 1964, Perry L. McCarty published a series of papers on anaerobic waste treatment that provided a comprehensive summary of the fundamentals of anaerobic treatment (McCarty, 1964a, 1964b, and 1964c). Over time, a large number of suspended and attached growth processes have been developed for the anaerobic treatment of wastewater.

Anaerobic treatment of wastewater involves the stabilization of organic matter, with a concurrent reduction in odors, pathogens, and the mass of solid organic matter that requires further processing. This is accomplished by the biological conversion of organics to methane and carbon dioxide in an oxygen-free or anaerobic environment (Parkin and Owen, 1986).

The main advantages of anaerobic treatment processes over aerobic processes are (McCarty, 1964a; Metcalf and Eddy et al., 2013):

- A high degree of waste stabilization is possible at high organic loads.
- Low production of waste biological sludge.
- Less energy is required.
- Low nutrient requirements.
- Methane gas produced is a useful source of fuel.
- Smaller reactor volume is required.
- No oxygen is required, so treatment rates are not limited by oxygen transfer rates.
- Rapid reactivation of biomass is possible with substrate addition, after long periods of starvation.

The anaerobic process has some disadvantages. They are:

- Relatively high temperature (35°C) is required for optimal operation.
- A longer start-up time is required to develop the necessary amount of biomass, due to the slow growth rate of methane-forming bacteria.
- It may be necessary to add alkalinity or other specific ions, depending on the characteristics of the wastewater.
- Process can be more susceptible to toxic substances.
- Odor production may be a problem.
- Biological nitrogen and phosphorus removal may not be possible.

This chapter will provide an overview of the process microbiology, followed by a discussion of factors that affect the process, and process kinetics. Anaerobic suspended and attached growth processes used for wastewater treatment will be discussed in detail in the later part of this chapter.

Anaerobic processes used for the treatment of sludge and biosolids will be discussed in Chapter 12.

11.2 PROCESS CHEMISTRY AND MICROBIOLOGY

Anaerobic waste treatment is a complex biological process involving various types of anaerobic and facultative bacteria. A four-step process can be used to describe the overall treatment. Although the bacteria are represented by separate groups, it is not possible to separate the metabolism of each group. They are interdependent. The anaerobic biotransformation process is illustrated in Figure 11.1.

Five groups of bacteria are thought to be involved, each deriving its energy from a limited number of biochemical reactions. They are the following (Novaes, 1986):

1. **Fermentative bacteria** – This group is responsible for the first two stages of anaerobic conversion, hydrolysis, and acidogenesis. Anaerobic species belonging to the family of *Streptococcus* and *Enterobacter*, and to the genera of *Clostridium eubacterium* are mainly found in this group.

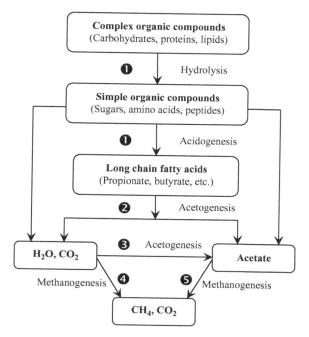

Figure 11.1 Metabolic steps involved in anaerobic biotransformation (adapted from McCarty and Smith, 1986). The numbers represent the different microbial groups.

2. **Hydrogen-producing acetogenic bacteria** – They catabolize sugars, alcohols, and organic acids to acetate and carbon dioxide. These include *Syntrophobacter wolinii* and *Syntrophomonus wolfei*.
3. **Hydrogen-consuming acetogenic or homoacetogenic bacteria** – These bacteria use hydrogen and carbon dioxide to produce acetate. They include the *Clostridium aceticum*, and *Butyribacterium methylotrophicum*, among others.
4. **Carbon-dioxide-reducing methanogens** – They utilize hydrogen and carbon dioxide to produce methane.
5. **Aceticlastic methanogens** – They cleave acetate to form methane and carbon dioxide.

The four steps of anaerobic biotransformation are:

1. Hydrolysis and liquefaction.
2. Fermentation or acidogenesis.
3. Hydrogen and acetic acid formation, or acetogenesis.
4. Methane formation or methanogenesis.

Step 1. Hydrolysis and liquefaction

The first step involves hydrolysis and liquefaction. Insoluble organics must first be solubilized before they are consumed. In addition, large soluble organic molecules must be diminished in size to facilitate transport across the cell membrane. The reactions are hydrolytic and catalyzed by enzymes such as amylase, proteinase, lipase, and nuclease. No waste stabilization takes place during this step, rather the organic matter is converted into a form that can be taken up by the microorganisms. Anaerobic digestion may be limited in the hydrolysis and liquefaction step if the waste contains large portions of refractory or non-biodegradable organic material which is not hydrolyzed by microorganisms. Particulate organic matter (lipids, polysaccharides, protein) is converted to soluble compounds and afterward hydrolyzed to simple monomers (fatty acids, monosaccharides, amino acids) in this step.

Step 2. Fermentation or acidogenesis

The simple monomers resulting from hydrolysis are used as carbon and energy sources by acid-producing bacteria. The oxidized end products of this step are primarily volatile fatty acids (VFAs), such as acetic, propionic, butyric, valeric, and caproic acid, together with the production of ammonia (NH_3), carbon dioxide (CO_2), hydrogen sulfide (H_2S), and other byproducts. This phase is referred to as acidogenesis or fermentation.

Step 3. Hydrogen and acetic acid formation, or acetogenesis

In the third step, VFAs and alcohols produced in the acidogenesis step are degraded primarily to acetic acid together with the production of CO_2 and H_2. In this conversion, the partial pressure of H_2 is an important factor. Free energy change associated with the conversion of propionate and butyrate to acetate and hydrogen, requires the hydrogen concentration to be low in the system ($H_2 < 10^{-4}$ atm) or the conversion will not take place (McCarty and Smith, 1986). Hydrogen is produced by the fermentative and hydrogen-producing acetogenic bacteria. Acetate is also produced by these groups in addition to the homoacetogenic bacteria.

Step 4. Methane formation or methanogenesis

Waste stabilization occurs in the third and final stage when acetic acid or acetate is converted to methane by the methanogenic bacteria. Approximately 72% of methane formed comes from acetate cleavage by aceticlastic methanogens (McCarty, 1964c). The proposed reaction is

$$CH_3COOH \rightarrow CH_4 + CO_2 \tag{11.1}$$

The remaining 28% results from the reduction of carbon dioxide (13% from propionic acid and 15 % from other intermediates, using hydrogen as an energy source by carbon dioxide-reducing methanogens, forming methane gas in the process:

$$CO_2 + H_2 \rightarrow CH_4 + H_2O \tag{11.2}$$

The following reactions describe the overall anaerobic biotransformation of acetic acid, propionic acid, butyric acid, ethanol, and acetone:

$$CH_3COOH \rightarrow CH_4 + CO_2 \tag{11.3}$$

$$4CH_3CH_2COOH + 2H_2O \rightarrow 7CH_4 + 5CO_2 \tag{11.4}$$

$$2CH_3CH_2CH_2COOH + 2H_2O \rightarrow 5CH_4 + 3CO_2 \tag{11.5}$$

$$2CH_3CH_2OH \rightarrow 3CH_4 + CO_2 \tag{11.6}$$

$$CH_3COCH_3 \rightarrow 2CH_4 + CO_2 \tag{11.7}$$

11.2.1 Syntrophic relationships

The rate-limiting step in the entire anaerobic process is the conversion of hydrogen to methane by CO_2-reducing methanogens. The hydrogen partial pressure must be maintained at an extremely low level to enable favorable thermodynamic conditions for the conversion of volatile acids and alcohols to acetate. Under standard conditions of 1 atm of hydrogen partial pressure, the free energy change is positive for this conversion and thus precludes it. The free energy change for conversion of propionate and butyrate to acetate and hydrogen does not become negative, until the hydrogen partial pressure decreases below 10^{-4} atm (Speece, 1983; McCarty and Smith, 1986). It is therefore obligatory for the hydrogen utilizing methanogens to utilize hydrogen rapidly, and maintain these extremely low hydrogen partial pressures in the system. Otherwise, higher volatile acids, such as propionic and butyric acids, will accumulate and waste stabilization will not occur.

11.3 METHANOGENIC BACTERIA

Methanogens are often considered to be the key class of microorganisms in anaerobic treatment. Methanogens are classified as *Archaea*, or *Archaebacteria*, and can be distinguished by the comparative cataloging of 16S rRNA sequences (Balch et al., 1979) as well as biochemical properties, morphology, and immunological analyses (Macario and Conway, 1988). They are obligate anaerobes with relatively slow reproduction rates since less energy is released in the reactions involved in the anaerobic stabilization of organic matter. This slow growth rate limits the rate at which the process can adjust to changing substrate loads, temperatures, and other environmental conditions.

A variation of methanogens is observed according to the following:

- Morphology: can be long or short rods, small or large cocci, numerous lancet, and spirillum shapes. The cell walls of methanogens are based on three major components: pseudomurien, protein, and heteropolysaccharide (Archer and Harris, 1986).
- Gram staining: all are Gram-negative.
- Growth temperature: some are thermophilic (55°–65°C) and some are mesophilic (30°–35°C) organisms.
- Generation time: The range is from 1.8 to 3.5 h (Dubach and Bachofen, 1985).

Methanogens can only use a small number of simple compounds which contain one or two carbons (Wose et al., 1978, Wose, 1987). The primary reactions of methane formation with their associated Gibbs free energy values are shown in Table 11.1. The methanogenic bacteria are dependent on other organisms for their substrates. Hence, a complex food web of anaerobes is required to convert most of the organic substrates to low molecular weight organic acids, CO_2, and hydrogen. The methanogens use the latter two of these products and eventually convert acetate to methane. It has been estimated that approximately 70% of the methane formed in nature is via acetate cleavage to methane and carbon dioxide. The optimum degradation

Table 11.1 Gibbs free energy values for selected methanogenic reactions

Reactants	Products	$G°$ (kJ/ mol CH_4)	Organisms
$4 H_2 + HCO_3^- + H^+$	$CH_4 + 3 H_2O$	-135	Most methanogens
$4 HCO_3^- + H^+ + H_2$	$CH_4 + 3 HCO_3^-$	-145	Most hydrogenotrophic methanogens
$4CO + 5H_2O$	$CH_4 + 3 HCO_3^- + 3 H^+$	-196	*Methanobacterium* and *Methanosarcina*
$2CH_3CH_2OH + HCO_3^-$	$2 CH_3COO^- + H^+ + CH_4 + H_2O$	-116	Some hydrogenotrophic methanogens
$CH_3COO^- + H_2O$	$CH_4 + HCO_3^-$	-31	*Methanosarcina* and *Methanothrix*
$4CH_3OH$	$3 CH_4 + HCO_3^- + H_2O + H^+$	-105	*Methanosarcina* and other Methylotrophic methanogens
$CH_3OH + H_2$	$CH_4 + H_2O$	-113	*Methanoshaera stadtmanii* and Methylotrophic methanogens

Source: Adapted from Wose (1987) and Thauer et al. (1977)

performance depends on a number of biochemical and physical interactions between methanogens and non-methanogens (Archer and Harris, 1986).

The majority of the species use hydrogen and carbon dioxide for both carbon and energy sources. Other substrates include formate, methanol, carbon monoxide, methylamines, and acetate. Three types of methanogenic bacteria have been identified that utilize acetate. They are *Methanosarcina* sp., *Methanothrix soehngenii*, and *Methanococcus mazei*. Formate is used by several genera, including *Methanobacterium*, *Methanogenium*, and *Methanospirillum* (Novaes, 1986, Daniels, 1984).

There is a variation in growth rate among different species of methanogens. Gujer and Zehnder (1983) evaluated growth kinetics for *Methanosarcina* and *Methanothrix* on acetate. *Methanosarcina* had a sharply increasing growth curve with a maximum specific growth rate (μ_{max}) of 0.3 d^{-1} and half-saturation coefficient (K_s) of 200 mg/L. *Methanothrix* had a flatter growth curve with a maximum specific growth rate of 0.1 d^{-1} and K_s of 30 mg/L. This meant that at low substrate concentrations, the *Methanothrix* outcompete the *Methanosarcina*. But at high substrate concentrations, the *Methanosarcina* predominate.

Even though the methanogens are the most important and sensitive microbial species in anaerobic treatment, a balance must be maintained between the acid-forming and hydrogen-forming bacteria and the methane formers, in order to achieve complete conversion of organic compounds to methane and carbon dioxide. The proper environmental conditions have to be maintained for growth and metabolism.

11.4 SULFATE-REDUCING BACTERIA

One group of bacteria often found in association with the methanogens is the *sulfate-reducing bacteria*. They produce hydrogen, acetate, and sulfides which are used by the methanogens. In sulfate-rich environments, the sulfate-reducing bacteria have a thermodynamic advantage over the methanogens (Thauer et al., 1977). The sulfate-reducers have lower K_s values for H_2 and acetate, as compared to the values for methanogens. The production of sulfide might inhibit methanogenesis, since the sulfate-reducing bacteria utilize hydrogen and acetic acid as energy sources, and outcompete methanogens for these substrates. Also, soluble hydrogen sulfide in excess of 200 mg/L (Parkin and Owen, 1986) are toxic to methanogens.

11.5 ENVIRONMENTAL REQUIREMENTS AND TOXICITY

Optimum environmental conditions are very important in the design and operation of anaerobic treatment processes. These conditions are usually dictated by the requirements of the methanogens, whose growth rate limits the process of waste stabilization. The following are important environmental factors affecting the process:

1. *Temperature* – Temperature is an important factor influencing anaerobic bacteria. There is a limited range of temperatures for optimum growth. Methane bacteria are active in two temperature zones, the mesophilic and the thermophilic ranges, and especially in the part of mesophilic range between 30°C and 35°C. The rates of degradation are slower at lower temperatures. The treatment process has to be operated at longer detention times, or the microbial population should increase in order to obtain the same degree of stabilization at lower temperatures.

   A rapid change of temperature is also detrimental to anaerobic treatment. Changing the temperature by a few degrees can cause an imbalance between the major bacterial populations, which can lead to process failure (Grady and Lim, 1980).

2. *pH* – pH is an important parameter affecting the enzymatic activity since a specific and narrow pH range is suitable for the activation of each enzyme. Anaerobic treatment processes operate best at a pH system of near neutrality. The pH has an effect on both acid-forming and methane-forming bacteria. The optimum pH for anaerobic treatment is in the range of 6.5–7.6 (McCarty, 1964a). If the pH drops below 6.3 or increases beyond 7.8, the rate of methanogenic activity reduces significantly. A sharp pH drop below 6.3 indicates that the

rate of organic acids production is faster than the rate of methane formation. For reactors treating a wastewater with a high concentration of protein, the buffering effect of ammonia released from amino acid fermentation can prevent the pH from dropping below the optimum range. On the other hand, a sharp pH increase above 7.8 can be due to a shift in NH_4^+ to NH_3, the toxic, unionized form of ammonia (Gomec et al., 2002). Buffers can also be added in the form of bicarbonates or hydroxides to maintain pH.

3. *Nutrients* – Nitrogen and phosphorus are the two major nutrients required for microbial growth and reproduction. In addition, sulfur, iron, cobalt, nickel, calcium, and some trace metals are necessary for the growth of methanogens. Sulfide is required by methanogens, even though it may adversely affect methane production by precipitating essential trace metals. It is toxic at concentrations above 100–150 mg/L of un-ionized hydrogen sulfide (Speece, 1983). Molybdenum, selenium, and tungsten have also been reported as trace metals used by methanogens.

4. *Toxic materials* – The methanogens are commonly considered to be the most sensitive to toxicity among all the microorganisms involved in the anaerobic conversion of organic matter to methane. However, acclimation to toxicity and reversibility of toxicity are frequently observed. Whether a substance is toxic to a biological system depends on the nature of the substance, its concentration, and its potential for acclimation. Changes in the concentration of the substance can change the classification of the substance from toxic to biodegradable. Table 11.2 presents a summary of concentrations of different cations at which they are reported to be stimulatory or inhibitory to the anaerobic process (McCarty, 1964c).

Control of toxicants is vital to the successful operation of an anaerobic process. Toxicity may be controlled by (1) dilution to reduce concentration

Table 11.2 Stimulatory and inhibitory concentrations of some compounds on anaerobic treatment

Substance	Stimulatory (mg/L)	Moderately inhibitory (mg/L)	Strongly inhibitory (mg/L)
Calcium	100–200	2500–4500	8000
Magnesium	75–150	1000–1500	3000
Potassium	200–400	2500–4500	12,000
Sodium	100–200	3500–5500	8000
Ammonia-nitrogen	50–1000	1500–3000	>3000

Source: Adapted from McCarty (1964c)

below the toxic threshold, (2) removal of toxic material from the feed, (3) removal by chemical precipitation, (4) neutralization, or (5) acclimation.

11.6 METHANE GAS PRODUCTION

11.6.1 Stoichiometry

A significant fraction of the COD removed in an anaerobic process is converted to methane. So the methane gas production can be estimated from the amount of COD that is biodegraded. The COD equivalence of methane can be determined from stoichiometry. The COD of methane is the amount of oxygen needed to completely oxidize methane to carbon dioxide and water as follows:

$$CH_4 + 2O_2 \rightarrow CO_2 + 2H_2O \qquad (11.8)$$

From the above equation, (2×32) or 64 g oxygen are required to oxidize one mole of methane. The volume occupied by one mole of gas at standard conditions of 0°C and 1 atm pressure (STP) is 22.4 L. So the methane equivalent of COD converted under anaerobic conditions is

$$\frac{22.4 \dfrac{L}{mol}}{64 g \dfrac{COD}{mol}} = 0.35 \frac{L\,CH_4}{g\,COD} \text{ or } 0.35 \frac{m^3\,CH_4}{kg\,COD} \qquad (11.9)$$

Equation (11.9) provides an estimate of the maximum amount of methane produced per unit of COD at STP conditions (Metcalf and Eddy et al., 2013).

The amount of methane gas produced at other temperature and pressure conditions can be determined by using the Universal gas law. This is demonstrated in Example 11.1.

Example 11.1: A wastewater treatment plant treats 2000 m³/d of high-strength wastewater in an anaerobic reactor operated at 35°C. The biodegradable soluble COD concentration of the wastewater is 3500 mg/L. Calculate the amount of methane gas that will be produced with 90% COD removal, and net biomass yield of 0.04 g VSS/g COD used. Assume, COD equivalent of VSS equals 1.42 kg COD/kg VSS. If the total gas contains 65% methane, calculate the total gas produced from the wastewater.

SOLUTION

Step 1. Conduct a steady state mass balance for the COD in the anaerobic reactor.

$$\text{Accumulation} = \frac{\text{Influent}}{\text{COD}} - \frac{\text{Effluent}}{\text{COD}} - \frac{\text{COD converted}}{\text{to new cells}}$$
$$- \frac{\text{COD converted}}{\text{to new methane}} \qquad (11.10)$$

Or, $0 = COD_{in} - COD_{out} - COD_{VSS} - COD_{methane}$ \qquad (11.11)

$COD_{in} = 2000\,m^3/d \times 3.5\,kg/m^3 = 7000\,kg/d$

$COD_{out} = (1 - 0.9)\,7000\,kg/d = 700\,kg/d$

$COD_{VSS} = 0.9 \times 7000\,kg\,COD/d \times 0.04\,kg\,VSS/kg\,COD$

$\qquad \times 1.42\,kg\,COD/kg\,VSS$

$\qquad = 357.84\,kg/d$

Using these values in equation (11.11), we obtain

$0 = 7000\,kg/d - 700\,kg/d - 357.84\,kg/d - COD_{methane}$

Or, $COD_{methane} = 5942.16\,kg/d$

Step 2. Determine volume (V) occupied by 1 mole of methane gas at 35°C.

From the Universal gas law, we have

$$V = \frac{nRT}{P} \qquad (11.12)$$

where

n = number of moles
R = universal gas constant = 0.082057 atm.L/mole.K
T = temperature, K
P = pressure, atm

Here, T = 273 + 35 = 308 K
n = 1 mol, and P = 1 atm

$$\text{Therefore, } V = \frac{1\,mol \times 0.082057\,atm.\dfrac{L}{mol}.K \times 308\,K}{1\,atm} = 25.27\,L$$

Step 3. Calculate the methane equivalent of COD converted.

The methane equivalent of COD converted under anaerobic conditions is

$$\frac{25.27\dfrac{L}{mol}}{64\,g\dfrac{COD}{mol}} = 0.395\,\frac{L\,CH_4}{g\,COD}\ or\ 0.395\,\frac{m^3 CH_4}{kg\,COD}$$

Step 4. Calculate the methane gas produced.

$$CH_4\text{produced} = 5942.16\,kg\,COD/d \times 0.395\,m^3 CH_4/kg\,COD$$
$$= 2347.15\,m^3/d$$

Step 5. Calculate the total gas produced.

Total gas contains 65% CH_4

$$\text{Total gas produced} = \frac{2347.15\dfrac{m^3}{d}}{0.65} = 3611\,\frac{m^3}{d}$$

11.6.2 Biochemical methane potential assay

The biological methane potential (BMP) assay measures the concentration of organic pollutants in a wastewater which can be anaerobically converted to methane, thus indicating waste stabilization. The BMP measures anaerobic biodegradability, and can be used to identify aerobic non-biodegradable components which are amenable to anaerobic biodegradation (Speece, 2008). It can be used to evaluate process efficiency.

The BMP test was developed by McCarty and his co-researchers as an indicator of the anaerobic pollution potential of a waste (Owen et al., 1979). Just as the BOD test is used to determine the aerobic pollution potential of a waste, so is the BMP test used as a correlative indicator in the anaerobic process. It has not been incorporated into the *Standard Methods* (AWWA et al., 2017), but it is widely used in practice.

In the BMP test, a sample of wastewater is placed in a serum bottle with an anaerobic inoculum. Care should be taken that the anaerobic inoculum or biomass is acclimated to the wastewater being tested. A small amount of nutrients is added to the bottle. The headspace is purged with 70:30 nitrogen:carbon dioxide gas to ensure anaerobic conditions and for pH control. The serum bottle is capped and incubated at 35°C for a period ranging from 30 to 60 d. A control with only the inoculum is also placed in the incubator. Gas production and composition are monitored at regular intervals. The gas volume produced is monitored by inserting a hypodermic needle connected to a calibrated fluid reservoir, through the bottle cap. Similar to the BOD test, a number of different sample volumes of the wastewater are used in the serum bottles. Average gas production should be similar. At 35°C, 395 mL of CH_4 production is equivalent to 1 g of COD used (Speece, 2008). This stoichiometric relationship can be used to calculate the COD reduction in the liquid phase.

11.6.3 Anaerobic toxicity assay (ATA)

The Anaerobic Toxicity Assay or ATA is used to measure the potential toxicity of a wastewater sample or compound to the anaerobic biomass. The procedure is similar to the BMP test, with the exception that excess substrate such as acetate is added initially to the serum bottles to avoid substrate limitation. If toxicity is present in the sample, it will be demonstrated by a reduced initial rate of gas production in proportion to the volume of wastewater added, as compared to the control. The test is run with a range of dilutions of the wastewater sample. The ATA was developed by McCarty and his co-researchers (Owen et al., 1979).

In the anaerobic biomass consortium, aceticlastic methanogens are the most sensitive to toxicity. For this reason, it is usually recommended to add acetate as the substrate, at about 1000 mg/L COD (Droste, 1997). More complex substrates such as glucose, ethanol, or others can be added to evaluate toxicity to other microorganisms in the consortia.

11.7 ANAEROBIC GROWTH KINETICS

The Monod model is the most widely used one among the models developed for the analysis of anaerobic growth kinetics. This model assumes that the rate of substrate utilization, and therefore the rate of biomass production, is limited by the rate of enzyme reactions involving the substrate. This has been described in detail in Chapter 8. The growth kinetics described in Section 8.2 are also applicable for anaerobic treatment reactors.

Table 11.3 shows the kinetic parameters for acetate utilization at various temperatures using batch, semi-continuous and continuous systems from various studies. Because of the different temperatures and different systems used, the values of the kinetic parameters from these studies show a wide range of variations.

The anaerobic process is stable when a sufficient methanogenic population exists in the reactor and sufficient time is available for VFA minimization and for methanogens to utilize H_2. The rate-limiting step is the conversion of VFAs by methanogenic organisms and not the fermentation of

Table 11.3 Kinetic coefficients for acetate utilization

Temp., °C	μ_{max}, d^{-1}	Y, $\dfrac{kg\,biomass}{kg\,COD}$	K_d, d^{-1}	K_s, $mgCOD/L$	Reference
37	0.11	0.023	ND	28	Zehnder et al. (1980)
35	0.34	0.04	0.015	165	Lawrence and McCarty (1969)
35	0.44	0.05	ND	250	Smith and Mah (1980)
30	0.24	0.054	0.037	356	Lawrence and McCarty (1969)
25	0.24	0.05	0.011	930	Lawrence and McCarty (1969)

Here, μ_{max} = maximum specific growth rate, Y = yield coefficient, K_d = decay coefficient, and K_s = half-saturation coefficient.

soluble substrates by acidogens. Therefore, most interest in anaerobic process design is given to methanogenic growth kinetics.

11.8 ANAEROBIC SUSPENDED GROWTH PROCESSES

Historically anaerobic treatment has been used for the stabilization of sludge and biosolids. Over the last 50 years, a lot of research and development have resulted in the application of anaerobic processes for wastewater treatment. Both suspended and attached growth processes are in use, especially for the treatment of high-strength wastewaters. Conventional anaerobic treatment using completely mixed reactors are used for the digestion of sludge, and are described in detail in Chapter 12. Some of the more common suspended growth processes used for wastewater treatment are described in this section.

11.8.1 Anaerobic contact process

The anaerobic contact process is similar to the activated sludge process in many aspects. The system consists of a completely mixed anaerobic reactor with gas collection, followed by a clarifier for solids–liquid separation. Part of the settled sludge is recycled to the reactor to increase the solids retention time (SRT). The SRT is usually greater than the HRT. By separating the HRT and the SRT, the reactor volume can be reduced. For anaerobic processes, the minimum SRT at 35°C is 4 d, with a recommended design SRT of 10–30 d. For the anaerobic contact process, the HRT ranges from 0.5 to 5 d with organic loading rates of 1–8 kg COD/m³.d (Metcalf and Eddy et al., 2013). The process flow diagram is illustrated in Figure 11.2(a).

The anaerobic sludge contains a large amount of entrained gases that are produced during anaerobic degradation. These gases can decrease the settleability of the sludge. Various methods are used to remove the gas bubbles

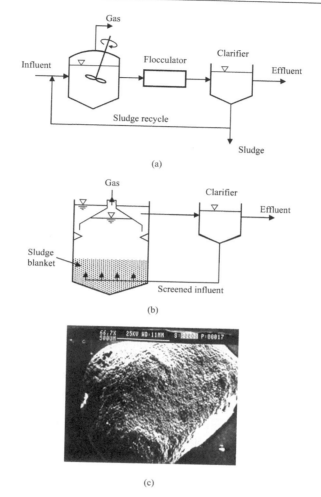

Figure 11.2 (a) Anaerobic contact process and (b) upflow anaerobic sludge blanket (UASB) process, (c) SEM image of granule formed in a UASB reactor. (Source: Courtesy of Somchai Dararat and Kannitha Krongthamchat).

from the sludge. These include vacuum degasification, inclined-plate separators, and chemical coagulation, among others.

11.8.2 Upflow anaerobic sludge blanket process

The upflow anaerobic sludge blanket (UASB) process was developed in the Netherlands by Lettinga and coworkers (Lettinga et al., 1980). This was one of the most important developments of anaerobic technology for the treatment of high-strength wastewaters. There are over 500 installations all over the world treating a wide range of industrial wastewaters (Metcalf and Eddy et al., 2013, Davis, 2011).

The UASB process is illustrated in Figure 11.2(b). The wastewater enters the reactor at the bottom and is distributed upward through a sludge blanket. Organic matter is degraded in the sludge blanket, after which the liquid effluent is discharged at the top. Gas production and evolution provide sufficient mixing in the sludge blanket. A quiescent zone above the sludge blanket is provided for solids settling. The liquid effluent is passed through a settling tank to collect solids that have escaped from the reactor. The collected solids are recycled back to the reactor. Critical design elements include the influent distribution system, gas–solids separator, and effluent withdrawal system.

The main characteristic of the UASB process is the formation of a dense granular sludge. The solids concentration can range from 50 to 100 g/L at the reactor bottom, to 5–40 g/L at the top of the sludge blanket. Several months may be required to form the granules, and seed is often supplied from other installations to accelerate the process. It was suggested that the UASB system promoted a selection between the sludge ingredients, such that lighter particles were washed out and heavier particles were retained. Growth was concentrated on these particles, which resulted in the formation of *granules* up to 5 mm in diameter (Hulshoff Pol et al., 1983). A typical granule is illustrated in Figure 11.2(c). Most of the organisms grow on the surface and in the interstices of the granules, while the core may contain inert extracellular material. A symbiotic relationship exists between the microbial consortia associated with granular sludge particles that is advantageous in enhancing biological activity. Very high specific activities have been observed, ranging from 2.2 to 2.3 kg COD/kg VSS.d. McCarty and Smith (1986) reported that reactors with granular sludge produced lower hydrogen partial pressures and more rapid hydrogen utilization, than reactors with dispersed sludge, resulting in increased efficiency. Granule development is influenced by wastewater characteristics, reactor geometry, upflow velocity, HRT, and organic loading rates. These are all important design considerations for the UASB process.

Volumetric loading rates can vary from 0.5 to 40 kg/m³.d (0.03–2.5 lb/ft³.d) for a UASB process (Droste, 1997). The HRT can vary from 6 to 14 h. Upflow velocities range from 0.8 to 3.0 m/h, depending on the type of wastewater and reactor height.

11.8.2.1 Design equations

The area of the reactor is given by

$$A = \frac{Q}{v} \tag{11.13}$$

where
  A = area of the reactor, m²
  Q = influent flow rate, m³/d
  v = design upflow superficial velocity, m/d

The required reactor volume depends on the organic loading rate and effective treatment volume. The effective treatment volume is the volume occupied by the sludge blanket and active biomass. An additional volume is provided between the sludge blanket and gas collection unit, where solids separation occurs. The nominal or effective liquid volume of the reactor is given by (Metcalf and Eddy et al., 2013),

$$V_n = \frac{QS_o}{L_{org}}$$
(11.14)

where
  V_n = effective or nominal liquid volume of the reactor, m³
  S_o = influent COD, kg COD/m³
  L_{org} = acceptable organic loading rate, kg COD/m³.d

The total liquid volume of the reactor exclusive of the gas storage area is given by

$$V_L = \frac{V_n}{E}$$
(11.15)

where
  V_L = total liquid volume of the reactor, m³
  E = effectiveness factor, representing the volume fraction occupied by sludge blanket, can vary from 0.8 to 0.9.

The reactor height (H_L) based on liquid volume is

$$H_L = \frac{V_L}{A}$$
(11.16)

So, the total height of the reactor is

$$H_T = H_L + H_G$$
(11.17)

where
  H_T = total reactor height, m
  H_G = reactor height corresponding to gas collection and storage volume, usually about 2.5–3m.

These concepts are illustrated in Example 11.2.

11.8.3 Expanded granular sludge bed (EGSB)

The expanded granular sludge bed (EGSB) process is a variation of the UASB process. It consists of two or more UASB reactors situated on top of each other. The EGSB system has been reported to successfully treat wastewaters with high lipid content, which cause foaming and scum, as well as handle organic loading rates 3–6 times greater than that of a conventional UASB system with similar efficiency (Vallinga et al., 1986).

> Example 11.2: Design a UASB reactor for treatment of a dairy wastewater at 35°C. The wastewater flow rate is 1500 m³/d with a soluble COD concentration of 3000 mg/L. Also, calculate the effluent soluble COD concentration and the reactor efficiency. The following parameters are given:
>
> SRT = 60 d
> Sludge blanket occupies 80% of liquid volume
> Height for gas collection = 2.5 m
> Upflow velocity = 1.5 m/h
> Design organic loading rate = 16 kg sCOD/m³.d
> Y = 0.08 kg VSS/kg COD
> k_d = 0.04 d⁻¹
> μ_{max} = 0.35 d⁻¹
> K_s = 160 mg sCOD/L

SOLUTION

Step 1. Determine the UASB reactor cross-sectional area and diameter based on the upflow velocity using equation 11.13.

$$v = 1.5\,\mathrm{m/h} \times 24\,\mathrm{h/d} = 36\,\mathrm{m/d}$$

$$A = \frac{Q}{v} = \frac{1500\,\dfrac{\mathrm{m^3}}{\mathrm{d}}}{36\,\dfrac{\mathrm{m}}{\mathrm{d}}} = 41.67\,\mathrm{m^2}$$

$$A = \frac{\pi D^2}{4} = 41.67\,\mathrm{m^2}$$

Or, $D = 7.28\,\mathrm{m} \cong 7.3\,\mathrm{m}$

Step 2. Calculate the liquid volume of the reactor using Equation (11.14).

$$V_n = \frac{QS_0}{L_{org}} = \frac{1500\,\dfrac{m^3}{d} \times 3\,\dfrac{kg}{m^3}}{16\,\dfrac{kg}{m^3}.d} = 281.25\,m^3$$

Step 3. Calculate the total liquid volume of the reactor using Equation (11.15).

$$V_L = \frac{V_n}{E} = \frac{281.25\,m^3}{0.8} = 351.56\,m^3$$

Step 4. Calculate liquid height using Equation (11.16).

$$H_L = \frac{V_L}{A} = \frac{351.56\,m^3}{41.67\,m^2} = 8.44\,m$$

Calculate the total height of the reactor using equation (11.17).

$$H_T = H_L + H_G = 8.44\,m + 2.5\,m = 10.94\,m \cong \mathbf{11\,m}$$

Therefore, UASB reactor height = 11m, and diameter = 7.3 m
Step 5. Calculate the effluent sCOD concentration using the kinetic coefficients and equation (8.46) (from Chapter 8).

$$S = \frac{K_s\left(1 + k_d\theta_c\right)}{\theta_c\left(\mu_{max} - k_d\right) - 1}$$

Or,

$$S = \frac{0.16\,\dfrac{kg}{m^3}\left(1 + 0.04\,d^{-1} \times 60\,d\right)}{60\,d\left(0.35 - 0.04\,d^{-1}\right) - 1} = 0.0309\,\frac{kg}{m^3} = 30.90\,\frac{mg}{L}$$

Step 6. Calculate the sCOD removal efficiency.

$$E = \frac{\left(3000 - 30.9\right)\dfrac{mg}{L}}{3000\,\dfrac{mg}{L}} \times 100\% = 98.97\% \approx 99\%$$

11.8.4 Anaerobic sequencing batch reactor

The anaerobic sequencing batch reactor (ASBR) was developed by Dague and co-researchers in the late 1980s at Iowa State University, in Ames, Iowa. It is a suspended growth process where biological conversions and solids–liquid separation all take place in the same reactor. Gas is collected on a continuous basis. One of the advantages of the process is the formation of a dense, granular sludge that has a high activity and settles well. The ASBR sequences through four steps as illustrated in Figure 11.3. They are (Sung and Dague, 1992):

1. Feed – a specific volume of the substrate is fed to the reactor at a specific strength. Reactor contents are usually mixed during feeding.

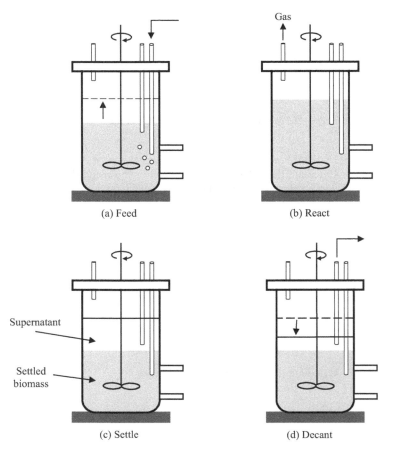

Figure 11.3 Operational steps of an anaerobic sequencing batch reactor.
Source: Adapted from Riffat, 1994.

2. React – The reactor contents are mixed intermittently to bring the substrate into close contact with the biomass. This is the most important step in the conversion of organic matter to biogas.
3. Settle – Mixing is turned off and the biomass is allowed to settle, leaving a layer of clear liquid at the top.
4. Decant – A specific volume of clear supernatant is decanted from the top. The volume decanted is usually equal to the volume fed in the first step.

These four steps constitute a *cycle* or *sequence*. The time for one sequence is called the *cycle length*. The ASBR is a very flexible system. The number of sequences per day may be varied, together with the time required for the various steps. The feeding and decanting times are short, while the time for the "react" step is the longest. Ideally, the react step should continue until the F/M ratio is quite low since a low F/M ratio is associated with improved flocculation and settling. The ASBR is capable of achieving a lower F/M ratio at the end of the react cycle than a similarly loaded CSTR, which was demonstrated by Sung and Dague (1992) as illustrated in Figure 11.4.

The time for settling depends on the settling characteristics of the biomass. HRT can vary from 6 to 24 h, while the SRT can range from 50 to 200 d. The ASBR has been demonstrated for the successful treatment of various types of high-strength wastewaters.

A number of variables influence the efficient operation of an ASBR. These include organic loading rate (OLR), HRT, SRT, and MLSS among others. The ratio of OLR to MLSS defines the F/M ratio, which is important in achieving efficient solids separation. The ASBR promotes granulation by imposing a selection pressure during the decant cycle. The decant process tends to wash out poorly settling flocs, so that the heavier, more rapidly

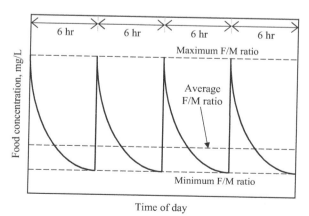

Figure 11.4 Typical variation of F/M ratio during ASBR operation.
Source: Adapted from Sung and Dague, 1992.

settling aggregates remain in the reactor. Reactor geometry, HRT, and OLR influence the size and characteristics of the granules. Settling velocities of 0.98–1.2 m/min were obtained for the granular sludge formed in the ASBR (Sung and Dague, 1992).

> **Example 11.3:** A laboratory-scale ASBR is operated at 35°C to treat a synthetic wastewater. The following operational parameters are given:
>
> Total liquid volume = 10 L
> Length of cycle = 6 h
> Feed phase = 15 min
> React phase = 300 min
> Settle phase = 30 min
> Decant phase = 15 min
> Volume fed/wasted per cycle = 2.5 L
>
> > i. Calculate the HRT for the given conditions.
> > ii. If the cycle length is increased to 8 h, what will be the new HRT of the system?
> > iii. If the cycle length remains the same, what can you do to increase the HRT?

SOLUTION

> i. Number of cycles per day = 24 h/cycle length = 24 h/6 h = 4
> The flow per day, Q = 2.5 L × 4 = 10 L/d

$$\text{HRT} = \frac{V}{Q} = \frac{10\,\text{L}}{10\dfrac{\text{L}}{\text{d}}} = 1\,\text{d}$$

> ii. Number of cycles per day = 24 h/cycle length = 24 h/8 h = 3

$$Q = 2.5\,\text{L} \times 3 = 7.5\,\text{L}$$

$$\text{HRT} = \frac{V}{Q} = \frac{10\,\text{L}}{7.5\dfrac{\text{L}}{\text{d}}} = 1.33\,\text{d}$$

> iii. If the cycle length remains the same, the HRT can be increased by reducing the volume fed/wasted per cycle.

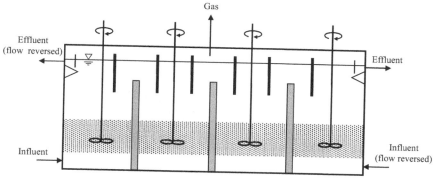

Figure 11.5 Anaerobic migrating blanket reactor (AMBR).

11.8.5 Anaerobic migrating blanket reactor

The anaerobic migrating blanket reactor (AMBR) consists of a number of compartments separated by over and under baffles, as illustrated in Figure 11.5. Mixing is provided in each compartment as the wastewater flows through. The sludge blanket in each compartment rises and falls with gas production and flow, and also moves through the reactor at a slow rate. After some time of operation, the influent feed port is changed to the effluent port, and vice versa. This helps to maintain a uniform sludge blanket across the reactor. Usually, the flow is reversed when a large quantity of solids accumulates in the last compartment. The AMBR was demonstrated to achieve high COD removal efficiencies at low temperatures of 15°C and 20°C in bench-scale tests with non-fat dry milk substrate.

11.9 ANAEROBIC ATTACHED GROWTH PROCESSES

Similar to the aerobic process, a media is used in this process on which the bacteria is allowed to attach itself and grow. Anaerobic conditions are maintained in the reactor for the conversion of organic matter to methane and other gases. Examples include: anaerobic filter or fixed-film reactor, and anaerobic rotating biological contactor (RBC), among others.

11.9.1 Anaerobic filter

An anaerobic filter is a column or reactor packed with highly porous material/medium. The wastewater usually passes through the reactor with vertical flow, either upflow or downflow, as illustrated in Figure 11.6. The microorganisms in the reactor attach to the porous inert medium or become

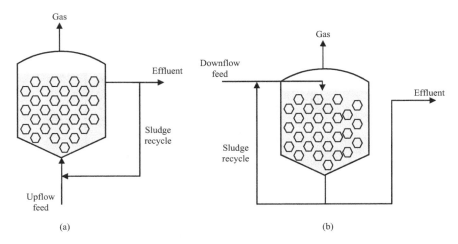

Figure 11.6 (a) Anaerobic upflow filter and (b) anaerobic downflow filter.

entrapped. The effluent gas flows upward through the support media and the gas produced is collected at the top. Anaerobic filters are also known as *fixed-film reactors* or *packed bed reactors*.

The first anaerobic filters, constructed by Young and McCarty (1969), were used to treat wastes of intermediate strength ranging from 6000 to 15,000 mg/L of COD, synthetic protein, carbohydrate, and volatile acid wastes at 25°C. The filters consisted of upflow reactors filled with small stones. The first full-scale anaerobic filter was described by Taylor and Burm (1972). The filters were operated in series to treat wheat starch wastes. The system accomplished up to 70 % COD reduction. After a shutdown period of 26 days, the filter was able to recover to maximum efficiency within 24 hours.

Anaerobic filters are capable of treating a wide variety of wastewaters at a high loading rate with a high rate of methane production. An anaerobic filter can switch from the treatment of one wastewater to another without adverse effects, and can operate at temperatures as low as 10°C (van den Berg, 1981). The anaerobic filter can effectively treat organic wastes in the presence of some toxic substances that are below a threshold level (Parkin and Speece, 1982). Effluent recycling can aid to reduce the toxic concentration and maintain a uniform pH through the filter. A very high SRT in excess of 100 d can be achieved. The effects of temperature and detention time can be minimized.

The disadvantages of the process include the inability to handle wastewaters with high suspended solids concentrations. The cost of packing or filter material is high. Clogging of the media can cause problems. Seeding is necessary for start-up, which may take from a few weeks to a few months to develop sufficient biomass for complete methanogenesis.

11.9.2 Anaerobic expanded bed reactor

The anaerobic expanded bed reactor (AEBR) is a variation of the upflow anaerobic filter. The packing material is usually silica sand with a diameter of 0.2–0.5 mm. The upflow velocity is designed to achieve about 20% expansion of media (Metcalf and Eddy et al., 2013). The AEBR process has been used mostly for the treatment of domestic wastewaters.

11.10 HYBRID PROCESSES

Hybrid processes are a combination of suspended and attached growth processes. Examples of hybrid processes are described below.

11.10.1 Anaerobic fluidized bed reactor

The AFBR consists of a reactor filled with a packing medium such as sand and operated at high upflow velocities to keep the media in suspension. Upflow velocities of 20 m/h may be used to provide 100% expansion of the packed bed (Metcalf and Eddy et al., 2013). The effluent is recycled to maintain a high upflow velocity. Reactor depth ranges from 4 to 6 m. The flow diagram of the process is similar to an upflow filter with effluent recycle. The AFBR is suitable for the treatment of wastewaters with mainly soluble COD and very low solids concentration. It can handle organic loading rates of 10–20 kg COD/m^3.d or higher, with greater than 90% removal, depending on the wastewater characteristics. Reactor biomass concentrations of 15–20 g/L can be established (Malina and Pohland, 1992).

Various types of packing materials can be used. These include sand, diatomaceous earth, resins, and activated carbon. Activated carbon is generally more expensive. But it is more efficient for the treatment of industrial and hazardous wastewaters. Granular activated carbon (GAC) can achieve a high biomass concentration due to its porous structure and can reduce toxicity and shock loads by adsorption.

11.10.2 Anaerobic membrane bioreactor

The anaerobic membrane bioreactor (AnMBR) system consists of an anaerobic bioreactor coupled with a membrane separation unit. The effluent from the bioreactor passes through the membrane unit, where the solids–liquid separation takes place. The liquid effluent or permeate is discharged, while the solids (concentrate) are recycled back to the reactor. The membrane bioreactor process is illustrated in Figure 11.7.

The major advantages of the AnMBR process are (i) high-quality effluent due to efficient solids capture, (ii) higher biomass concentration in the

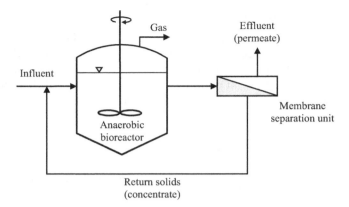

Figure 11.7 Anaerobic membrane bioreactor process.

reactor which results in higher COD loadings and smaller reactor size, and (iii) higher SRTs are achieved in the reactor due to solids recycle. Disadvantages of the process include the high cost of membranes and the potential for membrane fouling. A lot of recent research has focused on the fabrication of membranes and the application of coatings that can reduce fouling problems. A more detailed discussion on membrane bioreactors is provided in Chapter 13.

PROBLEMS

11.1 Compare aerobic and anaerobic processes in terms of growth rate, sludge production, energy use, and energy generation.

11.2 Name the scientist who discovered anaerobic life.

11.3 List three advantages and three disadvantages of the anaerobic treatment process.

11.4 Briefly describe the four steps of anaerobic biotransformation. What groups of bacteria are involved in each step?

11.5 List the factors that are important for anaerobic treatment. What temperature and pH ranges are best for the process?

11.6 Show with the help of a chemical equation and calculations that 1 kg COD can produce 0.35 m^3 CH_4 gas at STP conditions.

11.7 Design a UASB reactor to treat wastewater at 30°C from a food processing plant. The wastewater flow rate is 500 m^3/d with a soluble COD concentration of 6000 mg/L. The design parameters are given below.

Reactor effectiveness factor $(E) = 0.85$
Upflow velocity = 1.2 m/h

Organic loading rate = 12 kg sCOD/m^3.d
Y = 0.08 kg VSS/kg COD
k_d = 0.03 d–1
μ_{max} = 0.25 d–1
K_s = 360 mg/L
Height for gas collection = 3 m

Using the given information, determine the following:

    i. The reactor area and diameter
    ii. The reactor liquid volume
    iii. Liquid depth and total height of the reactor
    iv. The average SRT (assuming 97% degradation of sCOD)

11.8   Determine the methane gas production rate (m^3/d) for the reactor from Problem 11.7. Assume COD equivalent of VSS equals 1.42 kg COD/kg VSS.

11.9   An anaerobic reactor is operated at 35°C with an SRT of 30 d. Suddenly, the methane gas production rate is decreased significantly. Explain the possible reason(s) that can cause the reduction of methane.

11.10 An anaerobic sequencing batch reactor (ASBR) is operated at an HRT of 1.5 d at 35°C. Calculate the volume to be wasted per cycle, for the following operational parameters:

Total liquid volume = 15 L
Feed phase = 30 min
React phase = 240 min
Settle phase = 60 min
Decant phase = 30 min

11.11 Briefly differentiate between anaerobic suspended and attached growth processes. Give two examples of each process.

11.12 What are the advantages and disadvantages of anaerobic fluidized bed reactors (AFBR) and anaerobic membrane bioreactors?

REFERENCES

Archer, D.B. and Harris, J.E. 1986. Methanogenic Bacteria and Methane Production in Various Habitats. In *Anaerobic Bacteria in Habitats Other Than Man.*, Edited by Barness, E. M., and Mead, G. C. Blackwell Scientific Publication, Hoboken, NJ, USA.

AWWA, WEF, and APHA. 2017. *Standard Methods for the Examination of Water and Wastewater.* 23rd edn. Edited by Baird, R. B., Eaton, A. D., and Rice, E. W. American Water Works Association, Denver, CO, USA.

Babbitt, H. E. and Schlenz, H.E. 1929. Results of Tests on Sewage Treatment. *University of Illinois Engineering Experiment Station Bulletin.*, Urbana, Illinois., 198.

Balch, W.E., Fox, G.E., Magrum, J., Woese, C.R. and Wolfe, R.S. 1979. Methanogens: Reevaluation of a Unique Biological Group. *Microbiology Reviews*, 43:260-296.

van den Berg, L. 1981. Effect of Type of Waste on Performance of Anaerobic Fixed-Film and Upflow Sludge Bed Reactors. *Proceedings of the 36th Industrial Waste Conference*, Purdue University, Indiana, USA.

Dague, R.R., McKinney, R.E. and Pfeffer, J.T. 1966. Anaerobic Activated Sludge. *Journal Water Pollution Control Federation*, 38 (2):220–226.

Daniels, L. 1984. Biological Methanogenesis: Physiological and Practical Aspects. *Trends in Biotechnology*, 2 (4):91–98.

Davis, M. 2011. *Water and Wastewater Engineering: Design Principles and Practice*. McGraw-Hill, Inc., New York, NY, USA.

Droste, R.L. 1997. *Theory and Practice of Water and Wastewater Treatment*. John Wiley & Sons, Inc. NJ, USA.

Dubach, A.C. and Bachofen, R. 1985. Methanogens: A Short Taxonomic Overview. *Experientia*, 41:441.

Fair, G.M. and Moore, E.W. 1932. Heat and Energy Relation in the Digestion of Sewage Solids, III. Effect of Temperature of Incubate Upon the Course of Digestion. *Sewage Works Journal*, 4:589.

Fuller, G.W. 1912. *Sewage Disposal*, McGraw-Hill, Inc., New York, NY, USA.

Gomec, C.Y., Kim, M., Ahn, Y., Speece, R.E. 2002. The Role of pH in Mesophilic Anaerobic Sludge Solubilization. *Journal of Environmental Science and Health.*, Part A, 37 (10):1871–1878.

Grady, C.P.L. Jr. and Lim, H.C. 1980. *Biological Wastewater Treatment*, Marcel Dekker Inc., New York, NY, USA.

Gujer, W. and Zehnder, A. J. B. 1983. Conversion Process in Anaerobic Digestion. *Water Science and Technology*, 15:127–167.

Heukelekian, H. 1933. Digestion of Solids Between the Thermophilic and Non-Thermophilic Range, *Sewage Works Journal*, 5 (5):757–762.

Heukelekian, H. 1958. Basic Principles of Sludge Digestion. *Biological Treatment of Sewage and Industrial Wastes.*, vol. II, Reinhold Publishing Corp., New York, NY, USA.

Hulshoff Pol, L.W., de Zeeuw, W.J., Velzeboer, C.T.M. and Lettinga, G. 1983. Granulation is UASB-Reactors. *Proceedings of the 66th Annual Conference of Water Environment Federation*, Anaheim, CA, USA.

Lawrence, A.W. and McCarty, P.L. (1969) "Kinetics of Methane Fermentation in Anaerobic Treatment." *Journal of Water Pollution Control Federation*, 41:R1–R17.

Lettinga, G., van Velsen, F.M., Hobma, S.W., de Zeeuw, W. and Klapwijk, A. 1980. Use of Upflow Sludge Blanket (USB) Reactor Concept for Biological Wastewater Treatment, Especially for Anaerobic Treatment. *Biotechnology and Bioengineering*, 22:699–734.

Macario, A.J.L. and Conway de Macario, E. 1988. Quantitative Immunological Analysis of the Methanogenic Flora of Digesters Reveals a Considerable Diversity. *Applied Environmental Microbiology*, 54:79.

Madigan, M., Bender, K. S., Buckley, D. H., Sattley, W. M., and Stahl, D. A. 2021. *Brock Biology of Microorganisms.* 16th edn., Pearson, New York, NY, USA.

Malina, J.F., and Pohland, F.G. 1992. Design of Anaerobic Processes for the Treatment of Industrial and Municipal Wastes, *Water Quality management Library*, vol. 7, Technomic Publishing Co., Lancastar, PA, USA.

McCarty, P.L. 1964a. Anaerobic Waste Treatment Fundamentals – Part One: Chemistry and Microbiology, *Public Works*, 95 (9):107–112.

McCarty, P.L. 1964b. Anaerobic Waste Treatment Fundamentals – Part Two: Environmental Requirements and Control. *Public Works*, 95 (9):123–126.

McCarty, P.L. 1964c. Anaerobic Waste Treatment Fundamentals – Part Three: Toxic Metaerials and Their Control. *Public Works*, 95 (9):91–94.

McCarty, P.L. 1981. One Hundred Years of Anaerobic Treatment. *Proceedings of Second International Conference on Anaerobic Digestion*, Travemunde, Germany.

McCarty, P.L., Jeris, J.S. and Murdoch, W. 1963. Individual Volatile Acids in Anaerobic Treatment. *Journal Water Pollution Control Federation*, 35:1501.

McCarty, P.L. and Smith, D.P. 1986. Anaerobic Wastewater Treatment. *Environmental Science and Technology*, 20:1200–1206.

Metcalf and Eddy, Tchobanoglous, G., Stensel, H., Tcuchihashi, R., and Burton, F. 2013. *Wastewater Engineering: Treatment and Resource Recovery*. 5th edn. McGraw-Hill, Inc., New York, NY, USA.

Novaes, R.F.V. 1986. Microbiology of Anaerobic Digestion. *Water Science and Technology*, 18 (12):1–14.

Owen, W.F., Stuckey, D.C., Healy, J.B., Young, L.Y., and McCarty, P.L. 1979. Bioassay for Monitoring Biochemical Methane Potential and Anaerobic Toxicity. *Water Research*, 13 (6):485–492.

Parkin, G.F. and Owen, W.F. 1986. Fundamentals of Anaerobic Digestion of Wastewater Sludges. *Journal of Environmental Engineering, ASCE*, 112 (5):867–920.

Parkin, G.F. and Speece, R.E. 1982. Attached Versus Suspended Growth Anaerobic Reactor: Response to Toxic Substances. *Proceedings of IAWPR Specialized Seminar on Anaerobic Treatment*, Copenhagen, Denmark.

Riffat, R. 1994. *Fundamental Studies of Anaerobic Biosorption in Wastewater Treatment*. Ph.D. Thesis, Iowa State University, Ames, Iowa.

Rudolfs, W. 1927. Effect of Temperature on Sewage Sludge Digestion. *Industrial and Engineering Chemistry*, 19 (1):241–243.

Sawyer, C.N., Howard, F.S. and Pershe, E.R. 1954. Scientific Basis for Liming of Digesters. *Sewage and Industrial Wastes*, 26:935.

Smith, M.R. and Mah, R.A. 1980. Acetate as Sole Carbon and Energy Source for Growth of Methanosarcina Strain 227. *Applied and Environmental Microbiology*, 39 (5):993–999.

Speece R.E. 1983. Anaerobic Biotechnology for Industrial Wastewater. *Environmental Science & Technology*, 17 (9):416A.

Speece, R.E. 2008. *Anaerobic Biotechnology and Odor/Corrosion Control for Municipalities and Industries*. Archae Press, Nashville, TN, USA.

Sung, S. and Dague, R.R. 1992. Fundamental Principles of the Anaerobic Sequencing Batch Reactor Process. *Proceedings of the 47th Purdue Industrial Waste Conference*, Lewis Publishers, Inc., Chelsea, Michigan. no. 423.

Taylor, D.W. and Burm, R.J. 1972. Full-Scale Anaerobic Filter Treatment of Wheat Starch Plant Waste. *American Institute of Chemical Engineers Symposium Series*, 69 (129):30.

Thauer, R.K., Jungermann, K. and Decker, K. 1977. Energy Conservation in Chemotrophic Anaerobic Bacteria. *Bacteriology Reviews*, 41(100).

Vallinga, S.H.J., Hack, P.J.F.M., and van der Vlught, A.J. 1986. New Type High Rate Anaerobic Reactor. *Proceedings of Anaerobic Treatment: A Grown-Up Technology*, 547–562.

Wose, C.R. 1987. Bacterial Evolution. *Microbiology Reviews*, 51:221.

Wose, C.R., Magrum, L.J., and Fox, G.E. 1978. Archaebacteria. *Journal of Molecular Evolution*, 11:245.

Young, J.C. and McCarty, P.L. 1969. The Anaerobic Filter for Waste Treatment. *Journal of Water Pollution Control Federation*, 41:R160.

Zehnder, A.J.B., Huser, B.A., Brock, T.D. and Wuhrmann, K. 1980. Characterization of an Acetate Decarboxylating Non-Hydrogen-Oxidizing Methane Bacteria. *Archives of Microbiology*, 124:1.

Chapter 12

Solids processing and disposal

12.1 INTRODUCTION

Solids that are generated from primary, secondary, and advanced waste-water treatment processes are called *sludge*. Sludge is usually in the form of liquid or semisolid liquid, which typically contains from 0.25% to 12% solids by weight. It is classified into the following categories: Primary sludge, secondary sludge, and sludge produced in the advanced treatment process. Primary sludge consists of settable solids carried in the raw wastewater; secondary sludge consists of biological solids as well as additional settleable solids. Sludge produced in advanced wastewater may include viruses, heavy metals, phosphorous, or nitrogen.

In general, municipal sludge consists of primary and waste activated sludge and must be treated to some extent before disposal. They contain various organics and inorganics, e.g. biomass produced by the biological conversion of organics, oil and grease, nutrients (nitrogen and phosphorus), heavy metals, synthetic organic compounds, and pathogens. Disposal of sludge represents up to 50% of the operating costs of a wastewater treatment plant (Appels et al., 2008).

Treated wastewater sludge, commonly referred to as *biosolids*, is the material produced as the ultimate byproduct of the processes used to treat municipal wastewater in wastewater treatment facilities. Biosolids are nutrient-rich organic materials. They can be used for soil enrichment and can supplement commercial fertilizers. Biosolids must meet strict regulations and quality standards before being applied to land. Approximately 8–9 million tons of biosolids are produced each year by municipal wastewater treatment facilities in the US (Hong et al., 2006). In 2003, about 60% of the biosolids were reused. The beneficial reuse of biosolids is expected to increase in the near future.

The first step in sludge handling is usually thickening. The purpose of thickening is to reduce the volume of sludge before further treatment. The main thickening methods used are Gravity thickening, floatation thickening, centrifugation, gravity-belt thickening, and rotary-drum thickening (Metcalf and Eddy et al., 2013).

DOI: 10.1201/9781003134374-12

The second step is sludge stabilization. The purpose of sludge stabilization is to reduce the organic matter content of the sludge, reduce pathogens, and eliminate offensive odors. The main sludge stabilization processes are alkaline stabilization, anaerobic digestion, aerobic digestion, and composting.

After stabilization, treated sludge or biosolids is usually dewatered, in order to reduce the volume further. Most widely used dewatering processes are centrifuge, belt-filter press, and sludge drying beds (Metcalf and Eddy et al., 2013).

Final disposal methods for biosolids are (i) landfilling, (ii) land application which is a disposal method with beneficial use, and (iii) incineration which is a total conversion of organic solids to oxidized end products of carbon dioxide, water, and ash. Incineration is usually applied to dewatered and untreated sludge. Figure 12.1 illustrates the various sludge treatment and disposal options.

Land application is the major municipal sludge and biosolids disposal method. Agricultural land application is a beneficial use of biosolids. In order to produce biosolids with a quality suitable for meeting the requirements for agricultural land application, both stabilization of sludge and pathogen reduction are of importance.

In this chapter, the various processes used for sludge thickening, stabilization, and disposal will be described in detail. Methods used for sludge treatment, such as anaerobic digestion processes will be emphasized. Energy generation from anaerobic digestion in the form of methane gas will be discussed.

12.2 CHARACTERISTICS OF MUNICIPAL SLUDGE

Considering conventional wastewater treatment, municipal sludge is generally comprised of primary sludge from primary sedimentation tanks, and secondary sludge from the secondary sedimentation tanks following biological treatment of wastewater. Primary sludge is composed of organic and inorganic particles coming from raw wastewater. It is influenced by the wastewater source and primary sedimentation tank operation. The secondary sludge which is also called waste activated sludge includes the excess microorganism cells from the biological treatment process. Typical properties of primary and secondary sludge are given in Table 12.1.

12.3 SLUDGE QUANTIFICATION

The mass and volume of sludge are important quantities that are used in the design of processes for sludge treatment and disposal. The quantity of sludge produced depends on the characteristics of the wastewater, the specific processes used for treatment, and their efficiencies.

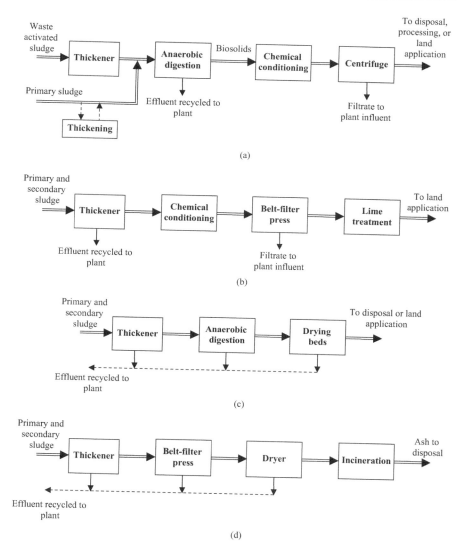

Figure 12.1 Flow diagrams for sludge treatment and disposal.

Source: Adapted from Metcalf and Eddy et al., 2013

Primary clarifiers typically remove 40–60% of the total influent solids. The mass of primary sludge can be calculated as

$$M_p = QX\left(E_p / 100\right) \tag{12.1}$$

Table 12.1 Typical properties of primary and secondary activated sludge

Parameter	Primary sludge	Secondary activated sludge
pH	5.5–8.0	6.6–8.0
Alkalinity (mg/L as $CaCO_3$)	600–1500	550–1200
Total solids % (TS)	4–9	0.6–1.2
Volatile solids (% of TS)	65–80	60–85
Protein (% of TS)	18–30	30–40
Fats and grease (% of TS)		
Ether soluble	5–30	–
Cellulose (% of TS)	8–16	–
Nitrogen (N, % of TS)	1.4–4.2	2.5–5.0
Phosphorus (P_2O_5, % of TS)	0.6–2.9	3–10
Organic acids (mg/L as HAc)	250–1800	1000
Energy content (kJ/kg TS)	24,000–28,000	18,000–23,000

Source: Adapted from U.S. EPA (1979) and Metcalf and Eddy et al. (2013).

where
  M_p = mass of primary sludge, kg/d
  Q = wastewater flow rate, m³/d
  X = total suspended solids in influent, kg/m³
  E_p = solids removal efficiency of primary clarifier, %

Secondary clarifiers following suspended growth processes are used for the thickening of sludge and clarification of effluent. The amount of sludge generated depends on the amount of new cells that are produced. This depends on the F/M ratio of the reactor, as well as organic loading rates and other factors. The mass of secondary sludge can be calculated as (Peavy et al., 1985)

$$M_s = Q(S_o - S)Y' \tag{12.2}$$

where
  M_s = mass of secondary sludge, kg/d
  Q = wastewater flow rate, m³/d
  S_o = influent BOD_5 concentration to secondary reactor, kg/m³
  S = effluent BOD_5 concentration from secondary reactor, kg/m³
  Y' = biomass conversion factor
    = fraction of BOD_5 converted to biomass, kg/kg

The value of Y' depends primarily on the F/M ratio of the biological reactor. Y' can be determined from Figure 12.2.
  The total sludge produced is given by

$$M_T = M_p + M_s \tag{12.3}$$

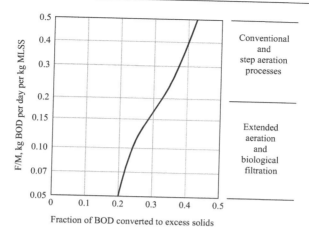

Figure 12.2 Typical variation of excess sludge production with F/M ratio. Actual quantities will vary from plant to plant.

Source: Adapted from Peavy et al., 1985

where
    M_T = Mass of total sludge produced, kg/d
    M_p and M_s are as defined previously.

Primary sludge is granular in nature and concentrated. Secondary sludge from activated sludge processes have a low solids content, and are light and flocculent in character. Sometimes primary and secondary sludge are mixed together prior to thickening to facilitate further treatment. Or, they may be thickened separately and then sent to digesters.

Solids content is usually determined on a mass/volume basis and expressed as a percent. For example, a 5% sludge contains 95% water by weight. The specific gravity of sludge is usually around 1.02–1.05. When the sludge contains less than 10% solids, the specific gravity of sludge can be assumed to be equal to that of water, or 1.00, without introducing significant error (Peavy et al., 1985). Each percent solids then corresponds to a solids concentration of 10,000 mg/L, or 1% solids = 10,000 mg/L.

The volume of sludge can be calculated using the following equation:

$$V = \frac{M}{SG_s\, \rho_w\, P_s} \tag{12.4}$$

where
    V = volume of sludge, m³/d
    M = mass of sludge, kg/d
    SG_s = specific gravity of sludge

P_s = percent solids expressed as a decimal

ρ_w = density of water = 1000 kg/m³

For a given solids content, the following relationship can be used for approximate calculations (Metcalf and Eddy et al., 2013):

$$\frac{V_1}{V_2} = \frac{P_2}{P_1} \tag{12.5}$$

where

V_1, V_2 = volumes of sludge

P_1, P_2 = percent of solids in V_1 and V_2, respectively

The calculation of sludge volumes is illustrated in Example 12.1.

Example 12.1: A conventional wastewater treatment plant treats 15,000 m³/d of municipal wastewater with a BOD₅ of 220 mg/L and suspended solids of 200 mg/L. The treatment consists of primary followed by secondary treatment. The effluent BOD₅ from the final clarifier is 20 mg/L. The following data are provided:

Primary clarifier: removal efficiency; SS = 55%, BOD₅ = 30%

water content = 95%, specific gravity = 1.04

Aeration tank: F/M = 0.33
Secondary clarifier: 2% solids in waste activated sludge, specific gravity = 1.02

a. Calculate the mass and volume of primary sludge.

b. Calculate the mass and volume of secondary sludge.

c. Calculate the total mass of primary and secondary sludge.

SOLUTION

Step 1. Calculate the mass of primary sludge.

$$SS = 200 \, mg / L = 0.2 \, kg / m^3$$

Calculate mass of primary sludge using Equation (12.1)

$$M_p = Q X \left(E_p / 100 \right)$$

$$= 15,000\,\text{m}^3/\text{d} \times 0.2\,\text{kg}/\text{m}^3 \times 0.55$$

$$= 1650\,\text{kg}/\text{d}$$

Step 2. Calculate volume of primary sludge.
Solids content = 100 − water content = 100 − 95% = 5%
  Use Equation (12.4) to calculate sludge volume

$$V_\text{p} = \frac{M}{SG_\text{s}\,\rho_\text{w}\,P_\text{s}}$$

$$= \frac{1650\,\text{kg/d}}{1.04 \times 1000\,\dfrac{\text{kg}}{\text{m}^3} \times 0.05}$$

$$= 31.73\,\text{m}^3/\text{d}$$

Step 3. Calculate mass of secondary sludge.
Primary clarifier removes 30% BOD_5
  Therefore, BOD_5 going to aeration tank = 220 mg/L (1 − 0.30)

$$\text{Or,}\, S_\text{o} = 154\,\text{mg}/\text{L} = 0.154\,\text{kg}/\text{m}^3$$

$$\text{Given,}\, S = 20\,\text{mg}/\text{L} = 0.02\,\text{kg}/\text{m}^3$$

For F/M of 0.33, biomass conversion factor $Y' = 0.38$ from Figure 12.2
Use Equation (12.2) to calculate mass of secondary sludge

$$M_\text{s} = Q(S_\text{o} - S)Y'$$

$$= 15,000\,\text{m}^3/\text{d}(0.154 - 0.02)\,\text{kg/m}^3 \times 0.38$$

$$= 763.80\,\text{kg/d}$$

Step 4. Calculate the volume of secondary sludge using Equation (12.4)

$$V_\text{s} = \frac{M}{SG_\text{s}\,\rho_\text{w}\,P_\text{s}}$$

$$= \frac{763.80\,\text{kg}/\text{d}}{1.02 \times 1000\,\dfrac{\text{kg}}{\text{m}^3} \times 0.02}$$

$$= 37.44\,\text{m}^3/\text{d}$$

Step 5. Calculate the total mass of primary and secondary sludge using Equation (12.3)

$$M_T = M_p + M_s$$

$$= 1650 + 763.8\,\text{kg/d}$$

$$= 2413.80\,\text{kg/d}$$

12.4 SLUDGE THICKENING

The objective of sludge thickening is to reduce the volume of sludge and increase the solids content. The sludge generated from primary, secondary, and tertiary treatment processes can have a wide range of solids concentrations and characteristics. Reducing the water content is advantageous for subsequent treatment processes. Volume reduction reduces pipe size, pumping cost, and tank sizes for further treatment.

All wastewater treatment plants use some method of sludge thickening. In small plants treating less than 4000 m³/d (less than 1 Mgal/d), thickening is accomplished in the primary clarifier, and/or sludge digestion units (Metcalf and Eddy et al., 2013). In larger plants, separate thickening processes are used. Examples of these are gravity thickener, dissolved air flotation (DAF), centrifugation, gravity-belt thickener, rotary-drum thickener, among others. The thickened sludge is pumped to a subsequent sludge stabilization process, while the liquid effluent is usually recycled to primary treatment. The thickeners have to be designed to meet peak demands and prevent septicity and odor problems during the thickening process. A number of the major sludge thickening processes are described in the following sections.

12.4.1 Gravity thickener

Gravity thickening is used for primary sludge, or a combination of primary and waste activated sludge. The design is similar to a secondary clarifier. The thickening function is the major design parameter and tanks deeper than secondary clarifiers are used. The surface area required for thickening

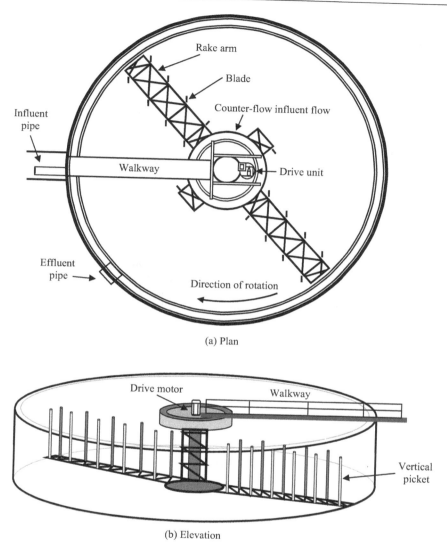

(a) Plan

(b) Elevation

Figure 12.3 Diagram of a typical gravity thickener: (a) plan and (b) elevation.

may be determined using the *solids flux analysis* or the *state point analysis* methods. A typical circular gravity thickener is illustrated in Figure 12.3. Dilute sludge is fed to a center feed well where it is allowed to settle. The sludge scraper mechanism can be in the form of vertical pickets or deep trusses. The scraper stirs the sludge gently, which helps to release the water trapped in the sludge and promotes compaction. The thickened sludge is pumped to digesters or dewatering processes, and storage space has to be provided for the sludge. The liquid effluent is recycled to the head works of the plant.

A sludge blanket is maintained at the bottom of the thickener to help in concentrating the sludge. Blanket depths can range from 0.5 to 2.5 m (2–8 ft), with shallower depths in warmer months. An operating variable is the *sludge volume ratio*, which is the volume of sludge blanket in the thickener divided by the volume of thickened sludge removed daily. The sludge volume ratio can range from 0.5 to 20 d. The solids loading rate ranges from 100 to 150 kg/m².d, with maximum hydraulic overflow rates of 15.5–21 m³/m².d for primary sludge. For a combined primary and waste activated sludge thickener, the solids loading rate ranges from 25 to 80 kg/m².d, with maximum hydraulic overflow rates of 6–12 m³/m².d (Metcalf and Eddy et al., 2013). High hydraulic loading can result in excess solids carryover in the effluent, while low hydraulic loadings can cause septic conditions and sludge floatation.

> **Example 12.2:** Consider the wastewater treatment process described in Example 12.1. The primary sludge is thickened in a gravity thickener. The thickener has a diameter of 4.5 m with a side water depth of 5 m. A sludge blanket of 1.2 m is maintained at the bottom. Primary sludge is applied at 31.73 m³/d with 5% solids to the thickener. An additional 270 m³/d of treated wastewater is applied to the thickener, to increase the overflow rate and improve odor control and thickening. The thickened sludge is withdrawn at 17 m³/d with 7% solids content. Calculate the following:
>
> a. Hydraulic overflow rate.
> b. Solids loading rate.
> c. Sludge volume ratio.
> d. Percent solids captured in thickener.
>
> **SOLUTION**
> Step 1. Calculate the surface area of the thickener.
>
> $$A_s = \frac{\pi}{4}(4.5)^2 = 15.90\,\text{m}^2$$
>
> Step 2. Calculate the hydraulic overflow rate.
>
> $$Q_{in} = (31.73 + 270)\,\text{m}^3/\text{d} = 301.73\,\text{m}^3/\text{d}$$
>
> $$Q_{thickened} = 17\,\text{m}^3/\text{d}$$

$$Q_{\text{effluent}} = Q_{\text{in}} - Q_{\text{thickened}} = 301.73 - 17 = 284.73\,\text{m}^3\,/\,\text{d}$$

$$\text{Overflow rate} = \frac{Q_{\text{effluent}}}{A_s} = \frac{284.73\,\text{m}^3/\text{d}}{15.90\,\text{m}^2} = \textbf{17.90}\,\textbf{m}^3\textbf{/m}^2\textbf{.d}$$

Step 3. Calculate the solids loading rate.

Primary sludge has 5% solids $= 50,000\,\text{mg}\,/\,\text{L} = 50\,\text{kg}\,/\,\text{m}^3$

$$\text{Mass of primary sludge solids} = 31.73\,\text{m}^3/\text{d} \times 50\,\text{kg/m}^3$$
$$= 1586.50\,\text{kg/d}$$

$$\text{Solids loading rate} = \frac{1586.50\,\text{kg}\,/\,\text{d}}{15.90\,\text{m}^2} = \textbf{99.78}\,\textbf{kg}\,/\,\textbf{m}^2\textbf{.d}$$

Step 4. Calculate the sludge volume ratio.

$$\text{Volume of sludge blanket in thickener} = 1.2\,\text{m} \times 15.90\,\text{m}^2$$
$$= 19.08\,\text{m}^3$$

$$\text{Sludge volume ratio} = \frac{\text{volume of sludge blanket}}{\text{rate of thickened sludge withdrawal}}$$

$$= \frac{19.08\,\text{m}^3}{17\,\text{m}^3/\text{d}} = \textbf{1.12}\,\textbf{d}$$

Step 5. Calculate the solids capture.

Mass of solids coming in $= 1586.50\,\text{kg}\,/\,\text{d}$

$$\text{Mass of solids in thickened sludge} = 17\,\text{m}^3/\text{d} \times 0.07 \times 1000\,\text{kg/m}^3$$
$$= 1190\,\text{kg/d}$$

Or, solids captured in thickener $= 1190\,\text{kg}\,/\,\text{d}$

$$\text{Solids capture} = \frac{1190\,\text{kg/d}}{1586.50\,\text{kg/d}} \times 100\% = \textbf{75}\%$$

12.4.2 Dissolved air flotation

DAF is used for thickening waste sludge from suspended growth processes, such as waste activated sludge. The process is especially suitable for thickening the light, flocculent sludge that is generated from the activated sludge process. It can also be used for the thickening of combined primary and waste activated sludge.

In this process, water or secondary effluent is aerated under a pressure of about 400 kPa. The super-saturated liquid is released at the bottom of the tank through which sludge is passed at atmospheric pressure. Fine air bubbles are released into the tank. The air bubbles attach themselves to the sludge particles, floating them up to the tank surface. The floating sludge is removed from the top with a skimmer, while the liquid is removed and recycled to the plant. Polymers can be added for sludge conditioning. The DAF system is illustrated in Figure 12.4.

Important factors that affect the design of DAF systems include air-to-solids ratio, hydraulic loading, polymer addition, and solids loading rate, among others (WEF, 1998). For waste activated sludge without polymer addition, solids loading rates ranging from 2 to 5 kg/m².h can produce thickened sludge with 3–5% solids. With polymer addition, the loading rate can be increased by 50–100%. Operational difficulties can arise when the solids loading rate exceeds 10 kg/m².h.

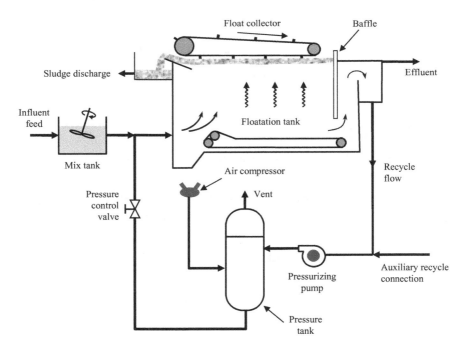

Figure 12.4 Typical dissolved air flotation system.

Example 12.3: Consider the wastewater treatment process described in Example 12.1. The secondary waste activated sludge is thickened in a dissolved air flotation process, shown in Figure 12.5. If the DAF process thickens the solids to 3.5%, calculate the volume of sludge thickened per day. Assume that the process captures 95% of the solids.

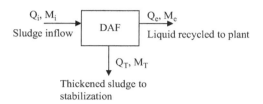

Figure 12.5 Dissolved air floatation process (for Example 12.3).

SOLUTION

$$\text{Mass}_{in} \text{ to DAF} = M_i = 763.80 \, \text{kg} \, / \, \text{d}$$

$$\text{Volume}_{in} \text{ to DAF} = Q_i = 37.44 \, \text{m}^3 \, / \, \text{d}$$

DAF process captures 95% solids.

$$\text{Therefore, } M_T = 0.95 \times M_i = 0.95 \times 763.80 \, \text{kg} \, / \, \text{d} = 725.61 \, \text{kg} \, / \, \text{d}$$

Assume, specific gravity of thickened sludge = specific gravity of water = 1.0
    Use Equation (12.4) to calculate the volume of thickened sludge

$$Q_T = \frac{M_T}{SG_s \, \rho_w \, P_s}$$

$$= \frac{725.61 \, \text{kg/d}}{1.0 \times 1000 \, \dfrac{\text{kg}}{\text{m}^3} \times 0.035}$$

$$= 20.73 \, \text{m}^3/\text{d}$$

12.4.3 Centrifugation

The process of centrifugation is used for both thickening and dewatering of sludge. The solid-bowl centrifuge is used mainly for the thickening of waste activated sludge. The basic principle involves the thickening of sludge by

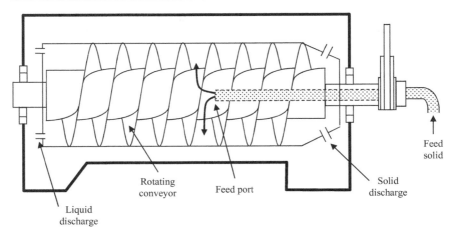

Figure 12.6 Diagram of a solid-bowl centrifuge.

the use of centrifugal forces. Thickened solids concentration of 4–6% can be achieved. Thickening can usually be achieved without polymer addition. Maintenance and power costs are high for the process.

The solid-bowl centrifuge consists of a long cylinder tapered at one end, which is mounted on a horizontal plane and rotates at a particular speed. Sludge flows into the cylinder and the solids concentrate on the periphery. An internal helical scroll rotates at a different speed, and moves the concentrated solids toward the tapered end, from where they are discharged. The liquid centrate is collected at the other end, and recycled to the plant. Figure 12.6 illustrates the solid-bowl centrifuge process.

12.5 SLUDGE STABILIZATION

Thickened sludge may be stabilized by various means at wastewater treatment plants. The commonly used methods for sludge stabilization are (1) alkaline stabilization, usually with lime, (2) anaerobic digestion, (3) aerobic digestion, and (4) composting. These are described in detail in the following sections. Not all plants practice sludge stabilization after thickening. Some plants dewater thickened sludge and then use lime stabilization prior to disposal. Other plants use anaerobic digestion to stabilize thickened sludge. This is followed by dewatering and final disposal. The selection of treatment methods depends on regulatory requirements for the final disposal of biosolids.

The objectives of sludge stabilization are the following (Metcalf and Eddy et al., 2013):

- Reduce pathogens.
- Eliminate offensive odors.
- Inhibit, reduce or eliminate the potential for putrefaction.

12.5.1 Alkaline stabilization

Quick lime or hydrated lime are added to the sludge for stabilization. Lime is added to raise the pH to 12 or higher. The alkaline environment inhibits pathogenic microorganisms, and significantly reduces or halts the bacterial decomposition of organic matter in the sludge. This prevents odor production and vector attraction. Health hazards are not a problem as long as the pH is maintained at this level. Lime can be used for (i) *pretreatment,* or (ii) *post-treatment* of sludge.

12.5.1.1 Chemical reactions

A variety of chemical reactions can occur depending on the characteristics and constituents of the sludge. Some of these are given below (Metcalf and Eddy et al., 2013; WEF, 1998):

$$\text{With calcium}: Ca^{2+} + 2HCO_3^- + CaO \rightarrow 2CaCO_3 + H_2O \tag{12.6}$$

$$\text{With phosphorus}: 2PO_4^{3-} + 6H^+ + 3CaO \rightarrow Ca_3(PO_4)_2 + 3H_2O \tag{12.7}$$

$$\text{With } CO_2 : CO_2 + CaO \rightarrow CaCO_3 \tag{12.8}$$

$$\text{With fats}: Fat + Ca(OH)_2 \rightarrow glycerol + fatty\ acids + CaCO_3 \tag{12.9}$$

$$\text{With acids}: RCOOH + CaO \rightarrow RCOOCaOH \tag{12.10}$$

Other reactions also take place with proteins, carbohydrates, and polymers. As the reactions progress, the pH can decrease due to the production of acids, etc. So excess lime is added. Ammonia is produced from amino acids, in addition to volatile off-gases which require collection and treatment for odor control.

When quicklime (CaO) is used, its reaction with water is exothermic, producing about 64 kJ/g.mol (2.75×10^4 BTU/lb.mol). The reaction of quicklime with CO_2 illustrated in Equation (12.8) is also exothermic, releasing approximately 180 kJ/g.mol (7.8×10^4 BTU/lb.mol) (US EPA, 1983).

12.5.1.2 Lime pretreatment

Pretreatment involves the application of lime to liquid sludge before dewatering. This requires more lime per unit weight of sludge. Lime pretreatment is used for direct application of sludge on land, or for conditioning and stabilization prior to dewatering. The design objective is to maintain the pH above 12 for about 2 h to ensure pathogen destruction, and to provide sufficient alkalinity to maintain the pH above 11 for several days. Excess lime is used to ensure the latter.

12.5.1.3 Lime post-treatment

In this process, hydrated lime or quicklime is applied to dewatered sludge. The advantages of post-treatment are that dry lime can be used, and there are no special requirements for dewatering. Scaling problems are eliminated. Adequate mixing is important to avoid the formation of pockets of putrescible material. The stabilized biosolids have a granular texture, can be stored for long periods, and are easily spread on land by a conventional manure spreader.

12.5.2 Anaerobic digestion

Anaerobic digestion is the traditional method for stabilization of municipal sludge, which results in volatile solids reduction, biogas production as an energy source, pathogen reduction, and reduced odor production. Anaerobic digestion processes are generally operated at mesophilic or thermophilic temperatures. Over the years, many process modifications have been developed. In addition to using single-stage digesters, two-phased digestion processes (staging the digestion process by adding a pretreatment step for acid production) or temperature-phased digestion processes (using mesophilic and thermophilic digestion) are also used. Thermal, mechanical, and chemical pretreatment options can be used as well before a mesophilic anaerobic digestion process.

The main advantages of anaerobic digestion over aerobic processes are reducing the energy need by eliminating the necessity of aeration, low nutrient requirements, energy production in the form of methane gas, and a lower amount of bacterial synthesis (Gomec et al., 2002). Energy in the form of methane can be recovered from the biological conversion of organic substrates. Sufficient digester gas can be produced to meet the energy requirements of digester heating and the operation of other plant processes. Another advantage is that anaerobic processes can handle higher volumetric organic loads compared to aerobic processes resulting in smaller reactor volumes. For these reasons, anaerobic digestion is the primary preferred method for the treatment of municipal sludge and high-strength organic wastes.

Anaerobic digestion has some disadvantages as well. Some of these disadvantages are longer start-up time required to develop the necessary amount of biomass due to the slow growth rate of methane-forming bacteria, the possible necessity of alkalinity and/or specific ion addition, and sensitivity to the adverse effect of lower temperatures on reaction rates.

The following are important factors that should be considered in the design of anaerobic digesters (Metcalf and Eddy et al., 2013):

- pH
- Temperature
- Alkalinity
- Presence of toxic compounds
- Bioavailability of nutrients
- Solids retention time
- Hydraulic retention time
- Volumetric loading of volatile solids.

Process description – Anaerobic digestion is composed of four major steps: (1) Hydrolysis, (2) acidogenesis, (3) acetogenesis, and (4) methanogenesis, as described previously in Chapter 11. In conventional single-stage anaerobic digestion of municipal sludge, all four steps take place in the same reactor. However, metabolic characteristics and growth rates of acid-producing- and methane-producing bacteria are different. Methanogens convert the end products (mainly H_2 and acetate) from previous steps to methane and CO_2, therefore maintaining a low partial pressure of H_2 and shifting the equilibrium of fermentation reactions toward the formation of more H_2 and acetate. When this balance is disturbed and methanogens do not utilize the H_2 formed by acidogens fast enough, accumulation of VFAs (volatile fatty acids) and a drop in pH is observed due to slow fermentation of propionate and butyrate resulting in digester failure (Metcalf and Eddy et al., 2013).

In order to maintain a favorable environment for this mixed culture of microorganisms, VFA production and utilization rates should be balanced. With short retention times, VFA production may exceed VFA utilization. The rate-limiting step is the conversion of VFAs by methanogenic organisms and not the fermentation of soluble substrates by acidogens. Digester upset can occur due to disturbance of the proper balance between acid and methane formers (Ghosh and Pohland, 1974).

pH is an important parameter affecting the enzymatic activity since a specific and narrow pH range is suitable for the activation of each enzyme. pH range in which the methanogens work efficiently is 6.7–7.4. A sharp pH drop below 6.3 indicates that the rate of organic acids production is faster than the rate of methane formation. On the other hand, a sharp pH increase above 7.8 can be due to a shift in NH_4^+ to NH_3, which is the toxic, union-ized form of ammonia (Gomec et al., 2002).

Buffering effect of ammonia released from amino acid fermentation can prevent the pH fall in anaerobic digesters. Primary sludge from domestic wastewater consists of high amounts of protein and detergent.

Alkalinity-generating cations like ammonium ions from protein degradation and sodium from soap degradation increase the alkalinity and pH.

The microbiology of the anaerobic treatment process is discussed in detail in Chapter 11. Factors affecting growth and toxicity are also provided. The discussion in the following sections will be focused on the design and operation of anaerobic digesters for the stabilization of sludge. These include (1) single-stage mesophilic digestion, (2) two-stage mesophilic digestion, (3) thermophilic anaerobic digestion, (4) temperature-phased anaerobic digestion (TPAD), (5) acid-gas phased digestion, (6) Enhanced Enzymic Hydrolysis™, and (7) Cambi™ process.

12.5.2.1 Single-stage mesophilic digestion

Single-stage mesophilic digesters can be standard-rate or high-rate digesters. Standard-rate digesters are used mainly by small plants processing less than 4000 m³/d, while high-rate digesters are used by larger wastewater treatment plants (Peavy et al., 1985). Digesters can have fixed covers or floating covers to adjust for variable volumes of sludge and gas production. Single-stage conventional floating cover digesters perform three functions: (1) Volatile solids destruction, (2) gravity thickening of digested sludge, and (3) storage of digested sludge (Hammer and Hammer, 2012). The optimum operating temperature is 35°C, with a range of 30–38°C.

Figure 12.7 illustrates a single-stage standard-rate mesophilic anaerobic digester (MAD). The sludge is fed continuously or at regular intervals to the

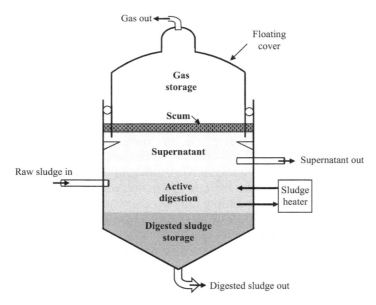

Figure 12.7 Diagram of single-stage standard-rate anaerobic digester.

digester. The temperature is maintained at 35°C by passing the sludge through a separate sludge heater. The sludge is mixed to some extent by pumping action to and from the sludge heater, in the zone of active digestion. A scum layer forms on top, with the supernatant liquid separating out from the solids. The supernatant is withdrawn and recycled to the plant. The total solids are reduced by 45–50%, and released as gas. The digested sludge is withdrawn from the bottom and transported to the dewatering processes. Solids concentration of digested sludge is 4–6%. The produced gas is collected, which consists of about 60–70% methane, 25–35% carbon dioxide, and trace amounts of other gases. The gas can be used for heating purposes.

High-rate digesters are completely mixed, and there is no separation of solids from the liquid. Mixing may be conducted by gas recirculation or draft tube mixers. The entire contents of the digester are transported to dewatering processes.

12.5.2.1.1 Design of digester

Anaerobic digesters can be designed based on the principles outlined in Chapter 8 for suspended growth processes. A number of empirical methods have also been used. These include methods based on volumetric loading rate of solids, solids retention time, volatile solids destruction, observed volume reduction, and loading factors based on population (Metcalf and Eddy et al., 2013).

The solids loading rate can range from 1.6 to 4.8 kg VSS/m^3.d for completely mixed high rate anaerobic digesters with a solids retention time (SRT) of 15–20 d (US EPA, 1979). For conventional digesters, the solids loading rate can range from 0.32 to 1.0 kg VSS/m^3.d with SRT values of 30–90 d. Volatile solids destruction of 55–65% can be achieved at SRT values of 15–30 d (WEF, 1998). In practice, the design SRT ranges from 10 to 20 d. McCarty (1964) observed that a minimum SRT of 4 d was required at 35°C to prevent washout of the methanogens. He suggested a design SRT of 10 d. Grady et al. (1999) proposed a lower SRT limit of 10 d to ensure an adequate factor of safety against washout. They observed that incremental changes in volatile solids destruction were relatively small for SRT values above 15 d at 35°C.

When population equivalent load is used to design digesters, typical values used are 0.17 m^3 (6 ft^3) tank volume per capita for digestion of primary and waste-activated sludge, and 0.11 m^3 (4 ft^3) per capita for digestion of trickling filter sludge (Hammer and Hammer, 2012).

When the characteristics of raw and digested sludge are known, the volume required for a single-stage standard-rate digester can be calculated from the following equation (Peavy et al., 1985; Hammer and Hammer, 2012):

$$V_S = \frac{V_1 + V_2}{2} t_1 + V_2 t_2$$

(12.11)

where
V_S = volume of standard-rate digester, m^3
V_1 = raw sludge loading rate, m^3/d
V_2 = digested sludge accumulation rate, m^3/d
t_1 = digestion period, d
t_2 = digested sludge storage period, d

For a single-stage high-rate digester, the volume can be calculated based on solids loading rates, or detention times, or any of the other empirical methods mentioned above. High-rate digesters are designed as completely mixed reactors without solids recycle. For a design digestion period, the volume can be calculated from the following:

$$V_H = V_1 t_1 \qquad (12.12)$$

where
V_H = volume of the high-rate digester, m^3
t_1 = digestion period or SRT, d

V_1 is as defined previously. The design of mesophilic digesters is illustrated in Example 12.4.

12.5.2.1.2 Gas production and use

Gas produced from anaerobic digestion usually contains about 65–70% methane, 25–30% carbon dioxide, and trace amounts of nitrogen, hydrogen, hydrogen sulfide, water vapor, and other gases. The volume of methane gas produced can be estimated from the feed concentrations and biomass produced. A number of mathematical relationships are available in the literature. Total gas production can be estimated from the amount of volatile solids reduction. Typical values range from 0.75 to 1.12 m^3/ kg VS destroyed (12–18 ft^3/lb VS destroyed). A first approximation of gas production can also be made from the population. For primary plants treating domestic wastewater, the gas production is about 15–22 m^3/1000 persons.d (0.6–0.8 ft^3/person.d), while for secondary plants the value is about 28 m^3/1000 persons.d (1.0 ft^3/person.d) (Metcalf and Eddy et al., 2013).
   Natural gas has a heating value of 37,300 kJ/m^3 (1000 BTU/ft^3). Pure methane gas at standard temperature and pressure (20°C and 1 atm) has a heating value of 35,800 kJ/m^3 (960 BTU/ft^3). Digester gas has about 65% methane, which has a heating value of approximately 22,400 kJ/m^3 (600 BTU/ft^3). Digester gas can be used as fuel for boilers and internal combustion engines. The electricity generated is then used for pumping wastewater, and heating digesters, among others. It can also be used in the cogeneration of electricity and steam.

12.5.2.1.3 Digester heating

Heat has to be provided to a digester to achieve the following: (1) Raise the temperature of feed sludge to temperature of digestion tank, (2) compensate for heat losses through the floor, walls, and cover of the digester, and (3) compensate for losses in the piping to the heat exchanger. The sludge is heated by transporting the sludge to an external heat exchanger and pumping it back to the digester.

In order to calculate the energy required to heat the incoming feed sludge, it is assumed that the specific heat of incoming feed sludge is equal to that of water (4.186 kJ/kg.K). The heat required to raise the temperature of the incoming sludge can be calculated using the following equation (Metcalf and Eddy et al., 2013; Davis, 2011):

$$q_r = M_D C_p (T_D - T_I)$$ (12.13)

where
q_r = heat required, kJ/d
M_D = mass of sludge fed to digester, kg/d
C_p = specific heat of water = 4.186 kJ/kg.K
T_D = digestion temperature, K
T_I = temperature of incoming feed sludge, K

The heat losses from the walls, floor, and cover of the digester can be calculated from the following equation:

$$q_l = U A \Delta T$$ (12.14)

where
q_l = heat loss, J/s or W
U = heat transfer coefficient, J/m².s.K or W/m².K
A = cross-sectional area of heat loss, m²
ΔT = temperature change across the surface, K

Typical heat transfer coefficients can be found in various sources (US EPA, 1979; Metcalf and Eddy et al., 2013). Some typical values are provided in Table 12.2.

12.5.2.2 Two-stage mesophilic digestion

In a two-stage mesophilic digestion process, the first tank is designed as a high-rate digester with mixing, while the second tank is used for dewatering and storage of digested sludge. Usually, the second tank is not heated or mixed. Most of the gas is generated in the first stage. Less than 10% of the

Table 12.2 Heat transfer coefficients for anaerobic digesters

Part of digester	U, W/m².K
Fixed concrete cover	
100 mm thick and covered with built-up roofing, no insulation	4.0–5.0
100 mm thick and covered, with 25-mm insulation	1.2–1.6
225 mm thick, no insulation	3.0–3.6
Fixed steel cover 6 mm thick	4.0–5.4
Floating cover	
35-mm wood deck, built-up roofing, no insulation	1.8–2.0
25-mm insulating board installed under roofing	0.9–1.0
Concrete floor	
300 mm thick in contact with dry soil	1.7
300 mm thick in contact with moist soil	2.85
Concrete walls above ground	
300 mm thick with insulation	0.6–0.8
300 mm thick without insulation	4.7–5.1
Concrete walls below ground	
Surrounded by dry soil	0.57–0.68
Surrounded by moist soil	1.1–1.4

Source: Adapted from U.S. EPA (1979) and Metcalf and Eddy (2013).

total gas is generated in the second stage. Both tanks are equipped with gas collection systems. A two-stage mesophilic digester system is illustrated in Figure 12.8.

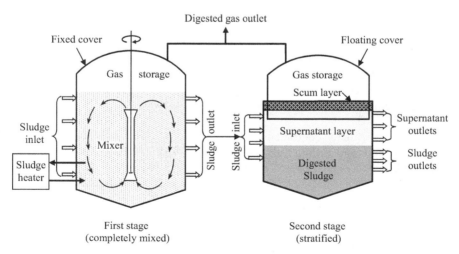

Figure 12.8 Diagram of a two-stage mesophilic digester.

Source: Adapted from Peavy et al., 1985

Example 12.4: A conventional wastewater treatment plant treats 30,000 m³/d of municipal wastewater with a BOD₅ of 240 mg/L and suspended solids of 200 mg/L. The effluent BOD₅ from the final clarifier is 15 mg/L. The flow diagram of the plant is given in Figure 12.9. The following data are provided:

Primary clarifier: removal efficiency; SS = 50%, BOD₅ = 35%

water content = 94%, specific gravity = 1.06

Aeration tank: F/M = 0.33, biomass conversion factor = 0.40

Final clarifier: 1.5% solids in waste activated sludge, specific gravity = 1.02

Flotation thickener: 96.5% water in thickened sludge

Anaerobic digestion: sludge is 74% organic, 55% reduction in VSS during digestion

solids content of digested sludge = 6.5%

a. Calculate the mass and volume of primary sludge.

b. Calculate the mass and volume of secondary sludge.

c. Calculate the total mass and volume of sludge entering the Blending tank.

d. Calculate the percentage of solids in the blended sludge.

e. Calculate the volume of a single-stage standard-rate digester for a digestion period of 25 days and a sludge storage period of 60 days.

f. Calculate the total volume of a two-stage digester, if the digestion period in the high-rate first stage is 10 days. The dewatering time is 5 days and the sludge storage time is 60 days in the second stage.

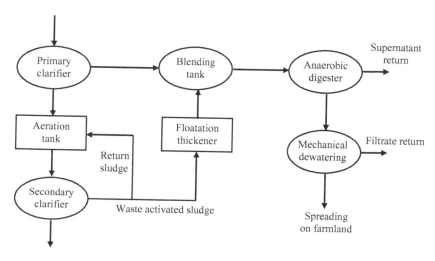

Figure 12.9 Flow diagram of wastewater treatment plant (for Example 12.4)

SOLUTION

Step 1. Calculate the mass and volume of primary sludge.

Mass of solids in influent $= 30,000\,\mathrm{m^3/d} \times 0.20\,\mathrm{kg/m^3} = 6000\,\mathrm{kg/d}$

With 50% removal, mass of primary sludge solids $M_p = 0.50 \times 6000\,\mathrm{kg/d}$
$$= 3000\,\mathrm{kg/d}$$

Use Equation (12.4) to calculate sludge volume.

$$V_p = \text{Volume of primary sludge} = \frac{3000\,\mathrm{kg/d}}{1000\,\dfrac{\mathrm{kg}}{\mathrm{m^3}} \times 1.06 \times (1-0.94)}$$

$$= 47.17\,\mathrm{m^3/d}$$

Step 2. Calculate mass and volume of secondary sludge.

$\mathrm{BOD_5}$ going to aeration tank $= (1-0.35) \times 240\,\mathrm{mg/L} = 156\,\mathrm{mg/L}$
$$= 0.156\,\mathrm{kg/m^3}$$

$\mathrm{BOD_5}$ consumed in aeration tank $= 156 - 15\,\mathrm{mg/L} = 141\,\mathrm{mg/L}$
$$= 0.141\,\mathrm{kg/m^3}$$

Use Equation (12.2) to calculate mass of secondary sludge solids

$M_s = \text{Mass of secondary sludge solids} = 0.4 \times 0.141\,\mathrm{kg/m^3} \times 30,000\,\mathrm{m^3/d}$

Or, $M_s = 1692\,\mathrm{kg/d}$

Use Equation (12.4) to calculate sludge volume

$$V_s = \text{Volume of secondary sludge} = \frac{1692\,\mathrm{kg/d}}{1000\,\dfrac{\mathrm{kg}}{\mathrm{m^3}} \times 1.02 \times 0.015}$$

$$= 110.59\,\mathrm{m^3/d}$$

Step 3. Calculate the volume of thickened sludge in Flotation thickener.

Mass of solids going to thickener $= 1692\,\mathrm{kg/d}$

Assume 100% capture of solids

Therefore, the mass of solids in thickened sludge $= 1692\,\mathrm{kg/d}$

$$\text{Volume of thickened sludge} = \frac{1692\,\mathrm{kg/d}}{1000\,\dfrac{\mathrm{kg}}{\mathrm{m^3}} \times 1.02 \times (1-0.965)}$$

$$= 52.44\,\mathrm{m^3/d}$$

Step 4. Calculate the mass and volume of sludge entering the Blending tank.

Mass of sludge solids entering the blending tank = primary solids + thickened secondary solids

$$M_B = 3000 + 1692\,kg/d = \mathbf{4692\,kg/d}$$

Volume of sludge entering the blending tank

$$V_B = 47.17 + 52.44 = \mathbf{99.61\,m^3/d}$$

Step 5. Calculate the percentage of solids (P_s) in blended sludge. Assume specific gravity of blended sludge = 1.0

$$V_B = \frac{M_B}{SG_s\,\rho_w\,(P_s/100)}$$

$$Or\ P_s = \frac{M_B}{SG_s\,\rho_w\,V_B}$$

$$Or, P_s = \frac{4692\,kg/d}{1.0\times 1000\dfrac{kg}{m^3}\times\left(\dfrac{1}{100}\right)\times 99.61\,m^3/d} = 4.71\%$$

Step 6. Single-stage digester design.

Raw sludge loading rate, $V_1 = 99.61\,m^3/d$

Mass of solids to digester = $4692\,kg/d$

Organic fraction = $4692\,kg/d \times 0.74 = 3472.08\,kg/d$

Inorganic fraction = $4692\,kg/d \times (1-0.74) = 1219.92\,kg/d$

Digestion reduces VSS or organic fraction by 55%

Therefore, organic fraction remaining = $3472.08\,kg/d \times (1-0.55)$
$$= 1562.44\,kg/d$$

Total mass remaining = $1219.92 + 1562.44\,kg/d = 2782.36\,kg/d$

Digested sludge accumulation rate, $V_2 = \dfrac{2782.36\,kg/d}{1000\dfrac{kg}{m^3}\times 1.0 \times 0.065}$
$$= 42.81\,m^3/d$$

Use Equation (12.11) to calculate single-stage digester volume.

$$V_S = \frac{V_1 + V_2}{2} t_1 + V_2 t_2$$

$$\text{Or } V_S = \frac{(99.61 + 42.81)\,m^3\,/\,d}{2} \times 25\,d + 42.81\frac{m^3}{d} \times 60\,d = 4348.85\,m^3$$

Step 6. Two-stage digester design.
Use Equation (12.12) to calculate the volume of first stage:

$$V_{1st\,stage} = V_1\,t_1 = 99.61\,m^3\,/\,d \times 10\,d = 996.10\,m^3$$

Use Equation (12.11) to calculate the volume of the second stage:

$$V_{2nd\,stage} = \frac{(99.61 + 42.81)\,m^3/d}{2} \times 5\,d + 42.81\frac{m^3}{d} \times 60\,d = 2924.65\,m^3$$

Therefore, total volume $= 996.10 + 2924.65\,m^3 = 3920.75\,m^3$

Note: The total volume required for the two-stage digester is less than that required for the single-stage digester. Total digestion time is also less for the two-stage digester.

Example 12.5: A single-stage mesophilic digester, shown in Figure 12.10, is used for sludge stabilization at the wastewater treatment plant mentioned in Example 12.4. The mass of sludge fed to the digester is 4872 kg/d. The temperature of the feed sludge is 12°C. Calculate the

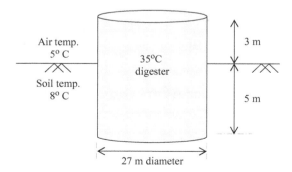

Figure 12.10 Single-stage mesophilic digester (for Example 12.5).

total heat that must be provided to maintain the digester temperature at 35°C, based on the data given below:

Digester: Diameter = 27 m
Total depth = 8 m, with depth below ground = 5 m
The digester has a fixed concrete cover with insulation, concrete floor and walls in contact with dry soil, and concrete walls above ground with insulation.
Temperature of soil surrounding digester = 8°C
Ambient air temperature in winter = 5°C

SOLUTION

Step 1. Calculate heat required to heat the feed sludge using Equation (12.13)

$$q_r = M_D\, C_p\left(T_D - T_I\right)$$

$$T_D = 273 + 35 = 308\,K$$

$$T_I = 273 + 12 = 285\,K$$

$$C_p = 4.186\,kJ\,/\,kg.K$$

$$\text{Therefore}, q_r = 4872\,kg/d \times 4.186\,kJ/kg.K\left(308\,K - 285\,K\right)$$
$$= 469{,}066.42\,kJ/d$$

Step 2. Calculate the surface area of the floor, walls, and cover.

$$\text{Floor area} = \pi\,/\,4 \times \left(27\right)^2 = 572.55\,m^2$$

$$\text{Area of fixed cover} = 572.55\,m^2$$

$$\text{Wall area above ground} = \pi \times 27m \times 3m = 254.47\,m^2$$

$$\text{Wall area below ground} = \pi \times 27m \times 5m = 424.12\,m^2$$

Step 3. Calculate the heat losses from the floor, cover, and walls using Equation (12.14) and heat transfer coefficients from Table 12.2.

$$q_1 = U\,A\,\Delta T$$

Heat loss from concrete floor with U = 1.7 W/m².K

$$q_{\text{floor}} = \left(1.7\,\text{W/m}^2.\text{K}\right)\left(572.55\,\text{m}^2\right)\left(308\,\text{K} - (273+8)\,\text{K}\right)$$

$$= 26,280.05\,\text{W or J/s}$$

$$= 26,280.05\,\text{J/s} \times 86400\,\text{s/d} \times 10^{-3}\,\text{kJ/J} = 2.27 \times 10^6\,\text{kJ/d}$$

Heat loss from concrete cover with U = 1.4 W/m².K

$$q_{\text{cover}} = \left(1.4\,\text{W/m}^2.\text{K}\right)\left(572.55\,\text{m}^2\right)\left(308\,\text{K} - (273+5)\,\text{K}\right)$$

$$= 24,047.10\,\text{W or J/s}$$

$$= 24,047.10\,\text{J/s} \times 86400\,\text{s/d} \times 10^{-3}\,\text{kJ/J} = 2.08 \times 10^6\,\text{kJ/d}$$

Heat loss from concrete wall below ground with U = 0.6 W/m².K

$$q_{\text{bg}} = \left(0.6\,\text{W/m}^2.\text{K}\right)\left(424.12\,\text{m}^2\right)\left(308\,\text{K} - (273+8)\,\text{K}\right)$$

$$= 6870.74\,\text{W or J/s}$$

$$= 6870.74\,\text{J/s} \times 86400\,\text{s/d} \times 10^{-3}\,\text{kJ/J} = 0.59 \times 10^6\,\text{kJ/d}$$

Heat loss from concrete wall above ground with U = 0.7 W/m².K

$$q_{\text{ag}} = \left(0.7\,\text{W/m}^2.\text{K}\right)\left(254.47\,\text{m}^2\right)\left(308\,\text{K} - (273+5)\,\text{K}\right)$$

$$= 5343.87\,\text{W or J/s}$$

$$= 5343.87\,\text{J/s} \times 86400\,\text{s/d} \times 10^{-3}\,\text{kJ/J} = 0.46 \times 10^6\,\text{kJ/d}$$

$$\text{Total heat loss} = \left(2.27 + 2.08 + 0.59 + 0.46\right) \times 10^6\,\text{kJ/d}$$

$$= 5.4 \times 10^6\,\text{kJ/d}$$

Step 4.Calculate total heat required for sludge and digester

$$q_{\text{Total}} = q_{\text{r}} + q_{\text{loss}} = 0.47 \times 10^6 + 5.4 \times 10^6\,\text{kJ/d} = 5.87 \times 10^6\,\text{kJ/d}$$

Example 12.6: Assume that 1 m³ of gas is produced per kg VS destroyed in the mesophilic digester given in Example 12.5. VSS fed to digester is 3605.28 kg/d. The heating value of the gas is 22,400 kJ/m³, with a methane content of 65%. The gas will be used to fuel a boiler, which will then be used to heat the digester. The efficiency of the boiler is 75%. Consider the treatment plant data from Examples 12.4 and 12.5. Will the power generated from the gas be sufficient to heat the digester?

SOLUTION

From Example 12.4,
VSS destroyed = 3605.28 kg/d × 0.55 = 1982.90 kg/d
Total gas produced = 1982.90 kg/d × 1 m³/kg = 1982.90 m³/d
Heating value of gas = 22,400 kJ/m³ × 1982.90 m³/d = 44.42 × 10⁶ kJ/d
Boiler efficiency = 75%
Heat generated by gas = 44.42 × 10⁶ kJ/d × 0.75 = 33.31 × 10⁶ kJ/d
From Example 12.5,
Total heat required for digester = 5.87 × 10⁶ kJ/d ≪ heat generated by gas in the boiler

The heat generated from the gas will be more than sufficient to maintain the digester heating requirements. Excess gas can be used for other purposes at the plant, e.g. provide energy for pumping, etc.

12.5.2.3 Thermophilic anaerobic digestion

Thermophilic digestion takes place at temperature ranges from 50°C to 60°C, with an optimum at 55°C. Advantages of thermophilic digestion when compared to mesophilic digestion are higher reaction rates of destruction of organic matter resulting in shorter retention times and therefore smaller reactor volumes, increased methane production due to increased solids destruction, improved dewatering characteristics of digested sludge, and higher destruction of pathogenic organisms. On the other hand, there are disadvantages associated with thermophilic digestion such as higher energy requirements for heating, poor supernatant quality, poor process stability, and increased odor problems (Metcalf and Eddy et al., 2013). Thermophilic digestion is seldom used as a single-stage digester. It is typically used as the first stage in a staged process. Increased pathogen destruction makes it desirable, especially when the digested biosolids are to be used for specific land applications.

One of the major drawbacks of thermophilic digestion is a higher sensitivity of this process to environmental changes, e.g. temperature, accumulation of intermediate products such as H_2, acetate, and propionate resulting in the ineffective conversion of VFAs to methane (van Lier et al., 1993).

Despite the higher substrate utilization and specific growth rates of thermophilic microorganisms when compared to mesophilic microorganisms, the yield of thermophilic bacteria per unit amount of substrate is lower. The lower yield of thermophilic microorganisms may be due to their higher energy requirement for maintenance or the specific molecular properties of enzyme reactions at thermophilic temperatures (Zeikus, 1979).

12.5.2.4 Temperature-phased anaerobic digestion (TPAD)

Temperature-phased anaerobic digestion (TPAD) is a two-stage digestion system consisting of a thermophilic stage as the first step followed by a mesophilic stage as the second step (Han et al., 1997). By combining the thermophilic and mesophilic digestion processes into one, TPAD offers the advantages of both while eliminating the problems associated with these systems when operated independently (Harikishan and Sung, 2003). The TPAD process is illustrated in Figure 12.11 (a).

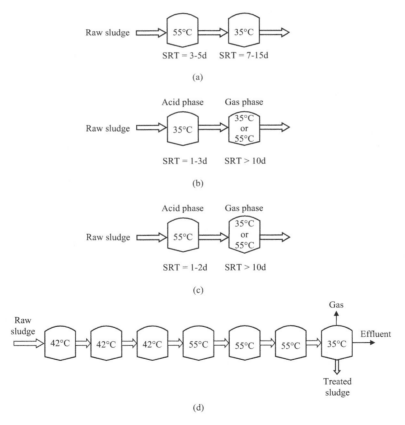

Figure 12.11 (a) TPAD process, (b) Acid-gas phased digestion with mesophilic acid phase, (c) acid-gas phased digestion with thermophilic acid phase, (d) Enhanced Enzymic Hydrolysis (EEH™) process.

The thermophilic step provides an increased rate of degradation of complex organics and improves pathogen reduction. On the other hand, the mesophilic stage is used as the polishing stage helping to diminish the drawbacks of the thermophilic stage such as poor process stability and poor effluent quality (Sung and Santha, 2003).

12.5.2.5 Acid-gas phased digestion

The acid-gas phased digestion system provides the separation of the two main stages of anaerobic biodegradation; hydrolysis/acidogenesis/acetogenesis and methanogenesis steps, increasing the process stability. In general, thermophilic temperatures are used for the acid-forming step, having the advantage of higher destruction rates of organic solids and increased destruction of pathogens. Hydraulic retention time has a considerable effect on the population levels of methanogens and the composition of fermentative products like VFAs (Fukushi et al., 2003). In addition to thermophilic temperatures, short retention times are adopted for the pretreatment step, in order to inhibit the methanogenic population and increase acid production. The second step is the methane-forming step, where neutral pH conditions and a longer SRT are provided for the growth of methane-forming bacteria and to maximize gas production.

One option for two-phase systems is employing thermophilic anaerobic digestion as the pretreatment step followed by mesophilic anaerobic digestion. Another option is to use a mesophilic acid-forming step followed by a mesophilic or thermophilic methane-forming step. These options are illustrated in Figures 12.11 (b) and (c).

Enhanced pathogen destruction in two-phase anaerobic digestion is thought to be a result of the combined effect of pH and acid concentration. A number of studies have been conducted to observe the separate and combined effects of pH and organic acid concentration on pathogen destruction (Fukushi et al., 2003, and Salsali et al., 2006).

12.5.2.6 Enhanced Enzymic Hydrolysis™

In January 2002, legislation was enacted in the UK that required pathogen reduction in municipal wastewater sludge for the first time. This new requirement led many utilities to search for methods to optimize their existing anaerobic digestion systems (Cumiskey, 2005), particularly mesophilic digesters, which included the majority of operating systems in the UK at that time. One such process was the Enhanced Enzymic Hydrolysis™ (EEH™) process developed by United Utilities (in the UK) in partnership with Monsal Limited.

The EEH™ process utilizes acid-phase digestion for hydrolysis of complex organic compounds and volatile fatty acid production, followed by batch thermophilic anaerobic digestion for pasteurization purpose, and continuous mesophilic anaerobic digestion for methane production and stabilization. The combination of 42°C and 55°C temperatures provide improved hydrolytic activities together with pasteurization to achieve required pathogen reduction (Werker et al., 2007). The enzyme hydrolysis step breaks down cell wall lipoprotein structures, enhancing the digestion process. The EEH™ process schematic is presented in Figure 12.11 (d).

The EEH™ process utilizes a novel plug-flow digester operation that provides the ideal condition for maximum production of digestive enzymes that are responsible for pathogen destruction and VFA production. This enables a pathogen destruction rate of 99.9999% (Le and Harrison, 2006). In this process, the sludge is pre-fermented at 42°C followed by pasteurization at 55°C, from where the sludge is transferred to a mesophilic digester. According to Werker et al. (2007), the EEH process has achieved 6 log *E. coli* removal and elimination of *Salmonella* and has enhanced volatile solids destruction by 10% at the Blackburn Wastewater Treatment Plant in the UK.

12.5.2.7 Cambi™ process

The Cambi™ process was developed in Norway in the 1990s. It is a patented sludge pretreatment process. The process has been installed in wastewater treatment plants in the US (DC Water, 2019a), Norway, Denmark, Japan, Ireland, Scotland, and England (Greater Vancouver Regional District, 2005). The process involves the oxidation of sludge under elevated temperature and pressure. Under these conditions, pathogens are destroyed and cell hydrolysis occurs releasing energy-rich compounds. Following hydrolysis, sludge is fed to an anaerobic digester where it readily breaks down, resulting in high volatile solids destruction (approximately 65%) and increased biogas production compared to conventional anaerobic digestion.

In the Cambi™ process, primary and secondary sludge is dewatered to approximately 17% solids before entering a *pulping* vessel. In the pulping vessel, the mixed sludge is heated to approximately 80°C, and then transferred to the thermal hydrolysis digester vessel, where it is heated to 160°C at a pressure of approximately 5.5 bar for 15–30 minutes. After digestion, the sludge is released to a *flash* tank which is at atmospheric pressure. The pressure drop between the digester and the flash tank causes cell lysis and a decrease in temperature to 100°C. A series of heat recovery and heat transfer systems are required to optimize the energy use of the process. Sludge in the flash tank is diluted with treated effluent to ensure that the solids concentration in the digester is not excessive. Figure 12.12 provides a flow diagram of the Cambi™ process. After thermal hydrolysis the viscosity of sludge is significantly reduced, thus allowing the digester to be operated at solids concentrations of about 9%. The digester sizes can be significantly reduced in the Cambi™ process.

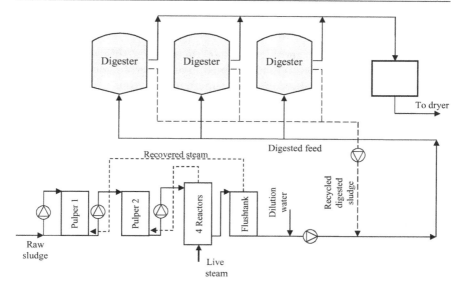

Figure 12.12 Flow diagram of Cambi™ process.

Blue Plains Advanced Wastewater Treatment Plant (AWTP) of DC Water has implemented the Cambi™ thermal hydrolysis process (THP) together with anaerobic digestion to produce Class A biosolids and generate energy. This is the first THP facility in North America and the largest in the world. The THP process pressure cooks the waste solids to generate combined heat and power, generating a net 10 MW of electricity (DC Water, 2019a and 2019b). The project has resulted in a 30% reduction in energy purchased from the grid, a 41% reduction in greenhouse gas production, and a 50% reduction in biosolids shipping costs (CDM Smith, 2021).

12.5.3 Aerobic digestion

Aerobic digestion is the biological conversion of organic matter in the presence of air, usually in an open-top reactor. Aerobic digestion is the oxidative microbial stabilization of sludge. It is based on the principle that when there are inadequate external substrates available, microorganisms will metabolize their own cellular mass, resulting in an overall reduction of volatile solids.

Aerobic digestion is similar to the activated-sludge process. Microorganisms start to consume their own protoplasm as an energy source as the supply of available substrate is depleted. When cell tissues become the energy source, microorganisms are said to be in the endogenous stage. Cell tissue is oxidized to carbon dioxide, water, and ammonia which are subsequently oxidized to nitrate; 20–25% of cell tissue is non-biodegradable, which remains after the digestion process as the final product.

Advantages of aerobic digestion are (Metcalf and Eddy et al., 2013):

- In a well-operated aerobic digester, the volatile solids reduction is approximately equal to that obtained in an anaerobic digester.
- Lower BOD concentration in the supernatant liquor.
- Production of an odorless biologically stable end product.
- Operation is relatively easy.
- Lower capital cost.
- Suitable for digesting nutrient-rich biosolids.

Disadvantages of aerobic digestion are:

- Higher power costs related to the supply of required oxygen.
- Digested solids have inferior mechanical dewatering characteristics.
- The process is significantly affected by temperature, location, tank geometry, concentration of feed, and type of mixing/aeration.

Some variations and combinations of aerobic and anaerobic digestion processes are presented in the following section.

12.5.3.1 Autothermal thermophilic aerobic digestion

Autothermal Thermophilic Aerobic Digestion (ATAD) is a solids treatment process where heat is released by the aerobic microbial degradation of organic matter (Layden, 2007). In ATAD, the heat released by the digestion process is the major heat source used to achieve the desired operating temperature. For the ATAD operation, feed sludge is typically thickened to 4–6% TS and volatile solids destruction provides heat production that results in autothermal conditions. Thermophilic temperatures between 55°C and 70°C can be achieved without external heat input by using the heat released from the microbial oxidation process. About 20,000 kJ of heat is produced per kg VS destroyed (Metcalf and Eddy et al., 2013). However, sometimes an outside heat source is required when the solids content of the raw sludge is not high enough to achieve the desired temperature.

The main advantages of the ATAD process are:

- Shorter retention times (5–6 d) can be used to achieve 30–50% VS destruction.
- Greater pathogen destruction
- Simplicity of operation.

The disadvantages include (i) odor production, (ii) lack of nitrification, and (iii) poor dewatering capabilities of digested sludge.

12.5.3.2 Dual digestion

The Dual digestion (DD) process involves the use of aerobic thermophilic digestion followed by anaerobic digestion. Typically, an ATAD process with a relatively short retention time is used as a pretreatment step to mesophilic anaerobic digestion. This is termed dual digestion (Ward et al., 1998; Zabranska et al., 2003). In the ATAD step, solids are pretreated by solubilization and partial acidification resulting in enhanced digestion together with improved pathogen destruction (Nosrati et al., 2007). The dual digestion process is illustrated in Figure 12.13.

Thermophilic aerobic digestion step through biological oxidation of volatile solids provides hydrolyzed and homogenized solids, which improve volatile solids destruction in the downstream anaerobic digester. ATAD is operated under oxygen limiting conditions that in conjunction with short HRT, result in the formation of VFAs through the fermentative metabolism of thermophilic bacteria (Borowski and Szopa, 2007). ATAD reactor provides a consistent feed with high VFA concentration to the anaerobic stage, which would perform as the methane-forming step. In addition, ammonification in the ATAD reactor produces a pH buffered feed to the anaerobic stage. In addition to enhancing the efficiency of the anaerobic digestion step, ATAD also provides better pathogen removal. In the short retention time, thermophilic phase, high levels VFAs, and ammonia produced results in a reduction in pathogenic bacteria.

A detailed laboratory-scale evaluation of the dual digestion process was conducted by Aynur et al. (2010, 2009, 2009, 2008). A dual digestion system consisting of ATAD followed by MAD (mesophilic anaerobic digestion) was operated and compared with other enhanced digestion processes. Effects of pretreatment HRTs, oxygen flow rates, and organic loading rates were evaluated. The ATAD process produced heat of 14,300 J/g VS removed from hydrolytic and acetogenic reactions without compromising overall methane yields, when the HRT was 2.5 days or lower and the total O_2 used

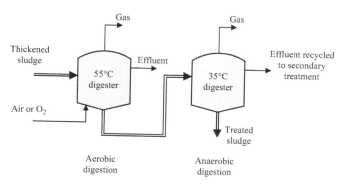

Figure 12.13 Schematic of the dual digestion process.

was 0.20 L O_2/g VS fed or lower. ATAD followed by TPAD was also evaluated by the researchers.

Tacoma Central Treatment Plant in Washington, USA, uses a DD set-up with a combination of ATAD followed by TPAD. ATAD is used as the first aerobic step at which pre-conditioning and Class A pasteurization are achieved. ATAD is followed by TPAD where volatile solids destruction and sludge stabilization takes place. The TPAD step consists of three anaerobic phases in a temperature-phased mode from thermophilic to high mesophilic to low mesophilic. Implementation of the TPAD system after ATAD resulted in the elimination of odor (Eschborn and Thompson, 2007).

12.5.4 Composting

The composting process involves biological degradation of the organic matter in sludge, to produce a stable end product. The process can be aerobic or anaerobic. In most cases, aerobic composting is used, as it enhances the decomposition of organic matter and results in the higher temperature necessary for pathogen destruction. It can be used for the stabilization of primary and waste activated sludge, as well as digested and dewatered sludge. The end product is a humus-like material that can be used as a fertilizer and soil conditioner.

Composting is carried out in the following steps (Metcalf and Eddy et al., 2013):

1. *Preprocessing* – where sludge is mixed with an organic amendment and/or a bulking agent. Commonly used amendments are sawdust, straw, recycled compost, which are used to reduce moisture content and increase air voids. A bulking agent such as wood chips is used to provide structural support and increase porosity.
2. *High-rate decomposition* – the compost pile is aerated by mechanical turning or air addition. The temperature first increases from ambient to about 40°C (mesophilic). As microbial degradation proceeds, the temperature further increases to the thermophilic range (50–70°C), where maximum degradation, stabilization, and pathogen destruction occurs. This takes place for 21–28 d.
3. *Recovery of bulking agent.*
4. *Curing* – this is the cooling period. As the temperature goes down, water of evaporation is released together with the completion of humic acid formation and pH stabilization. The curing period can last for 30 d or longer.
5. *Post-processing* – non-biodegradable materials are screened and removed. Grinding is used for size reduction of the finished product.

Commonly used methods of composting include the *aerated static pile* and *windrow* systems. A windrow system is illustrated in Figure 12.14 (a). In-vessel composting systems are also available from manufacturers, where

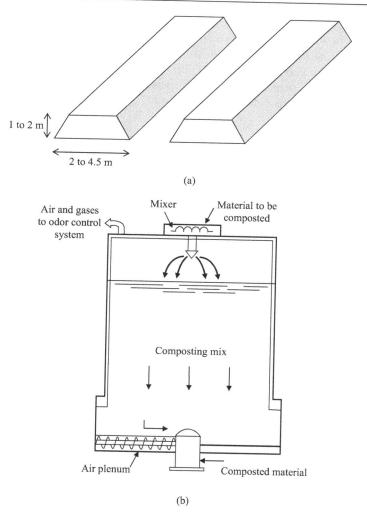

Figure 12.14 Composting systems: (a) Windrow system and (b) in-vessel composter.

the composting takes place in an enclosed vessel or reactor. These are used for small-scale applications, have better odor control, faster throughput, and small area requirement. An in-vessel composting system is illustrated in Figure 12.14 (b).

12.6 CONDITIONING OF BIOSOLIDS

Chemicals such as polymers are used to improve the dewatering characteristics of biosolids. Conditioning is used ahead of mechanical dewatering systems, such as belt-filter presses, centrifuges, etc. Chemicals such as ferric chloride, lime, alum, and polymers are added to the biosolids. These cause coagulation of the solids and release of absorbed water. The moisture content can be reduced from above 90% to a range of 65–85%. Besides chemical conditioning, other methods such as heat treatment, or freeze-thaw are also used to a limited extent at some plants.

12.7 BIOSOLIDS DEWATERING

The process of dewatering is used to reduce the volume of treated sludge or biosolids by reducing the water content. Dewatering is a physical unit operation. Dewatered sludge is easier to handle and transport for final disposal. Volume reduction reduces transportation costs. Dewatering is usually required prior to composting, incineration, and landfilling. A number of dewatering methods are used at various treatment plants. These include centrifugation, belt-filter press, recessed-plate filter press, drying beds, and lagoons. Heat drying is also used in some installations. Some of these methods are described in detail in the following sections.

12.7.1 Centrifugation

Centrifugation is a popular method for dewatering biosolids. Centrifugation is used for thickening as well as dewatering stabilized sludge. The *solid-bowl centrifuge* used for sludge thickening has been described in section 12.4.3. The *high-solids centrifuge* used for dewatering biosolids is presented in this section.

12.7.1.1 High-solids centrifuge

The high-solids centrifuge is a modification of the solid-bowl centrifuge. It has a longer bowl length, lower differential bowl speed to increase residence time, and a modified scroll to provide a pressing action at the end. Polymer application is required. Solids content ranging from 10% to 35% may be achieved. The centrate is usually high in suspended solids, which can pose problems when it is recycled back to the plant. Chemical conditioning or increased residence time in the centrifuge can be used to increase solids capture, and reduce solids load in the centrate. The power costs are high.

There are a number of advantages of the centrifugation process. They are:

- Low initial cost.
- Smaller footprint compared to other dewatering processes.
- High solids concentration in dewatered cake.
- Low odor generation as the unit is enclosed.

12.7.2 Belt-filter press

The belt-filter press is one of the most widely used dewatering equipment in the US. It is a continuous-feed process that can be used for most types of municipal wastewater sludges and biosolids. Dewatered cake solids content ranges from 12% to 32%, depending on the feed solids content and sludge characteristics. The process uses chemical conditioning, gravity drainage, and mechanical pressure to dewater biosolids.

A belt-filter press is illustrated in Figure 12.15. In the first stage, a polymer is added to condition the solids. In the second stage, conditioned sludge is applied to the upper belt of the gravity drainage zone. Water drains by gravity through the porous belt and is collected. About half of the water is removed by gravity. In the third stage, the sludge drops onto the lower belt where it is squeezed between opposing porous belts. This is followed by a high-pressure section where the sludge is subjected to shearing forces as the belts pass through a series of rollers. In the end, scraper blades remove the dewatered sludge cake from the belts.

Sludge loading rates range from 90 to 680 kg/m.h based on characteristics and concentration of feed (Metcalf and Eddy et al., 2013). The belt size varies from 0.5 to 3.5 m in width. Belt sizes of 2 m are commonly used for municipal sludge dewatering operations. Belt speeds can vary from 1.0 to 2.5 m/min (Davis, 2011).

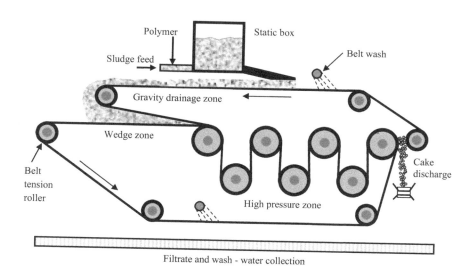

Figure 12.15 Belt-filter press.

Source: Adapted from Hammer and Hammer, 2012

12.7.3 Drying beds

The use of drying beds is a popular method for dewatering digested biosolids and unthickened primary and waste activated sludge. The advantages of the method are low cost, low maintenance, and high solids content in the dried cake. The disadvantages include large land requirements, odor problems, rodent problems, the impact of weather, and labor-intensive dried product removal. There are a number of different types of drying beds. These include conventional sand beds, paved beds, artificial media beds, solar drying beds, and vacuum-assisted beds.

12.8 DISPOSAL OF BIOSOLIDS

After thickening, stabilization, and dewatering, comes the disposal of the treated sludge and biosolids. The disposal can be by incineration or by land application and landfilling. The selection of disposal method depends on regulatory requirements and the degree of treatment received by the sludge. Some of the common methods are described in the following sections.

12.8.1 Incineration

Incineration is the complete combustion of organic matter in the sludge to end products such as carbon dioxide, water, and ash. Dewatered, untreated sludge is used as feed to the incinerator. The product ash has to be disposed of in an appropriate manner depending on whether it contains hazardous materials. The generated gases are passed through scrubbers and other air pollution control devices to remove air pollutants of concern, before releasing them into the atmosphere. Examples of combustion processes include multiple hearth incineration, fluidized bed incineration, and co-incineration with municipal solid waste.

The advantages of incineration are (Metcalf and Eddy et al., 2013):

- Maximum volume reduction is achieved.
- Pathogens and toxic compounds are destroyed.
- Potential for energy recovery.

The disadvantages include the following:

- High capital and operating costs.
- Hazardous waste may be produced as a by-product.
- Emission of air pollutants is a major concern.

12.8.2 Land disposal methods

Biosolids may be disposed of on land in a number of ways. These include land application, landfilling, and beneficial reuse. The land application can be on (1) agricultural land, (2) forest land, (3) disturbed land, and (4) dedicated land disposal sites. Recycling biosolids through land application has several advantages (US EPA, 2000):

- Biosolids provide essential nutrients, such as nitrogen and phosphorus to plants. It also contains other micronutrients, e.g. nickel, zinc, and copper. It can serve as an alternative to chemical fertilizers.
- The nutrients in biosolids are present in organic form. As such, they are released slowly to the plants and are less susceptible to runoff.
- Biosolids improve soil texture and water holding capacity. They can enhance root growth and increase the drought tolerance of vegetation.
- Biosolids can be used to stabilize and re-vegetate lands impaired by mining, dredging, and construction activities, as well as by fires and landslides.
- Biosolids are used in silviculture to increase forest productivity by accelerating tree growth, especially on marginally productive soil.

The selection of disposal method and site is dictated by local and state regulations and the degree of treatment received by the sludge. The regulatory requirements for the disposal of biosolids in the US are presented in the following section.

12.9 BIOSOLIDS DISPOSAL REGULATIONS IN THE US

On February 19, 1993, the United States Environmental Protection Agency under the authority of the Clean Water Act, promulgated risk-based regulations for the use and disposal of sewage sludge (US EPA, 1993). The regulations were for sludge from municipal wastewater treatment plants that were applied on land, sold or given away for use in home gardens, and incinerated. The regulations pertaining to land application are known as 40 CFR Part 503, as published in the Federal Register as the Code of Federal Regulations (CFR), and will be discussed in this section. The regulations are applicable to all treatment plants that use land applications for the final disposal of biosolids (e-CFR, 2021). The regulations are self-implementing, i.e. permits are not required by the plants. But failure to conform to the regulations is considered to be a violation of the law. The frequency of monitoring and reporting requirements are provided in detail.

The Part 503 regulations specify a number of methods for sludge stabilization, which include digestion, composting, and lime stabilization, among others (e-CFR, 2021). Maximum concentrations of metals that cannot be exceeded are given as *Ceiling Concentrations*. In addition, *Cumulative Pollutant Loading Rates* for eight metals are established, which may not be exceeded at land application sites. A third set of metals criteria, known as *Pollutant Concentrations* are provided. If these concentrations are not exceeded in the biosolids, then the cumulative pollutant loading rates do not have to be monitored.

The Part 503 rule defines two main types of biosolids according to the level of pathogen reduction: Class A and Class B. *Class A* biosolids can be applied to land without any restrictions. Class A biosolids have pathogens below detection limits, most stringent metal limits, and vector attraction standards. The term "exceptional quality biosolids" is also used for Class A. Class B biosolids have lesser treatment requirements. *Class B* biosolids can be applied on land, but are subject to restrictions with regard to public access to the site, livestock grazing, and crop harvesting schedules.

According to the 40 CFR Part 503 standards for Class A biosolids, fecal coliform indicator levels of less than 1000 MPN/gram TS should be achieved, or *Salmonella* sp. bacteria levels should be below detection limits (3 MPN/4 g TS) after treatment. Enteric viruses and viable helminth ova should each be less than 1 per 4 g TS (e-CFR, 2021).

The pathogen requirement for Class B biosolids is a fecal coliform concentration below 2×10^6 MPN/g TS, or that the sludge is treated in a process to significantly reduce pathogens (PSRP). As a point of reference, the concentration of fecal coliforms in undigested sludge is approximately 10^8 MPN/g TS, and *Salmonella* sp. concentration is approximately 2×10^3 MPN/g TS (US EPA, 1994).

12.9.1 Class A biosolids

According to US EPA (1993), there are six pathogen reduction alternatives by which sludge treatment processes can be considered to produce Class A biosolids. Class A biosolids can be applied on land immediately without any restrictions. The pathogen reduction alternatives are the following:

Alternative 1: Thermally Treated Sewage Sludge – This alternative may be used when the pathogen reduction process uses specific time-temperature regimes to reduce pathogens. Required time-temperature regimes must be met as well as the following requirement: Either fecal coliform densities should be below 1000 MPN/g TS (dry weight basis), or *Salmonella* sp. bacteria should be below detection limits (3 MPN/4 g TS).

Alternative 2: Sewage Sludge Treated in a High pH-High Temperature Process – High temperature-high pH process involves elevating the pH to greater than 12 and maintaining the pH for more than 72 h, or keeping the temperature above 52°C for at least 12 h at pH greater than 12, or air drying to over 50% solids after the 72 h period of elevated pH is necessary.

Alternative 3: Sewage Sludge Treated in Other Processes – This alternative applies to sewage sludge treated by processes that do not meet the process conditions required by Alternatives 1 and 2. The process must be demonstrated to reduce the density of enteric viruses and helminth ova in the sewage sludge to less than 1 PFU/4 g TS, in both cases.

Alternative 4: Sewage Sludge Treated in Unknown Processes – For this alternative, demonstration of the process is not necessary. Instead, the density of enteric viruses and helminth ova in the biosolids must be less than 1 PFU/4 g TS. In addition, as for all Class A biosolids, the sewage sludge must meet the FC or *Salmonella* sp. limits.

Alternative 5: Use of PFRP – Biosolids are considered to be Class A if they have been treated in one of the Processes to Further Reduce Pathogens (PFRP) as listed by US EPA (1993). These are composting, heat drying, heat treatment, thermophilic aerobic digestion, beta ray irradiation, gamma-ray irradiation, and pasteurization. In addition, products must meet the Class A fecal coliform or *Salmonella* sp. requirements.

Alternative 6: Use of Process Equivalent to PFRP – One of the alternatives to achieve Class A biosolids is to use a process equivalent to a PFRP, as determined by the permitting authority. In addition, products of all equivalent processes must meet the Class A fecal coliform or *Salmonella* sp. requirements.

12.9.1.1 Processes to further reduce pathogens (PFRP)

The PFRPs defined by US EPA are listed in Appendix B of 40 CFR Part 503 (Federal Register, 2010; e-CFR, 2021). These are outlined below.

i. Composting – maintain the temperature at 55°C or higher for 3 d for static aerated pile or in-vessel composting method. With windrow composting, maintain the temperature at 55°C or higher for 15 d or longer.
ii. Heat drying – at 80°C by direct or indirect contact with hot gases.
iii. Heat treatment – at 180°C or higher for 30 min.
iv. Thermophilic aerobic digestion – at 55–60°C with an SRT of 10 d.
v. Beta ray irradiation – at 1.0 mega rad at room temperature.
vi. Gamma-ray irradiation – at 1.0 mega rad at room temperature.
vii. Pasteurization – at 70°C or higher for 30 min or longer.

12.9.2 Class B biosolids

There are three pathogen reduction alternatives by which sludge treatment processes can be considered to produce Class B biosolids (US EPA, 1993). In addition to the pathogen reduction alternative, the biosolids must also meet one of the vector attraction reduction requirements. Site restrictions are placed for land application of Class B biosolids, with respect to crop harvesting, animal grazing, and public access to the site where the biosolids are applied. The details of these are available in the Federal Register (2010; e-CFR, 2021).

The pathogen reduction alternatives for the production of Class B biosolids are the following:

Alternative 1: Monitoring of Indicator Organisms – Seven samples of biosolids should be collected each time for monitoring. The geometric mean of the density of fecal coliforms in those samples should be less than 2×10^6 MPN/g TS.

Alternative 2: Use of PSRP – The sludge has to be treated by a PSRP (Processes to Significantly Reduce Pathogens). These include aerobic digestion, anaerobic digestion, air drying, composting, and lime stabilization.

Alternative 3: Use of Processes Equivalent to PSRP – Class B biosolids can be produced using a process that is equivalent to PSRP, as determined by the permitting authority.

12.9.2.1 Processes to significantly reduce pathogens (PSRP)

The PSRPs defined by US EPA are listed in Appendix B of 40 CFR Part 503 (Federal Register, 2010; e-CFR, 2021). These are presented below.

i. Aerobic digestion – at 20°C with SRT of 40 d, or at 15°C with SRT of 60 d.
ii. Anaerobic digestion – at 35–55°C with SRT of 15 d, or at 20°C with SRT of 60 d.
iii. Air drying – sludge should be dried in sand beds, or paved or unpaved beds for at least 3 months, when the ambient temperature is above 0°C.
iv. Composting – at 40°C or higher for 5 d. For 4 h during the 5 d period, the temperature in the compost pile should exceed 55°C.
v. Lime stabilization – sufficient lime should be added to raise the pH of the sludge to 12 after 2 h of contact.

PROBLEMS

12.1 Define the term *biosolids*. List the four steps of sludge processing and disposal. Give examples of each step.
12.2 Differentiate between primary and secondary sludge.
12.3 A municipal wastewater treatment plant processes an average flow of 15,000 m³/d with 200 mg/L of BOD_5 and 320 mg/L of suspended solids. The peak flow is 1.5 times the average flow. Determine the mass of BOD_5 and solids (kg/day) that are removed as sludge from the primary clarifier for average and peak flow conditions. Assume a reasonable efficiency for the primary clarifier and the same peak flow factor for BOD_5 and suspended solids.

12.4 A secondary treatment plant treats the wastewater from a community of 6000 people. The wastewater has an influent BOD_5 of 0.17 lb/capita-day and suspended solids of 0.20 lb/capita-day. The plant achieves an overall 90% BOD_5 removal. The primary clarifier removes 35% BOD_5 and 65% suspended solids. The primary sludge has 4% dry solids, a specific gravity of 1.03. The secondary sludge has 5% dry solids, a specific gravity of 1.01, and the biological solids produced is 0.35 kg/kg BOD_5 removed in the filter. Determine the volume of primary and secondary sludge produced per day.

12.5 A wastewater treatment plant consists of primary treatment followed by an activated-sludge secondary system. The plant processes 7000 m^3/d of wastewater with 180 mg/L of BOD_5 and 150 mg/L of suspended solids. The primary sludge contains 500 kg of dry solids per day with 4.5% solids content. The plant produces 760 kg of total sludge (primary and secondary) per day. Assume 30% BOD removal efficiency of the primary clarifier and F/M ratio of 0.2 for the aeration tank. Determine the following,
   a. Primary sludge volume and solids removal efficiency of the primary clarifier.
   b. Influent and effluent BOD_5 of the secondary activated sludge system.
   c. Volume of secondary sludge with 1% solids content.

12.6 Why it is necessary to thicken the primary and secondary sludge before further processing? Give examples of different sludge thickening processes. Describe the most common methods available for volume reduction of sludge.

12.7 In Problem 12.5, the primary and secondary sludges are mixed in a gravity thickener and sent for further treatment. The thickened sludge contains 4% solids. Calculate the percent volume reduction in the gravity thickener.

12.8 What are the objectives of sludge stabilization? List the common methods used to stabilize the sludge before disposal.

12.9 Briefly describe the advantages and disadvantages of anaerobic digestion. What factors should be considered for the design of an anaerobic digester?

12.10 The thickened sludge from Problem 12.7 is to be digested in a standard-rate mesophilic anaerobic digester (MAD). The sludge has 68% organic content and approximately 60% of the organic fraction is digested after a 30-day period. The digested sludge has a solids content of 6% and is stored for 90 days. Calculate the volume of the standard-rate digester.

12.11 Rework Problem 12.10 for a two-stage digester system employing a mixed, heated first stage with a digestion period of 10 days and second stage with a thickening period of 3 days. Determine the volume of the first-stage and second-stage digesters, and compare the total volume to that of the single-stage digester in Problem 12.10.

12.12 What are the advantages and disadvantages of aerobic digestion? Give examples of different types of aerobic digestion processes.

12.13 What is *composting*? Explain the steps of the composting process.

12.14 List the processes by which the water content of the sludge is reduced. Which one is the most widely used for dewatering? Briefly describe the process.

12.15 Name and describe the most common methods of sludge disposal. What is the basic difference between class A and class B biosolids?

12.16 Describe a process that can be used to produce Class A biosolids. Name a wastewater treatment plant that produces Class A biosolids.

12.17 An activated sludge plant has an anaerobic digester that serves a population of 50,000 people. Incoming sludge to the digester is a mixture of primary and secondary sludge that is equivalent to 0.15 kg/cap.d. The digester operating temperature is 35°C, average ambient winter and summer temperatures are 12°C and 20°C, respectively. The specific heat of water $C_p = 4.186$ kJ/kg.K.

    a. The incoming sludge to the digester has 70% volatile matter; 60% of the volatiles are destroyed during digestion after a 30 d period. The incoming sludge has a solids content of 4.5%. The digested sludge has a solids content of 6% and is stored for a period of 60 days. Calculate the volume of a standard rate digester.

    b. Calculate the heat required to raise the temperature of the incoming sludge to digestion temperature.

    c. Assume 0.9 m³ of gas is produced per kg VS destroyed. The heating value of the gas containing methane is 22,400 kJ/m³. Calculate the fuel value of the sludge gas in kJ/d. Will this be sufficient to heat up the incoming sludge?

12.18 Name two gases that are produced in significant quantities in anaerobic digesters.

12.19 Describe the TPAD process. How can we combine the benefits of the TPAD process with the acid-gas-phased digestion process?

12.20 A wastewater treatment plant consists of primary treatment followed by secondary treatment, and sludge digestion. The raw wastewater flow rate is 15,600 m³/d with 300 mg/L suspended solids and 250 mg/L BOD_5. The primary clarifier removes 30% BOD_5 and 60% suspended solids from raw wastewater. The secondary underflow contains 1200 kg/d of biological solids. There are two options for transporting the primary and secondary sludges to the digester. They are:

Option 1: Transport the primary and secondary sludges in separate pipelines to the digester. The primary sludge has a solids concentration of 4.5%, and the secondary sludge has a solids concentration of 0.5%.

Option 2: Mix the primary and secondary sludges in a mixing tank, before discharging into the digester. The mixed sludge has a solids concentration of 3.5%.

Answer the following:

a. Draw flow diagrams for the two treatment options.
b. Calculate the volume required for option 1, for treatment in a high-rate digester, which requires 12 days for complete digestion.
c. Calculate the volume required for option 2, for treatment in a high-rate digester, which requires 12 days for complete digestion.
d. Comment on which one is the better option and why.

REFERENCES

Appels, L., Baeyens, J., and Degreve, J., Dewil, R. 2008. Principles and Potential of the Anaerobic Digestion of Waste-Activated Sludge. *Progress in Energy and Combustion Science*, 34:755–781.

Aynur, S., Ahmed, F., Riffat, R., and Murthy, S. 2009. Evaluation of Two Dual Digestion Processes: Autothermal Thermophilic Aerobic Pretreatment followed by TPAD versus MAD. *Proceedings of 82nd Annual Conference of Water Environment Federation, WEFTEC '09*, Orlando, FL, USA.

Aynur, S., Chatterjee, S., Wilson, C., Riffat, R., Novak, J., Higgins, M., Abu-Orf, M., Eschborn, R., Le, S. and Murthy, S. 2008. Bench scale Evaluation of Two Enhanced Digestion Processes: Enhanced Enzymic Hydrolysis and Dual Digestion. *Proceedings of 81st Annual Conference of Water Environment Federation, WEFTEC '08*, Chicago, IL, USA.

Aynur, S. K., Dohale, S., Dumit, M., Riffat, R., Abu-Orf, M., and Murthy, S. 2009. Efficiency of Autothermal Thermophilic Aerobic Digestion under Two Different Oxygen Flowrates. *Proceedings of Water Environment Federation Residuals and Biosolids Conference*, Portland, OR, USA.

Aynur, S., Riffat, R., and Murthy, S. 2010. Effect of Hydraulic Retention Time on Pretreatment of Blended Municipal Sludge, *Water Science & Technology*, 64 (4):967–973.

Borowski, S., and Szopa, J. S. 2007. Experiences with Dual Digestion of Municipal Sewage Sludge. *Bioresource Technology*, 98:1199–1207.

CDM Smith. 2021. Driving Net-Zero at DC Water. https://cdmsmith.com/en/Client-Solutions/Projects/DC-Water-Driving-Net-Zero

Cumiskey, A. 2005. UK Leads the Way in Advanced Digestion Technology, *Water and Wastewater International*, 20 (4):35–36.

Davis, M. 2011. *Water and Wastewater Engineering: Design Principles and Practice*. McGraw-Hill, Inc., New York, NY, USA.

DC Water. 2019a. DC Water's Thermal Hydrolysis and Anaerobic Digester Project. https://www.dcwater.com/sites/default/files/documents/BioenergyFacility.pdf

DC Water. 2019b. *Blue Plains Advanced Wastewater Treatment Plant Brochure*. DC Water and Sewer Authority, Washington, DC, USA.

e-CFR. 2021. *Electronic Code of Federal Regulations*, Title 40, Part 503, Federal Government, USA, https://www.ecfr.gov/cgi-bin/text-idx?SID=5f51a56877

904ba1dccadd419dc305a8&mc=true&tpl=/ecfrbrowse/Title40/40cfr503_ main_02.tpl

Eschborn, R. and Thompson, D. 2007. The Tagro story - How the City of Tacoma, Washington Went Beyond Public Acceptance to Achieve the Biosolids Program Words We'd Like to Hear: Sold Out. WEF/AWWA Joint Residuals and Biosolids Management Conference.

Federal Register. 2010. *Code of Federal Regulations*, Title 40. Federal Government, USA.

Fukushi, K., Babel, S., Burakrai, S. 2003. Survival of Salmonella spp. in a Simulated Acid-Phase Anaerobic Digester Treating Sewage Sludge. *Bioresource Technology*, 86:53–57.

Ghosh, S. and Pohland, F.G. 1974. Kinetics of Substrate Assimilation and Product Formation in Anaerobic Digestion. *Journal of Water Pollution Control Federation*, 46:748.

Gomec, C. Y., Kim, M., Ahn, Y., Speece, R.E. 2002. The Role of pH in Mesophilic Anaerobic Sludge Solubilization. *Journal of Environmental Science and Health*, Part A, 37 (10):1871–1878.

Grady, C. P. L. Jr., Daigger, G. T. and Lim, H. C. 1999. *Biological Wastewater Treatment*. Marcel Dekker, NY, USA.

Greater Vancouver Regional District. 2005. *Review of Alternative Technologies for Biosolids Treatment.*Greater Vancouver Regional District, Burnaby, British Columbia, Canada.

Hammer, M. J. and Hammer, M. J. Jr. 2012. *Water and Wastewater Technology*. 7th edn. Pearson-Prentice Hall, Inc., Hoboken, NJ, USA.

Han, Y., Sung, S., Dague, R. R. 1997. Temperature-Phased Anaerobic Digestion of Wastewater Sludges. *Water Science & Technology*, 36 (6–7):367–374.

Harikishan, S., and Sung, S. 2003. Cattle Waste Treatment and Class A Biosolid Production using Temperature-phased Anaerobic Digester. *Advances in Environmental Research*, 7:701–706.

Hong, S. M., Park, J.K., Teeradej, N., Lee, Y.O., Cho, Y.K., Park, C. H. 2006. Pretreatment of Sludge with Microwaves for Pathogen Destruction and Improved Anaerobic Digestion Performance. *Water Environment Research*, 78 (1):76–83.

Layden, N. M. 2007. An Evaluation of Autothermal Thermophilic Aerobic Digestion (ATAD) of Municipal Sludge in Ireland. *Journal of Environmental Engineering Science*, 6:19–29.

Le, M. S. and Harrison, D. 2006. Application of Enzymic Hydrolysis Technology for VFA Production and Agricultural Recycling of Sludge. *Proceedings of the IWA Conference*, Moscow, Russia.

McCarty, P. L. 1964. Anaerobic Waste Treatment Fundamentals. *Public Works*, 95:9–12.

Metcalf and Eddy, Tchobanoglous, G., Stensel, H., Tcuchihashi, R., and Burton, F. 2013. *Wastewater Engineering: Treatment and Resource Recovery*. 5th edn. McGraw-Hill, Inc., New York, NY, USA.

Nosrati, M., Sreekishnan, T.R., Mukhopadhyay, S.N. 2007. Energy Audit, Solids Reduction, and Pathogen Inactivation in Secondary Sludges During Batch Thermophilic Aerobic Digestion Process. *Journal of Environmental Engineering*, 133 (5):477–484.

Peavy, H. S., Rowe, D. R., and Tchobanoglous, G. 1985. *Environmental Engineering*, McGraw-Hill, Inc., New York, NY, USA.

Salsali, H.R., Parker, W.J., Sattar, S.A. 2006. Impact of Concentration, Temperature, and pH on Inactivation of Salmonella spp. by Volatile Fatty Acids in Anaerobic Digestion. *Canadian Journal of Microbiology*, 52:279–286.

Sung, S., and Santha, H. 2003. Performance of Temperature-Phased Anaerobic Digestion (TPAD) System Treating Dairy Cattle Wastes. *Water Research*, 37:1628–1636.

US EPA. 1979. *Process Design Manual of Sludge Treatment and Disposal*. United States Environmental Protection Agency, Washington, DC, USA.

US EPA. 1983. *Process Design Manual for Land Application of Municipal Sludge*. EPA 625/1-83-016, United States Environmental Protection Agency, Washington, DC, USA.

US EPA. 1993. *Standards for the Use or Disposal of Sewage Sludge*: 40 CFR Part 257, 403 and 503, (58-FR-9248) Environmental Protection Agency, Office of Water, Washington, DC, USA.

US EPA. 1994. *A Plain English Guide to the EPA Part 503 Biosolids Rule*. EPA/832/R-93/003, United States Environmental Protection Agency, Office of Wastewater Management, Washington, DC, USA.

US EPA. 2000. *Biosolids Technology Fact Sheet: Land Application of Biosolids*. EPA 832-F-00-064, Environmental Protection Agency, Office of Water, Washington, DC, USA.

Van Lier, J. B., Grolle, K.C.F., Frijiters, C.T.M.J., Stams, A.J.M. and Lettinga, G. 1993. Effects of Acetate, Propionate, and Butyrate on Thermophilic Anaerobic Degradation of Propionate by Methanogenic Sludge and Defined Cultures. *Applied Environmental Microbiology*, 43:227–235.

Ward, A., Stensel, H.D., Ferguson, J.F., Ma, G., Hummel, S. 1998. Effect of Autothermal Treatment on Anaerobic Digestion in the Dual Digestion Process. *Water Science & Technology*, 38 (8–9):435–442.

WEF. 1998. *Design of Municipal Wastewater Treatment Plants. Manual of Practice 8*. 4th edn. Water Environment Federation, Alexandria, VA, USA.

Werker, A.G., Carlsson, M., Morgan-Sagastume, F., Le, M.S., and Harrison, D. 2007. Full Scale Demonstration and Assessment of Enzymic Hydrolysis Pre-Treatment for Mesophilic Anerobic Digestion of Municipal Wastewater Treatment Sludge. *Proceedings of the 80th Annual Conference of Water Environment Federation, WEFTEC '07*, San Diego, California USA.

Zabranska, J., Dohanyos, M., Jenicek, P., Ruzicikova, H., Vranova, A. 2003. Efficiency of Autothermal Aerobic Digestion of Municipal Wastewater Sludge in Removing Salmonella spp. and Indicator Bacteria. *Water Science and Technology*, 47 (3):151–156.

Zeikus, J.G. 1979. Thermophilic Bacteria: Ecology, Physiology, and Technology. *Enzyme and Microbial Technology* 1:243–251.

Chapter 13

Advanced treatment processes

13.1 INTRODUCTION

When the effluent from secondary treatment does not meet regulatory requirements for discharge, additional treatment may be needed to reduce the levels of specific contaminants. This is usually termed *advanced treatment or tertiary treatment*. Advanced treatment processes are used for the removal of nutrients such as nitrogen and phosphorus, removal of residual total suspended solids (TSS), removal of specific heavy metals or inorganics, and removal of emerging contaminants of concern, among others. Advanced treatment processes may be incorporated within the primary or secondary treatment units, e.g. for biological nutrient removal (BNR), or they may be added separately following secondary treatment, e.g. for wastewater reclamation. These include chemical precipitation, carbon adsorption, granular media filtration, and membrane filtration, among others. A number of these processes will be discussed in detail in this chapter, with regard to the types of contaminants that are removed by the processes. The focus of this chapter will be the removal of nitrogen, phosphorus, suspended and dissolved solids, and other inorganics from wastewater.

13.2 NITROGEN REMOVAL

Excess nutrients, such as nitrogen and phosphorus can cause *eutrophication* problems in water bodies as described previously in Section 3.5. Effluent discharge limits on nutrients can necessitate the use of advanced processes for the removal of nitrogen and phosphorus from the treatment plant effluent. Wastewater treatment plants that use biological processes for nutrient removal are commonly known as BNR plants. The most commonly used method of nitrogen removal is *biological nitrification–denitrification*. This can be accomplished by a number of different suspended and attached growth processes. Some of these are described in the following sections. Emerging technologies that use *biological deammonification* are

DOI: 10.1201/9781003134374-13

also presented. Physico-chemical processes can be used for nitrogen removal. One example of this is air stripping, which is also discussed.

13.2.1 Biological nitrogen removal

Nitrogen compounds are formed in domestic wastewater from the biodegradation of proteins and urea discharged in body waste. The organic nitrogen compounds are further converted to the aqueous ammonium ion (NH_4^+) or gaseous free ammonia (NH_3). These two species together are called *ammonia-nitrogen* (NH_4-N), and remain in equilibrium according to the following relationship:

$$NH_4^+ \rightleftharpoons NH_3 + H^+ \qquad\qquad (13.1)$$

The pH and temperature affect the relative concentrations of the two species in water, as illustrated in Figure 13.1.

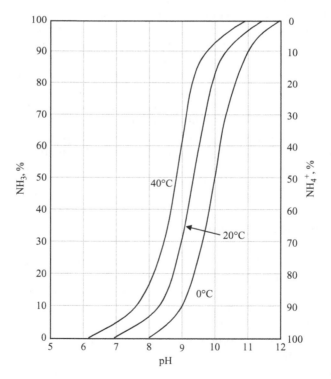

Figure 13.1 Relative distribution of ammonia and ammonium ion in water according to pH and temperature.

Source: Adapted from EPA, 1977

The removal of ammonia-nitrogen or ammonia from water is carried out by biological (i) *nitrification–denitrification* process, (ii) *nitritation–denitritation* process, and (iii) *deammonification* process. Descriptions of each process are provided in the following sections.

13.2.1.1 Nitrification–denitrification

Nitrification involves the conversion of ammonia to nitrates, while denitrification involves the conversion of the nitrates to nitrogen gas which is released to the atmosphere. The overall nitrification–denitrification process is illustrated in Figure 13.2. The conditions and process requirements for nitrification and denitrification are very different from one another as described below.

13.2.1.1.1 Nitrification stoichiometry

Nitrification is a two-step process where ammonia is oxidized to nitrite (NO_2^-) in the first step, and the nitrite is further oxidized to nitrate (NO_3^-) in the second step, as described previously in Section 3.2. Aerobic autotrophic bacteria carry out these reactions as shown below (Metcalf and Eddy et al., 2013):

$$2NH_4^+ + 3O_2 \xrightarrow{\text{nitrosomonas}} 2NO_2^- + 4H^+ + 2H_2O + \text{energy} \qquad (13.2)$$

$$2NO_2^- + O_2 \xrightarrow{\text{nitrobacter}} 2NO_3^- + \text{energy} \qquad (13.3)$$

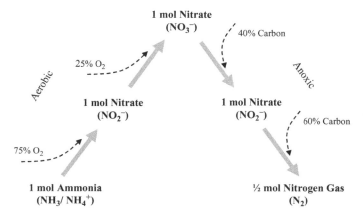

Figure 13.2 Schematic of the nitrification–denitrification process.

Source: Adapted from Murthy, 2011

The total oxidation reaction is

$$NH_4^+ + 2O_2 \rightarrow NO_3^- + 2H^+ + H_2O + energy \qquad (13.4)$$

From Equation (13.4), the oxygen required for total oxidation of ammonia is 4.57 g O_2/g N oxidized, with 3.43 g O_2/g N used for nitrite production and 1.14 g O_2/g N used for nitrate production. When cell synthesis is considered, the amount of oxygen is less than 4.57 g O_2/g N, as a portion of the ammonia is assimilated into cell tissue.

Neglecting cell tissue, the amount of alkalinity required to carry out the oxidation reaction given in Equation (13.4) is

$$NH_4^+ + 2HCO_3^- + 2O_2 \rightarrow NO_3^- + 2CO_2 + 3H_2O \qquad (13.5)$$

From Equation (13.5), 7.14 g of alkalinity as $CaCO_3$ is required for each gram of ammonia-nitrogen (as N) converted.

The biomass synthesis reaction is given by

$$NH_4^+ + 4CO_2 + HCO_3^- + H_2O \rightarrow C_5H_7O_2N + 5O_2 \qquad (13.6)$$

where $C_5H_7O_2N$ represents synthesized bacterial cells.

13.2.1.1.2 Nitrification kinetics

The rate-limiting step in nitrification is the conversion of ammonia to nitrite, as represented by Equation (13.2). This is true for systems operated below 28°C. The design of nitrification systems operated below 28°C are based on saturation kinetics of ammonia oxidation as shown below (Metcalf and Eddy et al., 2013), with the assumption that excess dissolved oxygen (DO) is available.

$$\mu_N = \left(\frac{\mu_{maxN}\, N}{K_N + N} \right) - k_{dN} \qquad (13.7)$$

where
μ_N = specific growth rate of nitrifying bacteria, d^{-1}
μ_{maxN} = maximum specific growth rate of nitrifying bacteria, d^{-1}
N = ammonia-nitrogen concentration, g/m^3
K_N = half saturation coefficient for ammonia-nitrogen, g/m^3
k_{dN} = endogenous decay coefficient for nitrifying bacteria, d^{-1}

Equation (13.7) is a form of the Monod model and can be applied to completely mixed activated sludge systems. Temperature, pH, and dissolved oxygen concentration are important parameters in nitrification kinetics. Temperature effects can be modeled by the van't Hoff–Arrhenius equation as shown in Equation (8.20). The maximum specific growth rate for nitrifiers varies from 0.25 to 0.77 d^{-1}, depending on site-specific conditions (Randall et al., 1992). The growth rate for nitrifying organisms is much lower than the corresponding values for heterotrophic organisms, requiring much longer Solids Retention Time (SRT) for the nitrification process. Typical design SRT values range from 10 to 20 d at 10°C, and from 4 to 7 d at 20°C. Above 28°C, both ammonia and nitrite oxidation kinetics have to be considered in the design.

Dissolved oxygen concentrations of 3–4 mg/L in the water can increase nitrification rates. To incorporate the effects of dissolved oxygen, Equation (13.7) can be modified as shown below:

$$\mu_N = \left(\frac{\mu_{maxN} \, N}{K_N + N} \right) \left(\frac{DO}{K_o + DO} \right) - k_{dN} \tag{13.8}$$

where
   DO = dissolved oxygen concentration, g/m^3
   K_o = half saturation coefficient for DO, g/m^3

All other terms are as defined previously. These kinetic models are best used to describe nitrification in systems at low to moderate organic loadings, but will usually over-predict rates in systems with high organic loadings.

13.2.1.1.3 Denitrification stoichiometry

The final step in biological nitrogen removal is denitrification, which involves the reduction of nitrate to nitric oxide (NO), nitrous oxide (N$_2$O), and nitrogen gas (N$_2$), and is carried out by a variety of heterotrophic and autotrophic bacteria. In wastewater processes, most are facultative anaerobes of the *Pseudomonas* species. The metabolic pathway of denitrification can be represented by the following equation:

$$NO_3^- \rightarrow NO_2^- \rightarrow NO \rightarrow N_2O \rightarrow N_2 \tag{13.9}$$

Denitrification can be considered as a two-step process, since nitrites may appear as an intermediate. This two-step process is termed "dissimilation." The first step represents a reduction of nitrate to nitrite, and the second step, a reduction of nitrite to nitrogen gas (McCarty et al., 1969). Nitric oxide gas (NO) is only an intermediate product, but nitrous oxide gas (N$_2$O) could be the final product of a few denitrifiers. In most cases, nitrogen gas (N$_2$) is the end product of denitrification.

Denitrification takes place in the presence of nitrate and the absence of oxygen. The dissolved oxygen level must be at or near zero, and a carbon supply must be available for the bacteria. The nitrate acts as an electron acceptor for organic or inorganic electron donors. Since a low carbon content is required for the previous nitrification step, additional carbon has to be added for the denitrification step. This can be added in the form of primary effluent wastewater which has BOD, or by adding an external carbon source such as methanol, ethanol, acetate, or glycerol. An external carbon source is used when it is desired to achieve very low levels of nitrogen in the effluent due to regulatory requirements.

When wastewater is the electron donor for denitrification, the reaction can be as follows (Metcalf and Eddy et al., 2013):

$$C_{10}H_{19}O_3N + 10NO_3^- \rightarrow 5N_2 + 10CO_2 + 3H_2O + NH_3 + 10OH^- \quad (13.10)$$

where $C_{10}H_{19}O_3N$ is a generalized formula for the wastewater.

The following reaction takes place with methanol as the electron donor:

$$5CH_3OH + 6NO_3^- \rightarrow 3N_2 + 5CO_2 + 7H_2O + 6OH^- \quad (13.11)$$

When ethanol is used as the external carbon source, the reaction is the following:

$$5CH_3CH_2OH + 12NO_3^- \rightarrow 6N_2 + 10CO_2 + 9H_2O + 12OH^- \quad (13.12)$$

The following reaction takes place with acetate as the electron donor:

$$5CH_3COOH + 8NO_3^- \rightarrow 4N_2 + 10CO_2 + 6H_2O + 8OH^- \quad (13.13)$$

In Equations (13.10)–(13.13), one equivalent of alkalinity (OH⁻) is produced per equivalent of NO_3-N reduced. This amounts to 3.57 g alkalinity as $CaCO_3$ per g nitrogen reduced. This indicates that about half of the alkalinity consumed in nitrification can be recovered in denitrification. The oxygen equivalents of nitrate and nitrite can be calculated as 2.86 g O_2/g NO_3-N and 1.71 g O_2/g NO_2-N, respectively.

The denitrification potential of wastewater is primarily determined as the stoichiometric ratio between the organic compound used and the nitrate, which is usually expressed as the COD/N or the BOD/N ratio. An important design parameter for the denitrification process is the amount of BOD or biodegradable soluble COD (bsCOD) required as the electron donor for nitrogen removal from wastewater. This can be estimated from the following relationship. Readers are referred to Metcalf and Eddy et al. (2013) for a complete derivation of the following:

$$g\,bsCOD/g\,NO_3 - N = \frac{2.86}{1 - 1.42\,Y_n} \qquad (13.14)$$

where

Y_n = net biomass yield, g VSS/g bsCOD
1.42 = oxygen equivalent of biomass, g O_2/g VSS
2.86 = oxygen equivalent of nitrate, g O_2/g NO_3-N

The net biomass yield can be calculated from the following equation (Metcalf and Eddy et al., 2013):

$$Y_n = \frac{Y}{1 + k_d\,\theta_c} \qquad (13.15)$$

where

Y = biomass yield for denitrifiers, g VSS/g bsCOD
k_d = decay coefficient for denitrifiers, d^{-1}
θ_c = anoxic SRT, d

13.2.1.1.4 Denitrification kinetics

The Monod model similar to Equation (13.7) can be developed for the denitrifying bacteria. The Specific Denitrification Rate (SDNR) or the rate of substrate utilization can be calculated from the Monod model and the concepts presented in Chapter 8, together with a term to account for the lower utilization rate in the anoxic zone.

$$r_{su} = -\frac{\mu_{max}\,S\,X\,\eta}{Y\,(K_S + S)} \qquad (13.16)$$

where

r_{su} = rate of substrate utilization, mg/L.d
η = fraction of denitrifying bacteria in the biomass.

All other terms are as defined previously. The value of η can range from 0.2 to 0.8 for pre-anoxic denitrification (Stensel and Horne, 2000). For postanoxic processes, η is 1.0, where the biomass is mainly denitrifying bacteria.

13.2.1.1.5 External carbon sources for denitrification

When a wastewater treatment plant requires significant nitrogen removal, the organic matter naturally present in the wastewater may be insufficient to achieve the required level of denitrification. This requires the addition of an external carbon source. Some external carbon sources include methanol, ethanol, acetic acid, glucose, glycerol, etc. In the last two decades, a significant amount of research has been conducted on the investigation of different carbon sources, their applications, kinetic parameters, and temperature effects. The impetus for this has been concerns of detrimental environmental effects of effluent nitrogen discharges to water bodies.

One example of this is the lowering of effluent limits to 3–4 mg/L of total N for wastewater treatment plants discharging into the Chesapeake Bay watershed in the eastern US. The Chesapeake Bay is the largest estuary in the US, encompassing six states including Delaware, Maryland, New York, Pennsylvania, Virginia, and West Virginia, and the District of Columbia. Both point and non-point sources have contributed excess nutrients to the Bay resulting in severe deterioration and impaired waters. As a result, stricter effluent limits have been imposed on the point sources which are mainly the wastewater treatment plants. Most of the treatment plants in that area use an external carbon source for denitrification. But sizing and operation of treatment units based on previously existing kinetic parameters have failed to produce the desired results, especially at low temperatures. That has provided the momentum for new research in methodology and determination of kinetic parameters for denitrification with external carbon sources. Understanding the kinetics and stoichiometry of the denitrifying organisms is of prime importance in designing and optimizing nitrogen removal processes.

Research on denitrification has been conducted for several decades now. Various researchers have measured kinetic parameters of denitrification for a variety of process configurations. In earlier studies, using methanol (MeOH) as the external carbon Stensel et al. (1973) reported maximum specific growth rates (μ_{maxDEN}) of 1.86 and 0.52 d^{-1} at 20°C and 10°C, respectively. Decay coefficients at these two temperatures were 0.04 and 0.05 d^{-1}, respectively. Recent researchers have observed lower growth rates for methanol utilizers in extensive studies conducted at a large number of treatment plants in the eastern US. Dold et al. (2008) observed a maximum specific growth rate of 1.3 d^{-1} at 20°C with an Arrhenius coefficient (θ) of 1.13, based on a decay rate (k_{dDEN}) of 0.04 d^{-1}. Nichols et al. (2007) obtained a maximum specific growth rate of 1.25 d^{-1} at 20°C with an Arrhenius coefficient of 1.13. The carbon-to-nitrogen ratio was approximately 4.73 mg MeOH COD/mg NO$_3$-N. The variation of μ_{max} with temperature is illustrated in Figure 13.3, with the dashed line representing the van't Hoff–Arrhenius model. Maximum specific growth rates of 1.0 and 0.5 d^{-1} at 19°C and 13°C, with an Arrhenius coefficient of 1.12 were observed by Mokhayeri et al. (2006). The low growth rate (similar to that of nitrifiers) indicated that systems should be designed based on a long enough anoxic SRT to ensure stable growth and avoid washout. This is

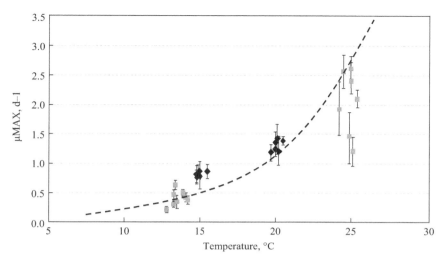

Figure 13.3 Variation of maximum specific growth rate of denitrifiers with temperature.

Source: Data from Nichols et al., 2007; and Hinojosa 2008.

exacerbated by the strong temperature dependency for plants operating at low temperatures.

Three external carbon sources – methanol, ethanol, and acetate – were evaluated for denitrification by Mokhayeri et al. (2009, 2008, 2007). At 13°C, the SDNRs for biomass grown on methanol, ethanol, and acetate were 10.1 mg NO_3-N/g VSS/h, 29.6 mg NO_3-N/g VSS/h, and 31.0 mg NO_3-N/g VSS/h, respectively, suggesting that acetate and ethanol were equally effective external carbon sources followed by much lower SDNR using methanol. The yield coefficients were observed to be 0.45 g/g, 0.53 g/g, and 0.66 g/g for methanol, ethanol, and acetate, respectively. Ethanol could be used with methanol biomass with similar rates as that of methanol. Additionally, methanol was rapidly acclimated to ethanol-grown biomass suggesting that the two substrates could be interchanged to grow respective populations with a minimum lag period. Methanol is used at a large number of treatment plants in the US because of its low cost. But the rates reduce significantly with methanol at cold temperatures. This may be overcome by using an alternate substrate to methanol in winter.

Studies conducted with glycerol as an external carbon source by Hinojosa et al. (2008) determined the following kinetic coefficients: maximum specific growth rate of 3.4 d^{-1} at 21°C, stoichiometric C:N ratio of 4.2 mg COD/mg NO_x-N, growth yield (Y_{DEN}) of 0.32 mg biomass COD/mg NO_x-N, SDNR of 1.4 mg NO_x-N/g VSS/h, and half saturation coefficient (K_{SDEN}) 5–8 mg COD/L.

Example 13.1: A municipal wastewater treatment plant is planning to upgrade to a nitrogen removal plant. It has successfully incorporated nitrification with BOD removal in the existing activated sludge process. The plant wants to add a separate denitrification system consisting of an anoxic tank followed by a clarifier. Design a suspended growth denitrification system for the plant using methanol as a carbon source. Calculate the tank volume and methanol dose required to achieve an effluent NO_3-N concentration of 3 mg/L. The following data are provided.

Wastewater effluent from nitrification system:
Flow rate = 3000 m^3/d, Temperature = 20°C
NO_3-N = 30 mg/L, TSS = 20 mg/L
Denitrification kinetic coefficients with methanol at 20°C:

$$\mu_{max} = 1.3\,d^{-1}, k_d = 0.04\,d^{-1}, K_s = 4\,mg\,bsCOD/L,$$
$$Y = 0.35\,kg\,VSS/kg\,bsCOD$$

COD equivalent of methanol = 1.5 kg COD/kg methanol
Denitrification tank: MLSS = 2500 mg/L, SRT = 6 d, HRT = 2 h
Clarifier: overflow rate = 24 m^3/m^2.d

SOLUTION
Step 1. Determine the tank volume based on HRT.

$$V = Q \times HRT = 3000\,m^3\,/\,d \times 2\,h \times \frac{1}{24\,h/d} = 250\,m^3$$

Use two tanks, with the volume of each tank = 250/2 = 125 m^3
Step 2. Determine methanol required for nitrate reduction.

$$NO_3 - N\,reduced = (30-3)mg\,/\,L = 27\,mg\,/\,L = 0.027\,kg\,/\,m^3$$

Calculate the net biomass yield using Equation (13.15)

$$Y_n = \frac{Y}{1 + k_d\,\theta_c}$$

$$or, Y_n = \frac{0.35\,kg\,/\,kg}{1 + (0.04\,d^{-1} \times 6\,d)} = 0.282\,kg\,/\,kg$$

Calculate the C:N ratio using Equation (13.14)

$$kg\,bsCOD/kg\,NO3 - N = \frac{2.86}{1 - 1.42\,Y_n} = \frac{2.86}{1 - (1.42 \times 0.282)} = 4.767\,kg/kg$$

Methanol required for nitrate reduction $= 4.767\,kg/kg \times 0.027\,kg/m^3$
$$= 0.129\,kg/m^3 \text{ as COD}$$

Daily methanol dose $= 3000\,m^3\,/\,d \times 0.129\,kg\,/\,m^3 = 387\,kg\,/\,d\text{ as COD}$

COD equivalent of methanol $= 1.5\,kg\,COD\,/\,kg\,methanol$

or, Daily methanol dose $= \dfrac{387\,kg\,COD\,/\,d}{1.5\,kg\,COD\,/\,kg\,methanol} = \mathbf{258\,kg\,/\,d}$

Step 3. Calculate the area of the clarifier based on the overflow rate.

$$A = \frac{Q}{v} = \frac{3000\,m^3\,/\,d}{24\,m\,/\,d} = \mathbf{125\,m^2}$$

Use two clarifiers.

Area of each clarifier $= 125\,/\,2 = 62.5\,m^2$

Diameter of each clarifier $= 8.92\,m \cong 9.0\,m$

13.2.1.1.6 Nitrification–denitrification processes

Biological nitrogen removal processes require an aerobic zone for nitrification, and an anoxic zone for denitrification. The processes may be broadly classified into three types: (1) *Preanoxic*, where an anoxic zone is followed by an aerobic zone; and (2) *Postanoxic*, where the aerobic zone is followed by an anoxic zone, and (3) a third type is where nitrification and denitrification occur in the same reactor/tank, e.g. SBR (Sequencing Batch Reactor). *Preanoxic* and *postanoxic* processes are illustrated in Figure 13.4. Suspended growth nitrogen removal processes can be further classified as (i) single-sludge system, where one clarifier is used for solids separation, though internal recirculation may be used between tanks (Figure 13.4), and (ii) two-sludge system, where the nitrification tank is followed by a clarifier, and the denitrification tank is followed by another clarifier. This is a post-anoxic process, usually where an external carbon source is added. This is illustrated in Figure 13.5 (a). The first tank may be used for combined BOD removal and nitrification.

Examples of preanoxic processes are the MLE (Modified Ludzack–Ettinger) process, step feed process, among others. Postanoxic processes include the oxidation ditch, etc. The Bardenpho process is a combination of

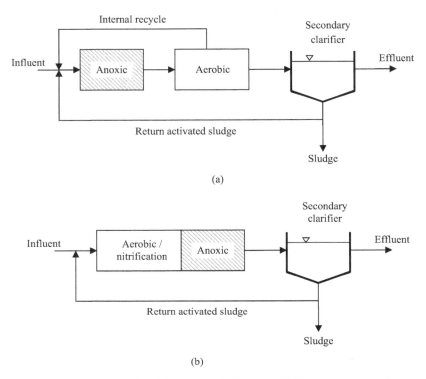

Figure 13.4 (a) Preanoxic Modified Lutzack–Ettinger (MLE) process and (b) post-anoxic process.

preanoxic and postanoxic processes. Membrane bioreactors can be used for nitrogen removal.

Attached growth processes are also used for nitrogen removal. Trickling filters and RBCs (Rotating Biological Contactors) can be used for both BOD removal and nitrification. Coarse media deep bed anaerobic filters have been used for denitrification for a long time. An external carbon such as methanol is usually added to the filter. The Moving Bed Biofilm Reactor (MBBR) process can be used for both nitrification and denitrification. Some of these processes are described in detail in the following sections.

Modified Ludzack–Ettinger (MLE) process – This is a widely used process that consists of the preanoxic system illustrated in Figure 13.4 (a). Wastewater flows into the anoxic zone and provides the carbon necessary for denitrification. The internal recycle was designed by Barnard (1973) to increase nitrate feed to the anoxic zone, as a modification of the original design. With sufficient influent BOD and anoxic contact time (2–4 h), average effluent NO_3-N concentrations of 4–7 mg/L can be achieved. Efficiency can be increased by dividing the anoxic zone into 3–4 stages in series.

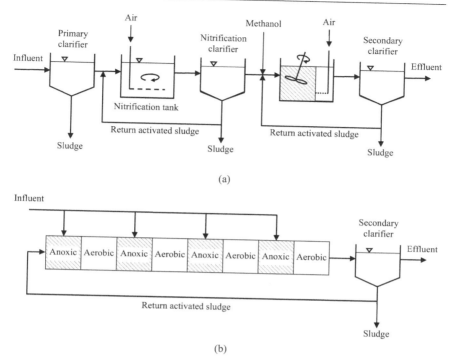

Figure 13.5 (a) Two-sludge or two-stage nitrification–denitrification system and (b) step feed process.

Step feed process – This is similar to the activated sludge step feed process, except that the tank is divided into anoxic and aerobic zones as shown in Figure 13.5 (b). The portion of flow going to the last anoxic/aerobic zone is critical since the nitrate produced in the last aerobic zone will not be reduced and will remain in the final effluent. Effluent NO_3-N concentrations less than 8 mg/L can be achieved (Metcalf and Eddy et al., 2013).

Bardenpho™ (four-stage) process – The Bardenpho™ process uses both preanoxic and postanoxic stages as illustrated in Figure 13.6 (a). The mixed liquor from the first aerobic zone is recycled to the preanoxic zone to provide nitrate for denitrification. The process was developed and applied at full scale in South Africa in the 1970s. Since then it has been used worldwide. It is capable of achieving both nitrogen and phosphorus removal. The name of the process is derived from the first three letters of the inventor's name, Barnard; denitrification and phosphorus (Barnard, 1974).

Oxidation ditch – This is a postanoxic process consisting of an aerobic zone followed by an anoxic zone in an oxidation ditch. Wastewater enters the aerobic zone close to the aerator. As it flows toward the anoxic zone, dissolved oxygen is used up by the microorganisms during the

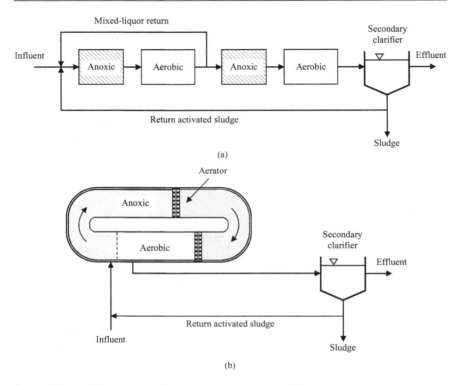

Figure 13.6 (a) Four-stage Bardenpho process and (b) oxidation ditch.

degradation of BOD and nitrification. Eventually when the dissolved oxygen is depleted, an anoxic zone is formed and denitrification occurs due to endogenous respiration by the bacteria using nitrate. The process is illustrated in Figure 13.6 (b). Large tank volumes and long SRTs have to be maintained.

MBBR (moving bed biofilm reactor) – The MBBR process can be used for BOD removal as well as nitrification and denitrification. This attached growth process has been described previously in Section 9.8. Figure 13.7 (a,b) illustrates the MBBR with two types of mixing options. Figure 13.7 (c) illustrates biofilm growth on an MBBR media. Full-scale plant applications in Norway have demonstrated high rates of nitrification and denitrification (Ødegaard, 2006, Andrettola et al., 2003).

A denitrification rate of 1.8 g N/m^2.d with methanol at 15°C was observed by Rusten et al. (1995) for an MBBR system. Aspegren et al. (1998) obtained maximum denitrification of 2.0 g N/m^2.d at 16°C using methanol for post-denitrification process in an MBBR system. Pilot plant MBBR studies at Blue Plains Advanced Wastewater Treatment Plant in Washington, DC, USA, produced specific denitrification rates ranging from

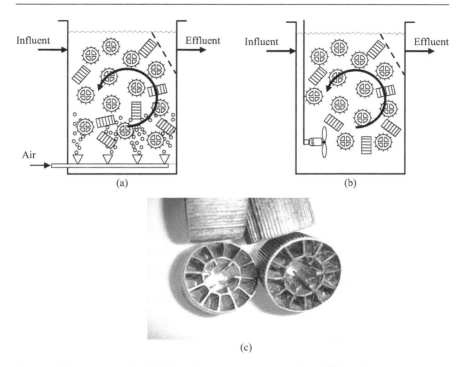

Figure 13.7 (a) Aerobic MBBR, (b) anoxic and aerobic MBBR. (Source: Adapted from Ødegaard, 2006), and (c) biofilm on MBBR media. Photo courtesy of Arbina Shrestha).

1.3 to 2.0 g NO_x-N/m^2-day, and stoichiometric C:N ratios of 4.6–5.8 mg COD/mg NO_x-N, for a temperature range of 13–18°C. Values of effective Ks_{NOx-N} were observed between 0.6 and 2.6 mg N/L (Shrestha et al., 2009, Peric et al., 2010).

13.2.1.2 Nitritation–denitritation

In the conventional nitrification–denitrification process, ammonia is converted to nitrites and then to nitrates, followed by reduction of nitrates again to nitrites and finally to nitrogen gas, as presented in Equations (13.2), (13.3), and (13.9). A number of researchers have investigated nitrogen removal by partial nitrification or *nitritation* (oxidation of ammonia to nitrite) followed by partial denitrification or *denitritation* (reduction of nitrite to nitrogen gas). This results in significant savings in oxygen demand, lower carbon requirement for denitrification, and a reduction in the amount of excess sludge produced (Ruiz et al., 2003; Ciudad et al., 2005). The

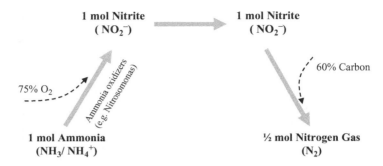

Figure 13.8 Schematic of the nitritation–denitritation process.

Source: Adapted from Murthy, 2011.

overall process is illustrated in Figure 13.8. Nitrite accumulation is obtained by optimizing dissolved oxygen, pH, and temperature. An example of this is the SHARON™ process described below.

13.2.1.2.1 SHARON™ process

The SHARON™ (Single reactor system for High Ammonium Removal Over Nitrite) process was developed in the Netherlands, initially to reduce the load of wastewater streams and side streams with high ammonium concentration. The reactor is designed to select for ammonium oxidizers by washing out nitrite oxidizers, using a short retention time of approximately 1 d, and a temperature above 30°C (van Dongen et al., 2001). Longer aerobic and shorter anoxic phases are used, with methanol addition in the anoxic phase. Compared to the conventional nitrification–denitrification process, the oxygen demand is reduced by 25% and equals 3.43 g O_2/g N, the carbon requirement is reduced by 40% and equals 2.4 g COD/g N (Mulder et al., 2001; Hellinga et al., 1998). There are a number of full-scale installations of the SHARON™ process in the Netherlands.

13.2.1.3 Deammonification

The term *deammonification* is used to describe an ammonium removal process that does not depend on the supply of organic matter (Hippen et al., 1997). It uses aerobic and anaerobic ammonium oxidizers to convert ammonium directly to nitrogen gas under oxygen-limited conditions. The ammonium reacts with nitrite acting as an electron acceptor to produce nitrogen gas. The anaerobic ammonium oxidizers or Anammox bacteria were discovered by Mulder et al. (1995) in a fluidized bed reactor. The Anammox bacteria belong to the phylum *Planctomycetales*. They have a very low growth rate of 0.072 d^{-1} with a mass doubling time of 11 d, which

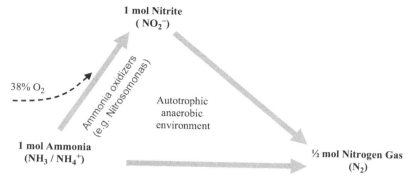

Figure 13.9 Schematic of deammonification.

Source: Adapted from Murthy, 2011.

can be an obstacle in process start-up (Jetten et al., 2001). The overall process is illustrated in Figure 13.9.

The following reactions are carried out by the Anammox bacteria (van Dongen et al., 2001).

Without cell synthesis:

$$NH_4^+ + NO_2^- \rightarrow N_2 + 2H_2O \tag{13.17}$$

With cell synthesis:

$$NH_4^+ + 1.32NO_2^- + 0.066HCO_3^- \rightarrow 1.02N_2 + 0.26NO_3^-$$
$$+ 2.03H_2O + 0.066CH_2O_{0.5}N_{0.15} \tag{13.18}$$

where $0.066CH_2O_{0.5}N_{0.15}$ indicates new cells.

The advantages of deammonification are zero oxygen demand, zero COD requirement, and low sludge production. An example of a deammonification process is the Anammox™ process.

13.2.1.3.1 Anammox™ process

The Anammox™ process was developed in the Netherlands in the late 1990s. The term Anammox is an abbreviation for anaerobic ammonium oxidation. The Anammox process is preceded by a nitritation process that converts half of the ammonium to nitrite, without subsequent conversion of nitrite to nitrate. The oxygen uptake based on initial ammonium concentration is 1.72 g O_2/g N, or 38% of the oxygen demand for oxidation of all the ammonium to nitrate (Gut, 2006). After this process, the Anammox process (Equation 13.17 and 13.18) follows without the addition of any organic material in a separate reactor (van Loosdrecht et al., 2004).

13.2.2 Physico-chemical process for nitrogen removal

Physico-chemical processes may be used for the removal of nitrogen and simultaneous ammonia recovery from wastewater. Air Strippers or Steam Strippers are used at a number of full-scale installations in Europe. In some cases, ammonia is recovered as an ammonium nitrate fertilizer product.

13.2.2.1 Air stripping

The process of air stripping involves the conversion of aqueous ammonium ion (NH_4^+) to gaseous ammonia (NH_3) and releasing it into the atmosphere. The equilibrium between the two species is outlined in Equation (13.1) and Figure 13.1. From Figure 13.1, above pH 11 or 11.5, more than 99% of ammonia will be present in the gaseous phase.

The process consists of pretreatment of the wastewater with lime to raise the pH above 11.5. Enough lime has to be added to precipitate the alkalinity and raise the pH to the desired level. Once the conversion to gaseous ammonia is complete, stripping or degasification is conducted. One of the most efficient reactors is the counter-current spray tower, illustrated in Figure 13.10 (Peavy et al., 1985). Large volumes of air are required. Packing material is provided to minimize film resistance to gas transfer, and to aid in the formation of liquid droplets. Air pollution control may be required for ammonia emissions. Another disadvantage is the reduction in efficiency at

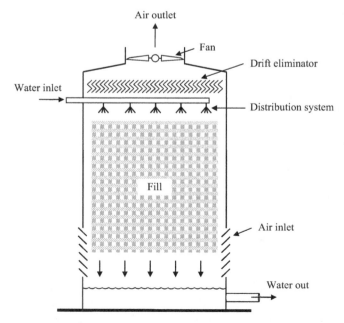

Figure 13.10 Counter-current spray tower for air stripping of ammonia.

cold temperatures. The process is economical when lime precipitation of phosphorus is also desired.

13.3 PHOSPHORUS REMOVAL

Phosphorus is contributed to wastewater mainly from human wastes, and from synthetic wastes such as detergents. The principal form of phosphorus in wastewater is orthophosphate, together with some polyphosphates and organically bound phosphorus. Polyphosphates originate from detergents and can be hydrolyzed to orthophosphates. Organically bound phosphorus comes from body and food wastes and is biologically degraded/converted to orthophosphates. The phosphorus or orthophosphates can be removed from wastewater by chemical or biological processes. These are described below.

13.3.1 Chemical precipitation

Chemical precipitation involves the addition of metallic coagulants or other chemicals to form insoluble compounds with phosphates and remove them by precipitation. Orthophosphates consist of the negative radicals PO_4^{3-}, HPO_4^{2-}, and $H_2PO_4^-$. These can form insoluble compounds with metallic cations, e.g. with iron or aluminum. The following reactions take place with chemical precipitation at acidic pH:

$$Fe^{3+} + PO_4^{3-} \rightarrow FePO_4 \tag{13.19}$$

$$Al^{3+} + PO_4^{3-} \rightarrow AlPO_4 \tag{13.20}$$

Coagulants such as ferric chloride or polyaluminum chlorides can be used for chemical precipitation.

Lime can be added to raise the pH to about 9.0 and form an insoluble complex with phosphates, as shown below:

$$Ca(OH)_2 + PO_4^{3-} \rightarrow Ca_5(OH)(PO_4)_3 + H_2O \tag{13.21}$$

Metallic coagulants and lime consume alkalinity from the wastewater. Thus, chemical dosages are usually 2–3 times greater than that predicted from stoichiometry. Coagulants may be added to the primary or secondary units for combined removal with solids and BOD, or separately in a tertiary unit. Application in primary clarifiers is beneficial since it results in enhanced clarification of BOD and solids. But polymers are required for flocculation (Peavy et al., 1985). Application in a tertiary unit results in the most efficient use of coagulants with the highest removal efficiency. It also has the highest capital cost and metal leakage. Tertiary units for phosphorus precipitation can be designed as flocculator-clarifiers, with in-line mixing of coagulants. Coagulation and flocculation are followed by settling to remove the precipitated compounds.

13.3.2 Biological phosphorus removal (BPR)

Biological phosphorus removal is accomplished by a group of bacteria collectively known as PAOs (*Phosphorus Accumulating Organisms*). The PAOs incorporate large amounts of phosphorus into cell biomass, which is subsequently removed from the process by sludge wasting. PAOs include *Acinetobacter, Arthrobacter, Aeromonas, Nocardia,* and *Pseudomonas* (Davis, 2020). Phosphorus content in PAOs can range from 0.2 to 0.3 g P/g VSS, while in ordinary heterotrophic bacteria it ranges from 0.01 to 0.02 g P/g VSS.

The basic biological phosphorus removal process consists of an anaerobic zone or tank followed by an aeration tank. The anaerobic zone is called a "selector," since it provides the favorable conditions for growth and proliferation of PAOs, with a short HRT of 0.5–1.0 h. A fraction of the biodegradable COD is fermented to acetate and consumed by the PAOs. They produce intracellular PHB (poly-hydroxy-butyrate) storage products and release orthophosphates. In the aerobic zone, PHB is metabolized for new cell synthesis. The energy released from PHB oxidation is used to form polyphosphate bonds in cell storage, leading to the removal of orthophosphates from solution and incorporation into polyphosphates within the bacterial cell (Metcalf and Eddy et al., 2013). Phosphorus is removed from the system when the biomass is wasted. A maximum specific growth rate of 0.95 d^{-1} was observed for PAOs by Barker and Dold (1997).

From stoichiometry, it is estimated that about 10 g of bsCOD is required to remove 1 g P by the biological storage mechanism. This value is based on the following assumptions (Metcalf and Eddy et al., 2013): (1) 1.06 g acetate/g bsCOD is produced in the anaerobic zone; (2) Cell yield is 0.3 g VSS/g acetate; and (3) cell phosphorus content of PAO is 0.3 g P/g VSS. In biological systems, other cations associated with polyphosphate storage, such as Ca, Mg, and K must also be available in sufficient quantities for efficient phosphorus removal. Municipal wastewaters usually have the cations in required quantities.

13.3.2.1 Selected processes for BPR

The following are descriptions of selected processes used for biological phosphorus removal. SBRs can also be used for combined nitrogen and phosphorus removal, where the reactor sequences through anaerobic/anoxic/aerobic phases.

13.3.2.2 Phoredox

The term *Phoredox* was used by Barnard (1975) to designate any BPR process with an anaerobic/aerobic sequence, as illustrated in Figure 13.11 (a). The anaerobic detention time is 0.5–1 h. Low-operating SRT is used to prevent nitrification in the aerobic zone. SRT values range from 2 to 3 d at 20°C, and 4 to 5 d at 10°C, to promote phosphorus removal without simultaneous nitrification (Grady et al., 2011). A variation of this process with multiple stages was patented as the A/O™ process.

13.3.2.3 A²O™ process

The A²O™ process is used for combined nitrogen and phosphorus removal. An anoxic zone is provided between the anaerobic and aerobic zones, as illustrated in Figure 13.11 (b). Anoxic zone detention time is about 1 h. Chemically bound oxygen in the form of nitrate is introduced into the anoxic zone by recirculating effluent from the aerobic zone. This reduces the amount of nitrate fed to the anaerobic zone in the return activated sludge.

13.3.2.4 Modified Bardenpho™ (five stage)

The five-stage Bardenpho™ process illustrated in Figure 13.11 (c) is used for combined carbon, nitrogen, and phosphorus removal. The anaerobic, anoxic, and aerobic stages provide phosphorus, nitrogen, and carbon removal. A second anoxic stage achieves additional denitrification using nitrate produced in the aerobic zone and endogenous organic carbon. The final aerobic stage is used to strip nitrogen gas from the solution and minimize the release of phosphorus in the final clarifier. Process SRT ranges from 10 to 20 d.

13.3.2.5 UCT process

The UCT process was developed at the University of Cape Town in South Africa, and hence its name. The standard UCT process is illustrated in Figure 13.11 (d). The introduction of nitrate to the anaerobic stage is avoided by recycling the activated sludge to the anoxic stage. This improves phosphorus uptake. Anoxic effluent recycle to the anaerobic stage results in increased organic utilization. The anaerobic detention time ranges from 1 to 2 h. The anaerobic recycle rate is usually two times the influent flow rate. The standard process was later modified to provide a second anoxic tank, after the first one. This improved nitrate removal for the process.

13.4 SOLIDS REMOVAL

The presence of excess solids in wastewater effluent can create problems in receiving bodies, depending on the type of solids (suspended or dissolved)

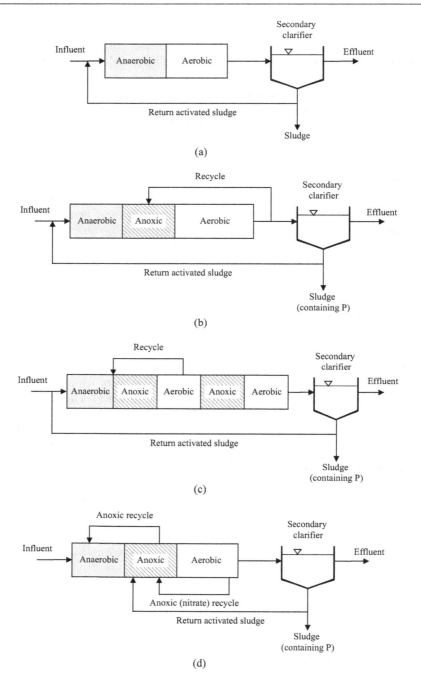

Figure 13.11 (a) Phoredox (A/O™) process, (b) A²O™ process, (c) modified Bardenpho™ process, (d) standard UCT process.

Source: Adapted from Metcalf and Eddy et al., 2013.

and its constituents. Regulatory requirements may also necessitate the use of tertiary treatment for further removal of solids of concern. Granular media filtration is used for the removal of TSS. Processes like membrane filtration, activated carbon adsorption, and ion exchange can be used for the removal of suspended and dissolved solids. Activated carbon adsorption is also used for the removal of odorous compounds that are produced at wastewater treatment facilities.

13.4.1 Granular media filtration

Granular media filtration has always been a part of conventional drinking water treatment. The use of the process in wastewater treatment has increased over the past few decades. It is used as a tertiary treatment to remove TSS from the secondary effluent. Filtration is used when the regulatory limit for effluent TSS is less than or equal to 10 mg/L (Davis, 2020). A simultaneous reduction in BOD is achieved, since a fraction of the TSS is biomass which contributes to the BOD. Deep bed filters are used for denitrification and solids removal. Filtration with chemical coagulation can be used for simultaneous solids and phosphorus removal.

Conventional filters that are used at municipal wastewater treatment plants are operated in a down-flow mode, mainly by gravity. There are some proprietary filters such as the (a) deep bed upflow continuous backwash filter, (b) pulsed bed filter, and (c) traveling bridge filter, that use additional methods of wastewater and airflows. Pressure filters and vacuum filters are used in industrial operations for wastewater treatment. The capital and operation costs of these are much higher than conventional filters.

A typical filter consists of a tank filled with granular media, with an underdrain system at the bottom. A layer of gravel is placed between the media and the underdrain, to prevent loss of media with the effluent. The wastewater enters at the top of the tank and flows down through the media. Solids from the wastewater are removed in the pores of the media by adsorption, diffusion, settling, and other mechanisms. The clarified effluent leaves through the underdrains, as illustrated in Figure 13.12. Over time as solids build up in the filter bed, the efficiency decreases and headloss increases. The filter is cleaned by *backwashing* when the headloss reaches a predetermined terminal limit. The wastewater flow is stopped, and clean water is passed through the filter bed via the underdrain system in a reverse direction at a high velocity to dislodge the collected solids and remove them from the bed.

Important design parameters include flow rates, bed depth, and media characteristics. Type of media and characteristics, such as porosity, effective size, uniformity coefficient, and specific gravity are carefully considered. Typical filtration rates range from 5 to 20 m/h with terminal headlosses of 2.4–3 m (Davis, 2020).

Different types of granular media can be used in a filter. Monomedia or single media filters are hardly used anymore, due to problems of clogging.

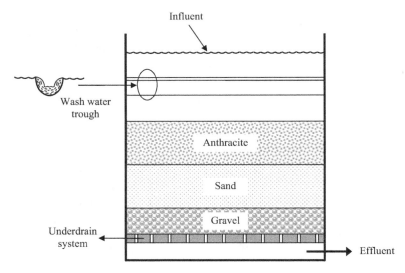

Figure 13.12 Conventional dual media filter.

Dual-media filters and multi-media filters are commonly used. Dual-media filters consist of a layer of anthracite at the top and a layer of silica sand at the bottom, as illustrated in Figure 13.12. Anthracite of a larger effective size and silica sand of a smaller effective size is used to make efficient use of the bed depth. Larger solids are trapped in the anthracite layers, while the smaller solids travel through the bed and are trapped in the lower layers of silica sand. Another advantage is that after backwashing, as the particles settle back in the tank based on their terminal settling velocities and Stokes' law (Equations 7.5 and 7.10), the silica particles with the higher specific gravity settle at the bottom, while the anthracite particles with the lower specific gravity settle on top. Thus, the original configuration is maintained with an intermixed layer in the middle. The ratio of depths of the anthracite and silica sand layer is typically 2:1 or 3:1.

Multi-media filters have a third layer of garnet sand at the bottom, with anthracite at the top and silica sand in the middle layer. This type of filter has a higher capital cost, but has a higher efficiency of solids removal, especially of smaller-sized particles. The garnet sand has a higher specific gravity compared to the other two media, and a smaller effective size is used. The ratio of depths of the anthracite, silica, and garnet sand layer is typically 5:5:1 or 6:5:1. Some of the characteristics of these three types of media are provided in Table 13.1.

13.4.2 Activated carbon adsorption

Activated carbon adsorption is used as a tertiary treatment for the removal of refractory organic compounds and other inorganics including sulfides,

Table 13.1 Characteristics of different media used in granular media filters

Characteristic	Anthracite	Silica sand	Garnet sand
Specific gravity	1.40–1.75	2.55–2.65	3.60–4.30
Porosity	0.55–0.60	0.40–0.45	0.42–0.55
Effective size, mm	1.0–2.0	0.4–0.8	0.2–0.6
Uniformity coefficient	1.4–1.8	1.3–1.8	1.5–1.8
Shape factor	0.4–0.6	0.7–0.8	0.6–0.8

Sources: Metcalf and Eddy et al., 2013; Cleasby and Logsdon, 1999.

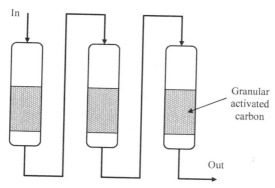

Figure 13.13 Downflow activated carbon contactor in series.

nitrogen, and heavy metals. Refractory organic compounds are resistant to biodegradation, and hence remain in the secondary effluent. When these include chemical contaminants of concern, activated carbon adsorption can be used to remove them. Carbon adsorption is also used when wastewater is reused. Pretreatment of wastewater by granular media filtration and chlorination is usually done prior to carbon adsorption, to improve process efficiency.

Powdered activated carbon (PAC) or granular activated carbon (GAC) is used. PAC can be added directly to the aeration tank, or to the secondary effluent. GAC is used in a column, that may be fixed bed or moving bed. Downflow columns are commonly used in tertiary treatment. Flow can be by gravity or pressure. A number of columns can be operated in series to increase removal efficiencies. Figure 13.13 illustrates a downflow carbon contactor operated in series. GAC columns can be cleaned by backwashing to some extent to limit the headloss, and reduce solids build-up. When the adsorption capacity is exhausted, the spent carbon column can be regenerated by heating under controlled conditions. Sometimes, the carbon column has to be disposed of as hazardous waste and replaced with a new one.

Typical design parameters include carbon size, bed depth, hydraulic loading rate, and empty bed contact time. Adsorption isotherms should be developed for the type of carbon and wastewater from bench-scale laboratory tests. These can be combined with available design parameters from WEF (1998) to design a particular carbon adsorption system.

13.4.3 Membrane filtration

Membrane filtration is used as tertiary treatment to remove solids from wastewater, especially when it is desired to use the effluent for aquifer recharge or indirect reuse. Membrane filtration is used as a pretreatment to remove particulate solids, prior to dissolved ion removal by reverse osmosis. Membrane processes such as *reverse osmosis (RO)* and *nanofiltration (NF)* are widely used for drinking water treatment. These processes operate at very high pressures, usually in excess of 500 kPa. Low-pressure *microfiltration (MF)* and *ultrafiltration (UF)* membranes are used in the tertiary treatment of wastewater. The operating range of pressure for wastewater treatment is from 70 to 200 kPa (Davis, 2020). MF membranes have a pore size ranging from 0.1 to 1.0 μm, while UF membrane pore size can range from 0.005 to 0.1 μm. They are used for the removal of particulates and microorganisms. Removal mechanisms include straining, adsorption, and cake filtration. Over time, particles build up on the membrane surface forming a cake, that increases the filtration efficiency of the membrane. Like granular media filters, MF and UF filters are cleaned by backwashing with water and/or air scouring. Over time, chemical cleaning agents have to be used.

Secondary effluent has to be pretreated before it is fed to membrane filters. The quality of feed water to MF/UF filters must at least meet the standards for secondary effluent, e.g. $BOD_5 \leq 30$ mg/L, TSS ≤ 30 mg/L, and fecal coliforms (FC) $\leq 200/100$ mL, in order to achieve high-quality effluent for potential reuse (WEF, 2006). Chemical coagulation, chlorination, and screening are some of the pretreatment options.

13.4.3.1 Fundamental equations

Membranes are organic polymers that are semipermeable to selected constituents. Commonly used membrane materials include polysulfone (PS), polyethersulfone (PES), polyvinylidene difluoride (PVDF), among others. For filtration, the membrane is placed in a tank. As wastewater flows through the tank under pressure, the membrane prevents the contaminants from flowing through it. As a result, a waste stream with concentrated contaminants (*reject* or *concentrate*), and a clarified product stream (*permeate*) are produced as effluent. This is illustrated in Figure 13.14. The rate at which the permeate flows through the membrane is known as the *flux*, expressed in units of mass/area.time. The reject or concentrate has to undergo further treatment. This has to be incorporated into the design of a complete treatment system.

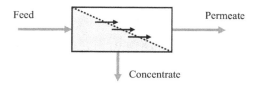

Figure 13.14 Membrane filtration.

For the membrane filter in Figure 13.14, from continuity, we can write

$$Q_F = Q_P + Q_C \tag{13.22}$$

where
  Q_F = flow rate of feed, m³/s
  Q_P = flow rate of permeate, m³/s
  Q_C = flow rate of concentrate, m³/s

The mass balance equation for the contaminant can be written as

$$Q_F\, C_F = Q_P\, C_P + Q_C\, C_C \tag{13.23}$$

where
  C_F = contaminant concentration in feed, kg/m³
  C_P = contaminant concentration in permeate, kg/m³
  C_C = contaminant concentration in concentrate, kg/m³

The rate of rejection (R) or removal efficiency is given by

$$R,\% = \frac{C_F - C_P}{C_F} \times 100\% \tag{13.24}$$

More advanced models have been developed to calculate the rejection based on particle diameter and pore size (MWH, 2005).

The volumetric flux of pure water across a clean membrane can be modeled using a modified form of Darcy's law (AWWA, 2005):

$$J = \frac{Q}{A} = \frac{\Delta P}{\mu\, R_m} \tag{13.25}$$

where
  J = volumetric flux through clean membrane, m³/m².h
  Q = flow rate of pure water, m³/h

A = surface area of the membrane, m²
ΔP = transmembrane pressure, kPa
μ = dynamic viscosity of water, kPa.h
R_m = membrane resistance coefficient, m⁻¹

Important design parameters include flux, rejection or removal efficiency, membrane resistance, transmembrane pressure, and temperature effects. A variety of models have been developed for membrane flux, as a function of time, or membrane thickness, or particle concentration, etc. These include the time-dependent models, the blocking filtration laws, and the cake filtration law, among others (MWH, 2005). However, pilot plant studies should be conducted to determine process parameters and removal efficiencies for a specific wastewater using a particular membrane filter.

13.4.3.2 Membrane fouling

The term *fouling* is used to denote the deposition and accumulation of particulates from the feed stream onto the membrane (Metcalf and Eddy et al., 2013). Fouling reduces the efficiency of the membrane. It can occur in three forms: (i) cake formation or build-up of constituents on the membrane surface, e.g. metal oxides, colloids, bacteria; (ii) scaling or chemical precipitation on the membrane, e.g. calcium carbonate, calcium sulfate, etc.; and (iii) damage to the membrane by acids, bases, or bacteria present in the feed.

There are a number of options available to control membrane fouling. The most important one is the pretreatment of feed water to remove the fouling compounds. Cartridge filters can be used to remove colloidal particles. Chemical conditioning of feed water is used to prevent chemical precipitation. *Reversible fouling* can be treated by backwashing the membrane. Chemical cleaning is used to remove scaling. When the membrane efficiency or desired flux rate cannot be recovered by the above methods, it is termed *irreversible fouling*. In that case, the damaged membrane has to be replaced with a new one.

Recent research has focused on developing membranes that are more resistant to fouling. The application of a coating of nanomaterials on the membrane has been investigated, to reduce fouling. Coating materials investigated include TiO_2 (titanium dioxide), Al_2 (alumina), silver, silica, iron, and magnesium-based nanoparticles, among others. In general, nanomaterials have improved the hydrophilicity, selectivity, conductivity, fouling resistance, and anti-viral properties of membranes (Su et al., 2012; Lu et al., 2009; Zodrow et al., 2007; Bae and Tak, 2005). But researchers have cautioned about the loss of nanomaterials with the permeate and emphasized the need for further study on their potential environmental and health effects (Kim and van der Bruggen, 2010).

13.4.3.3 Membrane configurations

A membrane unit comprised of the membranes, pressure support structure for the membranes, together with feed inlet and permeate/retentate outlet ports is called a *module*. The main types of modules used for wastewater treatment are (1) hollow fiber, (2) tubular, and (3) spiral-wound. The hollow fiber membrane module is the most common, where a bundle of hollow fibers is placed inside a pressure vessel. The fibers have an outside diameter of 0.5–2.0 mm, and a wall thickness of 0.07–0.60 mm (WEF, 2006). Each vessel contains a bundle of hundreds to thousands of hollow fibers. A hollow fiber membrane module is illustrated in Figure 13.15. In a tubular module, the membrane is cast on the inside of a support tube. A number of these tubes are then placed in a pressure vessel. The spiral-wound module consists of flat membrane sheets separated by flexible spacers, rolled into a circle, and placed in a pressure vessel. Membranes can also be pressure-driven or vacuum-driven.

Four different process configurations are used with hollow fiber membrane modules, depending on the direction of flow of feed water and retentate. They are (1) inside-out (dead-end), (2) inside-out (cross-flow), (3) outside-in (dead-end), and (4) outside-in (cross-flow) (Davis, 2020). In the inside-out configuration, the feed water flows into the hollow membrane, and the permeate passes out through the membrane to the outside. In the outside-in configuration, feed water flows against the walls of the membrane, and the permeate is collected inside. In the dead-end or direct feed mode, all of the feed water passes through the membrane. In the cross-flow mode, feed water is pumped tangentially to the membrane. Water that does not pass through the membrane is recirculated through the membrane after blending with feed water.

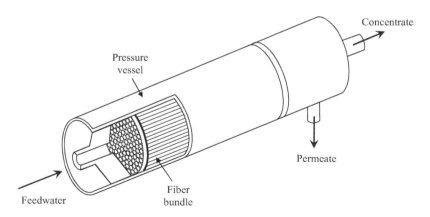

Figure 13.15 Hollow fiber membrane module.

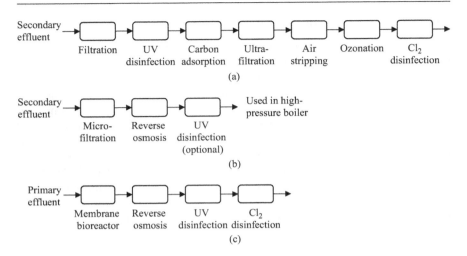

Figure 13.16 Flow diagrams for advanced treatment options.

13.4.4 Process flow diagrams

Tertiary treatment options can be used for advanced treatment of wastewater for the removal of specific constituents and solids. Advanced treatment is required when the wastewater is to be reused for aquifer recharge, in high-pressure boilers, or for indirect reuse. Process flow diagrams are provided in Figure 13.16, incorporating a number of the processes discussed in previous sections.

PROBLEMS

13.1 When are advanced or tertiary treatment processes used at a treatment plant? List the pollutants that are removed through these processes.

13.2 What are the sources of nitrogen and phosphorus in wastewater? List the forms of nitrogen and phosphorus present in wastewater.

13.3 Define the three biological nitrogen removal processes. Draw the schematic diagram of each of the processes.

13.4 Provide some examples of external carbon sources. Why are they used for biological denitrification processes? What is the limitation of using methanol as an external carbon source?

13.5 List three types of nitrification–denitrification processes. Give an example of each type of process.

13.6 A municipal wastewater treatment plant has an activated sludge process for combined BOD removal and nitrification, followed by a denitrification system consisting of an anoxic tank and clarifier. The plant used methanol as an external carbon source for denitrification. Effluent from the activated sludge process has a flow rate of 3500 m³/day and NO_3-N concentration of 22 mg/L. The denitrification tank has an SRT of 6 days and the kinetic coefficients with methanol at 20°C are as follows:

$\mu_{max} = 1.5$ d^{-1}
$k_d = 0.03$ d^{-1}
$K_s = 6.5$ bsCOD/L
$Y = 0.35$ kg VSS/kg bsCOD

COD equivalent of methanol is 1.5 kg COD/kg methanol. If the daily dosage of methanol is 300 kg/day, calculate the effluent NO_3-N concentration from the denitrification tank.

13.7 Why is lime pretreatment necessary for the air stripping process? What are the disadvantages of this process?

13.8 Describe the process of chemical precipitation for the removal of phosphorus. Why are the actual chemical requirements higher than stoichiometric requirements?

13.9 What chemicals can be used to enhance biological phosphorus removal? Give an example of a process used for biological phosphorus removal.

13.10 Why is filtration used in wastewater treatment? List the advantages of using dual-media or multi-media filters.

13.11 What is membrane fouling? Why is it of concern and how can it be controlled? What are the advantages and disadvantages of using nano-materials on membranes?

13.12 A municipal wastewater with a TDS concentration of 3000 g/m^3 is to be treated using membrane filtration. For regulatory requirements, the product water is to have a TDS of no more than 200 g/m^3. Estimate the rejection rate and the concentration of the concentrate stream. Assume 90% of the water is recovered by the system.

13.13 A pilot membrane filtration plant is set up to determine the operational parameters of a novel type of membrane. The flow rate of pure water through a 20 cm^2 membrane is 0.5 mL/min. If the transmembrane pressure is 2500 kPa, calculate the membrane resistance coefficient.

REFERENCES

Aspegren, H., Nyberg, U., Andersson, B., Gotthardsson, S., and Jansen, J. 1998. Post Denitrification in a Moving Bed Biofilm Reactor Process. *Water Science and Technology*, 38 (1):31–38.

Andrettola, G.; Foladori, P.; Gatti, G.; Nardelli, P.; Pettena, M. and Ragazzi, M. 2003. Upgrading of a Small Overload Activated Sludge Plant Using a MBBR System, *Journal of Environmental Science and Health*, A38 (10):2317–2338.

AWWA. 2005. *Microfiltration and Ultrafiltration Membranes for Drinking Water*, AWWA Manual M53, American Water Works Association, Denver, CO, USA.

Bae, T., and Tak, T. 2005. Preparation of Tio2 Self-Assembled Polymeric Nanocomposite Membranes and Examination of Their Fouling Mitigation Effects in a Membrane Bioreactor System, *Journal of Membrane Science*, 266 (1–2):1–5.

Barker, P. S., and Dold, P. L. 1997. General Model for Biological Nutrient Removal in Activated Sludge Systems: Model Presentation. *Water Environment Research*, 69 (5):969–984.

Barnard, J. 1973. Biological Denitrification. *Water Pollution Control* (G.B.), 72 (6):705–720.

Barnard, J. 1974. Cut P and N Without Chemicals. *Water and Wastes Engineering*, 11:41–44.

Barnard, J. 1975. Biological Nutrient Removal Without the Addition of Chemicals. *Water Research*, 9:485–490.

Ciudad, G., Robilar, O., Muñoz, P., Ruiz, G., Chamy, R., Vergara, C. and Jeison, D. 2005. Partial Nitrification of High Ammonia Concentration Wastewater as a Part of a Shortcut Biological Nitrogen Removal Process, *Process Biochemistry*, 40:1715–1719.

Cleasby, J. L. and Logsdon, G. S. 1999. Granular bed and precoat filtration. in *Water Quality and Treatment*, 5th ed., R. D. Letterman (ed.), American Water Works Association, McGraw-Hill, New York, NY, USA, p. 8.1–8.99.

Davis, M. 2020. *Water and Wastewater Engineering: Design Principles and Practice.* 2nd edn. McGraw-Hill, Inc., New York, NY, USA.

Dold, P., Takacs, I., Mokhayeri, Y., Nichols, A., Hinojosa, J., Riffat, R., Bott, C., Bailey, W. and Murthy S. 2008. Denitrification with Carbon Addition – Kinetic Considerations. *Water Environment Research*, 80 (5):417–427.

EPA. 1977. *Process Design Manual for Nitrogen Control.* United States Environmental Protection Agency, Technology Transfer Office, Washington, DC, USA.

Grady, C.P.L. Jr., Daigger, G., Love, N. and Filipe, C. 2011. *Biological Wastewater Treatment*, 3rd edn. CRC Press, Boca Raton, FL, USA.

Gut, L. 2006. *Assessment of a Partial Nitritation/Annamox System for Nitrogen Removal.* Licentiate Thesis, Silesian University of Technology, KTH, Sweden, ISBN: 91-7178-167-6.

Hellinga, C., Schellen, A. A. J. C., Mulder, J. W., van Loosdrecht, M. C. M. and Heijnen, J. J. 1998. The SHARON Process: An Innovative Method for Nitrogen Removal from Ammonium-Rich Wastewater. *Water Science and Technology*, 37 (9):135–142.

Hinojosa, J. 2008. *A Study of Denitrification Kinetics Using Methanol And Glycerol As External Carbon Sources.* Master's Thesis, George Washington University, Washington, DC, USA,

Hinojosa, J., Riffat, R., Fink, S., Murthy, S., Selock, K., Bott, C., Wimmer, R., Dold, P., and Takacs, I. 2008. Estimating The Kinetics and Stoichiometry of Heterotrophic Denitrifying Bacteria With Glycerol as an External Carbon Source. *Proceedings of 81st Annual Conference of Water Environment Federation, WEFTEC '08*, Chicago, IL, USA.

Hippen, A., Rosenwinkel, K. H., Baumgarten, G. and Seyfried, C. F. 1997. Aerobic Deamonification: A New Experience in the Treatment of Wastewaters. *Water Science and Technology*, 35 (10):111–120.

Jetten, M. S. M., Wagner, M., Fuerst, J., van Loosdrecht, M. C. M., Kuenen, G. and Strous, M. 2001. Microbiology and Application Of Anaerobic Ammonium

Oxidation ('Anammox') Process. *Current Opinion in Biotechnology*, 12:283–288.

Kim, J., and van der Bruggen, B. 2010. The Use of Nanoparticles in Polymeric and Ceramic Membrane Structures: Review of Manufacturing Procedures and Performance Improvement for Water Treatment. *Environmental Pollution*, 158 (7):2335–2349.

Lu, Y., Yu, S., and Meng, L. 2009. Preparation of Poly(Vinylidene Fluoride) Ultrafiltration Membrane Modified By Nano-Sized Alumina and Its Antifouling Performance. *Harbin Gongye Daxue Xuebao/Journal of Harbin Institute of Technology*, 41 (10):64–69.

McCarty, P. L., Beck, L., and Amant, P. S. 1969. Biological Denitrification of Wastewaters by Addition of Organic Materials. *Proceedings of the 24th Industrial Waste Conference*, Purdue University, IN, USA. p. 1271–1285.

Metcalf and Eddy, Tchobanoglous, G., Stensel, H., Tcuchihashi, R., and Burton, F. 2013. *Wastewater Engineering: Treatment and Resource Recovery*. 5th edn. McGraw-Hill, Inc., New York, NY, USA.

Mokhayeri, Y., R. Riffat, S. Murthy, W. Bailey, I. Takacs, and C. Bott. 2009. Balancing Yield, Kinetics and Cost for Three External Carbon Sources used for Suspended Growth Post Denitrification. *Water, Science and Technology*, 60 (10):2485–2491.

Mokhayeri, Y., Riffat, R., Takacs, I., Dold, P., Bott, C., Hinojosa, J., Bailey, W., and Murthy, S. 2008. Characterizing Denitrification Kinetics at Cold Temperature using Various Carbon Sources in Lab-Scale Sequencing Batch Reactors. *Water Science and Technology*, 58 (1):233–238.

Mokhayeri, Y., Murthy, S., Riffat, R. Bott, C., Dold, P., and Takacs, I. 2007. Denitrification Kinetics using External Carbon Sources: Methanol, Ethanol and Acetate. *Proceedings of 80th Annual Conference of Water Environment Federation*, WEFTEC '07, San Diego, CA, USA.

Mokhayeri, Y., Nichols, A., Murthy, S., Riffat, R., Dold, P. and Takacs, I. 2006. Examining the Influence of Substrates and Temperature on Maximum Specific Growth Rate of Denitrifiers. *Water Science and Technology*, 54 (8):155–162.

Mulder, M. W., van Loosdrecht, M. C. M., Hellinga, C. and Kempen, R. 2001. Full-scale Application of the SHARON Process for Treatment of Rejection Water of Digested Sludge Dewatering. *Water Science and Technology*, 43 (11):127–134.

Mulder, A., van der Graaf, A. A., Robertson, L. A. and Kuenen, J. G. 1995. Anaerobic Ammonium Oxidation Discovered in a Denitrifying Fluidized Bed Reactor. *FEMS Microbiology and Ecology*, 16 (3):177–184.

Murthy, S. 2011. Personal communication at George Washington University, Washington, DC, USA.

MWH. 2005. *Water Treatment: Principles and Design*, John Wiley & Sons, Hoboken, NJ, USA, p. 882–894.

Nichols, A., Hinojosa, J., Riffat, R., Dold, P., Takacs, I., Bott, C., Bailey, W., and Murthy, S. 2007. Maximum Methanol Utilizer Growth Rate: Impact of Temperature on Denitrification. *Proceedings of 80th Annual Conference of Water Environment Federation*, WEFTEC '07, San Diego, CA, USA.

Ødegaard, H. 2006. Innovations in Wastewater treatment: the Moving Bed Biofilm Process. *Water Science and Technology*, 53 (9):17–33.

Peavy, H. S., Rowe, D. R., and Tchobanoglous, G. 1985. *Environmental Engineering*, McGraw-Hill, Inc., New York, NY, USA.

Peric, M., Shrestha, A., Riffat, R., Stinson, B. and Murthy, S. 2010. Observation of Self Regulation of Biofilm Thickness in Denitrifying MBBR, *Proceedings of the Annual Conference and Exhibition of Water Environment Federation, WEFTEC '10*, New Orleans, LA, USA.

Randall, C. W., Barnard, J. L., and Stensel, H. D. 1992. Design and Retrofit of Wastewater Treatment Plants for Biological Nutrient Removal. *Water Quality Management Library*, vol. 5, Technomic Publishing Co., Lancaster, PA, USA.

Ruiz, G., Jeison, D. and Chamy, R. 2003. Nitrification With High Nitrite Accumulation for the Treatment of Wastewater With High Ammonia Concentration. *Water Research*, 37 (6):1371–1377.

Rusten, B., Hem, L.J. and Ødegaard, H. 1995. Nitrogen Removal from Dilute Wastewater in Cold Climate using Moving-Bed Biofilm Reactors. *Water Environment Research*, 67 (1):65–74.

Shrestha, A., Riffat, R., Bott, C., Takacs, I., Stinson, B., Peric, M., Neupane, D., and Murthy, S. 2009. Process Consideration to Achieve Nitrogen Removal in a Moving Bed Biofilm Reactor. *Proceedings of 82nd Annual Conference of Water Environment Federation, WEFTEC '09*, Orlando, FL, USA.

Stensel, H. D., and Horne, G. 2000. Evaluation of Denitrification Kinetics at Wastewater Treatment Facilities. *Research Symposium Proceedings, 73rd Annual Water Environment Federation Conference*, Anaheim, CA, USA.

Stensel, H. D.; Loehr, R. C.; Lawrence, A. W. 1973. Biological Kinetics of Suspended-Growth Denitrification. *Journal of Water Pollution Control Federation*, 45:249.

Su, Y., Huang, C., Pan, J. R., Hsieh, W., and Chu, M. 2012. Fouling Mitigation by TiO2 Composite Membrane in Membrane Bioreactors, *Journal of Environmental Engineering (ASCE)*, 138 (3):344–350.

van Dongen, L. G. J. M., Jetten, M. S. M. and van Loosdrecht, M. C. M. 2001. The Sharon-Anammox Process for Treatment of Ammonium Rich Wastewater. *Water Science and Technology*, 44 (1):153–160.

van Loosdrecht, M.C. M., Hao, X., Jetten, M.S.M. and Abma, W. 2004. Use of Anammox in Urban Wastewater Treatment, *Water Science and Technology: Water Supply*, 4 (1):87–94.

WEF. 2006. *Membrane Systems for Wastewater Treatment*, Water Environment Federation, Alexandria, VA, USA.

WEF. 1998. *Design of Municipal Wastewater Treatment Plants*, 4th ed., Water Environment Federation, Manual of practice 8, Alexandria, VA, USA.

Zodrow, K., Brunet, L., Mahendra, S., Li, D., Zhang, A., Li, Q., and Alvarez, P. J. 2007. Polysulfone Ultrafiltration Membranes Impregnated With Silver Nanoparticles Show Improved Biofouling Resistance and Virus Removal, *Polymers for Advanced Technologies*, 18 (7):562–568.

Chapter 14

Resource recovery and sustainability

14.1 INTRODUCTION

In recent decades, wastewater has been evolving from *waste* to valuable resource. Wastewater can help dampen the effects of water shortages through water reclamation. It also provides a means for energy and nutrient recovery, which can reduce dependency on non-renewable resources. Resource recovery from wastewater treatment facilities in the form of energy, reusable water, biosolids, and other resources, such as nutrients, represents an economic and financial benefit that contributes to the sustainability of water supply and sanitation systems, as well as the water utilities. According to the World Bank report authored by Rodriguez et al. (2020), one of the key advantages of adopting circular economy principles in the processing of wastewater is that resource recovery and reuse can transform sanitation from a costly service to one that is self-sustaining and adds value to the economy. If financial returns can cover operation and maintenance costs partially or fully, improved wastewater management will result in a double value proposition.

14.2 SUSTAINABLE DESIGN PRINCIPLES

With the establishment of the Environment Protection Agency (EPA) in 1972, a shared societal value and growing environmental societal movement emerged in the US. In order to achieve the EPA mission to protect human health and the environment, the US has made tremendous progress through research, development, and technological improvements. However, modern-day environmental challenges are more complex as a higher level of understanding is required of the clear linkages between society, economy, and the environment. These three pillars of sustainability must be considered simultaneously in any decision-making scenario.

Sustainable development is defined by the Brundtland Commission as the development which meets the needs of the present without compromising the ability of the future to meet its needs (Mihelcic and Zimmerman, 2014).

DOI: 10.1201/9781003134374-14

From an environmental engineering perspective, sustainable wastewater engineering can be defined as the design of a system to treat wastewater with minimal adverse impact on societal conditions, human health, and environment, and ensuring the use of natural resources without diminishing the quality of life for the present and future generations.

Life cycle thinking is necessary in every aspect of the design of a wastewater treatment system, from raw materials to process design and end-of-life management. Wastewater treated with conventional treatment processes and discharged into water bodies is not sufficient. The potential for various end-of-life strategies like resource recovery, reclamation, and reuse are important to minimize environmental impacts and reduce energy demand and carbon emissions.

The potential benefits and impacts over the lifetime of the materials and design must be considered to develop and implement solutions. Resources, such as nutrients, energy, or water recovery from wastewater can help achieve the sustainable design principles, without adverse impacts to historically harmed environments and societies, while maintaining sufficient natural resources for the current and future generations. Water reclamation can free scarce freshwater resources for other uses or preservation. In addition, by-products and end products of wastewater treatment can be of value for agriculture and energy generation, making wastewater treatment plants more environmentally and financially sustainable.

14.3 NUTRIENT RECOVERY

Biosolids produced from the treatment of wastewater sludge are major sources of nutrients such as nitrogen, phosphorus, iron, calcium, magnesium, and other micronutrients (chromium, selenium, zinc, copper, etc.) which are essential for plant growth. Biosolids can be used as fertilizer after proper stabilization. Multiple nutrients found in biosolids are also essential components of animal feed. A significant amount of nutrients are present in the wastewater in soluble form after secondary treatment and ends up in the return flow (also known as sidestream) from sludge digestion and dewatering processes. Sidestream treatment is often economical due to the relatively low volume and high concentration of nutrients present. The sidestream flow is about 1% of the forward flow; however, sidestream return accounts for 15–30% of the total nitrogen load on the process (Husnain et al., 2015). The recovery of nutrients from mainstream and sidestream can help achieve lower nitrogen and phosphorus concentrations in the receiving water bodies and meet the NPDES permit levels for nutrient discharge.

Nutrients can be recovered via urine source separation, composting toilets, or reclaimed water with nutrients, and reused to fertilize plants (Asano et al., 2007; Anand and Apul, 2014). There are innovative ways to produce fertilizer directly from urine utilizing urine-diverting toilets. For urine source separation, volume reduction can be achieved via evaporation, drying, or struvite precipitation to facilitate transportation. Pilot projects have been investigated or are underway around the world, including projects in Australia, Austria, China, Germany, the Netherlands, Sweden, Switzerland, and the US (Mitchell et al., 2013; Noe-Hayes et al., 2015).

There are numerous technologies available to recover phosphorus from wastewater and treatment byproducts (e.g. digester supernatant, sludge, etc.) including crystallization, precipitation, wet chemical processes, and thermo-chemical processes. 50–94% of the phosphorus can be recovered from the liquid phase (Rahman et al., 2014). A large number of crystallization and precipitation processes have been developed or commercialized including PhoStrip, PRISA, DHV Crystalactor, CSIR, Kurita, Ostara, Phosnix, Berliner Verfahren, and FIX-Phos (Mehta et al., 2015).

14.3.1 Nutrient applications

Organic nitrogen and phosphorus recovered from the liquid stream, or present in the Class A and Class B biosolids can be used for agricultural and non-agricultural land applications. The major beneficial uses of biosolids include (i) agricultural land application, (ii) non-agricultural land application, (iii) energy recovery and generation, and (iv) commercial uses (Metcalf and Eddy et al., 2013).

Biosolids applied to the land provide sufficient nitrogen supply for crop growth. Such recycling of biosolids provides economic benefits to the farmers and environmental benefits from reduced use of chemical fertilizer. It improves the soil structure, water holding capacity, water filtration, and soil aeration, and facilitates plant growth. Organic matter present in the biosolids helps the soil to retain calcium, magnesium, and potassium, and also improves biological diversity of the soil.

Class A and Class B biosolids produced from anaerobic digestion or lime stabilization are generally applied to the land as cake or granular pellets after dewatering, or as compost materials. In the US, the land application of biosolids and disposal are regulated under Code of Federal Regulations (CFR), 40 CFR Part 503. The regulations provide guidelines for management practices, pathogen and vector attraction reduction alternatives, as well as requirements for monitoring, record keeping, and reporting. Biosolids production, land disposal methods, and regulations are discussed in detail in Chapter 12.

Biosolids have been widely used for non-agricultural land applications, such as landscaping, land reclamation, forest crops, and reclamation of mining sites. Biosolids are typically combined with mine soil to increase the organic content, soil nutrient level, control pH, and metal content, and are used for the repair of land damaged by surface mining, coal refuse piles, and abandoned mine sites. Reclaiming and improving disturbed soils are achieved with improved physical property and water holding capacity of biosolids. Reclamation of disturbed land such as superfund sites has been successful (Henry and Brown, 1997). For landscaping, biosolids composts are widely used as soil conditioners. Application of biosolids in the forest-land is also practiced, although there are difficulties associated with spreading in heavily forested areas. Forest utilization has been practiced extensively in the US, and biosolids application has been recognized as being beneficial to forest growth (WEF, 2010).

14.4 ENERGY RECOVERY

The organic matter present in sludge and biosolids are significant sources of renewable energy. Energy generation and recovery from biosolids can be achieved in a properly engineered and controlled environment. Wastewater can be considered as a source of different types of energy: Electrical energy from bio-electrochemical wastewater treatment processes, low-head hydro-electric energy, biogas from anaerobic digestion, renewable fuels from residual sludge processing, and heat energy (Cecconet et al., 2020).

Fuels can be derived from wastewater in gaseous (biogas, synthetic gas or syngas, etc.), liquid, and solid forms. Sludge from the wastewater can be processed to produce methane gas from anaerobic digestion, thermally oxidized sludge, or produce syngas through gasification and/or pyrolysis, and produce oil and liquid fuel (WERF, 2008).

Producing methane-rich biogas through anaerobic digestion is the most popular and common way of producing energy from biosolids. Many plants are using combined heat and power (CHP) recovery from solids stabilization and dewatered sludge. The details of the anaerobic digestion process with gas production are discussed in Chapter 12.

Thermal oxidation of biosolids is practiced in locations with limited disposal sites, and less stringent regulations. The combustion process produces flue gas, from which the energy can be recovered for combustion air pre-heating and other energy needs, or electricity production (Metcalf and Eddy et al., 2013). The process has the potential of producing other resources as well. Ash can be used for cement or asphalt making, or phosphorus recovery. Gasification is a process that has been in practice since the 1800s to generate flue gas (also called syngas) from coal. Although the heating value of syngas is significantly lower compared to that of biogas generated from anaerobic digestion, there is increased interest and innovation to apply this technology for the processing of biosolids. Pyrolysis is a similar technology, which produces charcoal, activated carbon, and methanol at high temperatures. The process is not widely used for municipal applications. Converting sludge to liquid or oil fuel is another option, although it requires high capital and operating costs.

14.4.1 Energy applications

The gaseous fuel derived from wastewater is commonly used to generate electricity with reciprocating engines, gas turbines, and micro-turbines, and is less commonly used to operate pumps and blowers. The exhaust heat from the combustion process is used to heat water in boilers for use in heating buildings, anaerobic digesters, or provide hot water supply.

Gas pretreatment is necessary for the utilization of biogas and syngas to produce power and heat. The constituents such as H_2S, siloxanes, CO_2, and moisture are of major concern due to their hazardous nature or detrimental impact on the combustion system. The emission of flue gas from the combustion of biogas can cause air pollution and may be subject to strict regulation.

Energy recovered from the thermal oxidation (or incineration) of solid fuel is mostly waste heat. Part of that heat is utilized to preheat the incoming sludge and air. The solids content in the dewatered sludge is an important parameter. Solids content greater than 22–28% is required. The syngas is generally utilized in conventional boilers or engine generators, or directly into the thermal oxidation process.

The methane generated from anaerobic digestion is used in the fuel cell system to generate electricity and waste heat. In the fuel cell, methane gas is used to generate hydrogen to produce direct current (DC), which is converted to alternating current (AC). Three types of fuel cells: Phosphoric acid fuel cell (PAFC), molten carbon fuel cell (MCFC), and solid-oxide fuel cell (SOFC) are considered promising for application at wastewater treatment facilities (EPA, 2006). Waste heat generated from the fuel cells can be recovered for other purposes.

14.5 WATER RECOVERY

The demand for water quality varies in different sectors, thus water reclamation and reuse technologies vary accordingly. There are a large number of advanced and innovative technologies available to process wastewater for beneficial reuse – including reuse as drinking water. Compared to other water sources such as desalinated seawater, harvested rainwater, or water from melted icebergs, treated municipal wastewater provides a climate-independent high-quality freshwater source that can be locally controlled (Bloetscher et al., 2005) and has significant potential for helping to meet future water needs (NRC, 2012). In order to address the clean water challenge in a sustainable way, a paradigm shift must occur in wastewater management that emphasizes wastewater as a renewable resource, and a careful assessment of resource recovery systems is essential (Guest et al., 2009).

When treated wastewater is considered as the source of supply for potable or non-potable reuse, a combination of advanced treatment processes is utilized. A comprehensive summary of advanced treatment options is provided in the Framework for DPR (Direct Potable Reuse) (AWWA, 2015), including the biological treatment for nitrate and nitrite removal, membrane filtration with microfiltration (MF) and ultrafiltration (UF), membrane desalination with reverse osmosis (RO) and nanofiltration (NF), ozone as disinfection and/or oxidizing agent, and advanced oxidation with ozone (O_3) or ultraviolet (UV) light in combination with hydrogen peroxide.

Treatment via sand filtration and membrane bioreactors is common practice. The US EPA recommends secondary treatment and disinfection for non-potable applications; filtration should be added when direct human contact is expected, or to irrigate food crops that can be consumed raw. Small-scale Nexus treatment units (Nexus, 2015) are three-stage hybrid processes comprised of floatation, filtration, and disinfection. The systems were installed in San Diego and Fresno, CA, and typical water use from public supply was reduced by 50–72%. The recovered water was used mostly for irrigation. The Phipps Center for Sustainable Landscapes in Pittsburgh, Pennsylvania, implemented an aerated septic tank with sand filter and UV disinfection. The recycled water was used for toilet flushing. The reuse of wastewater, along with harvesting and treating rainwater, fulfilled 24% of their water demand (Hasik et al., 2017).

14.5.1 Reclaimed water applications

Water reuse applications include non-potable reuse, groundwater recharge, and DPR. Non-potable reuse is the most prevalent application of reclaimed water. The reuse application mainly consists of irrigation, toilet flushing, landscaping, and industrial processes like cooling and cleaning.

Singapore is the global hub for pioneering novel water reuse techniques for both potable and non-potable applications. They have been practicing potable and non-potable water reuse for more than two decades. DPR is practiced in many parts of the world, however, the psychological "yuck factor" is often the deterrent when direct potable use is considered. With increased water scarcity in the US and many parts of the world, the potential for DPR has expanded and innovative technologies are being developed. However, public engagement is critical for the success of potable reuse of wastewater. Transparent communication and building public confidence are essential initial steps. Project success requires sufficient time to create rapport, build trust, and establish effective communication strategies with the public and media. The successful implementation and ongoing success of potable reuse projects rely on educating and encouraging local communities in order to gain public support (AWWA, 2016).

Regulations for potable reuse have not been developed at the federal level in the US. As a result, state and local agencies are responsible for setting potable reuse standards. The majority of states have established regulations or guidelines for water reuse, and state regulations are expected to be regularly updated and developed as states gain experience and confidence in water reuse. Although few states have established regulations or guidelines specifically for potable reuse, most will still consider potable reuse projects on a case-by-case basis.

14.6 EXAMPLES OF BEST PRACTICES

Resource recovery from wastewater facilities in the form of energy, reusable water, biosolids, and other resources, such as nutrients, represents economic and financial benefits that contribute to the sustainability of water supply and sanitation systems, and the water utilities operating them. In recent years, many wastewater treatment plants from all over the world are moving toward energy self-sufficiency with a combination of energy-efficient strategies and energy harvesting from wastewater. Many plants are recovering the biosolids and reclaimed water for beneficial reuse for sustainable development. Below are some examples from the US, Europe, and other parts of the world.

Example 1: DC Water, Washington, DC, USA

DC Water's Blue Plains Advanced Wastewater Treatment Plant is the largest plant of its kind in the world, treating 384 MGD on average and over one billion gal/d at peak flow. DC Water is the first facility in North America to utilize CAMBI's thermal hydrolysis process for energy recovery and biosolids production. Four digesters (3.8 Mgal each) convert the organic matter to biogas and Class A biosolids in the highly efficient system. The cleaned biogas operates three 5 Megawatt (MW) turbines to produce enough power to run one-third of the Blue Plains plant. The biosolids produced from the digesters meet or exceed all EPA standards for Class A biosolids, and are used in rural and urban settings for landscaping, restoration, gardening, and tree planting. 200,000 tons of biosolids are processed per year. The project has resulted in a 30% reduction in energy purchased from the grid, a 41% reduction in greenhouse gas production, and a 50% reduction in biosolids shipping costs (CDM Smith, 2021). The project will reduce greenhouse gas emissions by 50,000 tons of CO_2 equivalent metric tons (DC Water, 2021).

Example 2: East Bay Municipal Utility District (EBMUD), California, USA

EBMUD's award-winning wastewater treatment protects San Francisco Bay and services 650,000 customers with a treatment capacity of 415 MGD. The plant is considered a *green factory* where biodegradable wastes in sewage, food scraps, and grease from local restaurants, plus waste streams from wineries and poultry farms are mixed together and treated in anaerobic digesters. The biogas is captured and used to generate energy to power the wastewater treatment plant. In 2012, EBMUD became the first wastewater treatment plant in North America to produce more energy onsite than was needed to run the facility. The plant sells excess energy back to the electrical grid to cut fossil fuel use and greenhouse gas emissions. Biogas production saves EBMUD approximately $3 million each year by reducing their electric power demand (EBMUD, 2021).

Example 3: Encina Water Pollution Control Facility (EWPCF), California, USA

The EWPCF treats wastewater sludge in anaerobic digesters to produce biogas, which is used in internal combustion engines to produce electricity and usable heat. Gas produced by the anaerobic digestion process is a mixture of approximately 60–65% methane, 35–40% carbon dioxide, and small quantities of nitrogen, hydrogen, and sulfur compounds. Over 80% of the electricity used by EWPCF is provided by the cogeneration facility. In addition, use of thermal energy from the cogeneration process is used to heat the digesters. As a result, the operation of the cogeneration facility saves the Encina Wastewater Authority about $2 million dollars each year in energy costs. In 2009, the plant implemented heat-drying technology to produce Class A biosolids. Class A biosolids have unrestricted use which provides more options for the application of the product, even generating revenue from the sale of *Pure Green* fertilizer. The total savings from biosolids recovery is about $2 million per year. In addition, EWPCF produces over five MGD of recycled water onsite. To produce recycled water, secondary treated effluent undergoes filtration (sand filtration or ultrafiltration) and chlorination. This water is used in the plant and reduces the cost of potable water. Recycled water uses include equipment washdown, cogeneration engine cooling, odor reduction, and landscape irrigation (EWPCF, 2021).

Example 4: Point Loma Wastewater Treatment Plant (PLWTP), California, USA

PLWTP treats approximately 175 MGD of wastewater generated in a 450 sq mile area by more than 2.2 million residents. The plant has eight anaerobic digesters that break down the organic sludge removed from the wastewater, producing methane gas that is collected, cleaned, and piped to an onsite gas utilization facility. The gas is used to provide space heating and cooling. The gas produced by the digesters also fuels two internal combustion reciprocating engines that run generators with a total capacity of 4.5 MW. Using the gas produced on-site, the Point Loma plant has not only become energy self-sufficient, but it is also able to deliver excess power to the grid and receive credit on other selected facilities' energy bills. Thermal energy produced by the generators is used to heat the plant's digesters. In addition, excess digester gas from the plant is treated to achieve natural gas standard, and injected into the utility natural gas pipeline for other uses. This was the first biogas cleaning and injecting into utility pipeline project in the state of California (PLWTP, 2021).

Example 5: West Point Treatment Plant, Kings County, Washington, USA

The West Point Treatment Plant in Washington state treats wastewater for about 1.5 million people. About 90 MGD of wastewater is treated during the dry months and 300 MGD during the rain/storm season. The anaerobic digesters at the plant produce 1.4 M ft^3/d of digester gas that is used to run two generators to produce 1.5–2 MW of electricity, which is sold to the local utility – Seattle City Light. West Point plant also uses digester gas as the primary fuel for their influent pumps, and exhaust heat from the engines is used to heat the digesters and the plant (WPTP, 2021).

Example 6: Sheboygan Regional Wastewater Treatment Facility, Wisconsin, USA

The Sheboygan Regional Wastewater Treatment Facility in Wisconsin is one of the few WWTPs in the world that can currently achieve nearly 100% energy self-sufficiency. The plant serves 68,000 people, with an average capacity of 37,854 m^3/d (10 MGD) and a peak design capacity of 221,824 m^3/d (58.6 MGD). The average annual electrical energy self-sufficiency is 80% (Sheboygan-WWTP, 2021).

Example 7: South Cross Bayou Water Reclamation Facility, Florida, USA

The South Cross Bayou Water Reclamation Facility (WRF) utilizes denitrification, filtration, and chlorine disinfection to recover water, energy, and nutrients. Reclaimed water is distributed to communities primarily for irrigation. The methane gas recovered from the digestion process partially fuels the dryer process for biosolids to produce fertilizer pellets (Diaz-Elsayed et al., 2019).

Example 8: Wolfgangsee-Ischl Wastewater Treatment Plant, Austria

The WWTP of Wolfgangsee-Ischl in Austria is a secondary treatment plant with a single-stage activated sludge system and anaerobic sludge digestion. Electrical energy at 20.6 kWh/PE per yr is produced from the digester gas. Surplus electric energy from the plant is delivered to the grid. In total, 19.2 kWh/PE per yr of electricity was consumed at the WWTP of which 11.5 kWh/PE per yr was used for aeration and mixing, and 7.7

kWh/PE per yr was used for all the other treatment steps and equipment. The Wolfgangsee-Ischl plant is a conventional activated sludge plant similar to thousands of others worldwide and there is no additional energy input – either by co-substrates fed into the anaerobic sludge digesters or by separate devices for electricity production. The main reason for the neutral-to-positive energy balance of this wastewater treatment plant is the longstanding and ongoing optimization of all mechanical equipment and optimal aeration control (Nowak et al., 2015).

Example 9: Rzeszów Wastewater Treatment Plant, Poland

In 2015, the Rzeszów WWTP was modernized and expanded to change the capacity of the facility to up to 54,500 m³/d and 398,000 PE. Anaerobic digestion with a CHP system was introduced. The energy produced from the generated biogas made it possible to meet 74.3% of the electrical and 95.5% of the heating requirements, in relation to the overall energy consumption of the wastewater treatment plant. The relation between total energy generation and consumption at Rzeszów WWTP had an average value of 83.6% (Maslon, 2017).

Example 10: Melbourne Water, Australia

Melbourne Water in Australia operates two regional wastewater treatment plants: The Western Treatment Plant (WTP) and Eastern Treatment Plant (ETP). The WTP uses land and grass filtration, and lagoon treatment. The three lagoon systems have 10 ponds each, staged from anaerobic to aerobic, treating about 600 million L (160 MG) of water. The WTP produces almost all of its electricity onsite from a 10 MW biogas-fueled power station. The ETP also uses generated biogas to produce a substantial amount of electricity for use in heating and cooling operations. The plant's seven generators run solely on biogas or with supplemental natural gas as needed (WERF, 2015).

Example 11: NEWater, Public Utility Board, Singapore

Singapore is considered the global hub for pioneering novel water reuse techniques for both potable and non-potable applications. Two decades ago, when membrane technology was not considered economically feasible, the reclaimed water produced by Singapore's Public Utility Board and marketed as *NEWater*, was of tremendous success, supplying 30% of Singapore's water demand (NEWater, 2021). The treatment system consisted of primary sedimentation/activated sludge/microfiltration (MF)/ultrafiltration (UF)/reverse osmosis (RO) and ultraviolet (UV) disinfection. The main technology behind the production of NEWater is pressure-driven RO, which allows particle removal first via size-exclusion by means of pores extending from 0.2 to 0.4 nm. At that pore dimension, permeate is solid-free, devoid of emerging pollutants, metals, salts, viruses, or other microbes (Ghernaout et al., 2019). The water is used for non-potable use and indirect potable use for reservoir recharge. There are five NEWater plants currently supplying up to 40% of Singapore's water needs. The goal is to meet up to 55% of the future water demand by 2060 (PUB, 2021).

Example 12: Baix Llobregat Wastewater Treatment Plant, Spain

The Baix Llobregat WWTP in Barcelona, Spain has a capacity of 2 million PE and treats approximately 270,000 m³/d of wastewater and 4000 m³/d of sludge. The treatment train is comprised of conventional activated sludge with nutrient removal. 30% of the secondary effluent undergoes tertiary treatment with UV and chlorine disinfection for agricultural reuse, and is also used to create a seawater intrusion barrier after additional treatment processes (Diaz-Elsayed et al., 2019).

REFERENCES

Anand, C. K., Apul, D. S. 2014. Composting Toilets as a Sustainable Alternative to Urban Sanitation – A Review. *Waste Management*, 34 (2):329–343.

Asano, T., Burton, F.L., Leverenz, H.L., Tsuchihashi, R., Tchobanoglous, G. 2007. *Water Reuse: Issues, Technologies, and Applications*. McGraw-Hill, New York, New York, NY.

AWWA. 2015. *Framework for Direct Potable Reuse*. http://www.awwa.org/resources-tools/water-knowledge/reuse.aspx

AWWA. 2016. *Potable Reuse 101: An Innovative and Sustainable Water Supply Solution*. American Water Works Association, Boulder, CO, USA.

Bloetscher, F., Englehardt, J.D., Chin, D. A., Rose, J.B., Tchobanoglous, G., Amy, V.P., Gokgoz, S. 2005. Comparative Assessment of Municipal Wastewater Disposal Methods in Southeast Florida. *Water Environment Research*. 77 (5):480–490.

CDM Smith. 2021. Driving Net-Zero at DC Water. https://cdmsmith.com/en/Client-Solutions/Projects/DC-Water-Driving-Net-Zero

Cecconet, D., Raček, J., Callegari, A. and Hlavínek, P. 2020. Energy Recovery from Wastewater: A Study on Heating and Cooling of a Multipurpose Building with Sewage-Reclaimed Heat Energy. *Sustainability*, 12, (1):116.

DC Water. 2021. *DC Water's Thermal Hydrolyis and Anaerobic Digester Project*. https://www.dcwater.com/sites/default/files/documents/BioenergyFacility.pdf, (accessed August 17, 2021)

Diaz-Elsayed, N., Rezaei, N., Guo, T., Mohebbi, S., and Zhang, Q. 2019. Wastewater-based Resource Recovery Technologies Across Scale: A Review. *Resources, Conservation and Recycling*, 145: 94–112.

EBMUD. 2021. East Bay Municipal District Website, http://www.ebmud.com/about-ebmud, (accessed August 11, 2021).

EPA. 2006. *Auxiliary and Supplemental Power Fact Sheet: Fuel Cells*, US Environmental Protection Agency, Washington, DC, USA.

EWPCF. 2021. Encina Water Authority website, https://www.encinajpa.com/environment-resource-management (accessed August 11 2021).

Ghernaout, D., Elboughdiri, N., and Alghamdi, A. 2019. Direct Potable Reuse: The Singapore NEWater Project as a Role Model. *OALib*, 6. 1–11. doi:10.4236/oalib.1105980.

Guest, J. S., Skerlos, S.J., Barnard, J.L., Beck, M.B., Daigger, G.T., Hilger, H., et al. 2009. A New Planning and Design Paradigm to Achieve Sustainable Resource Recovery from Wastewater. *Environmental Science and Technology*, 43: 6126–6130.

Hasik, V., Anderson, N.E., Collinge, W.O., Thiel, C.L., Khanna, V., Wirick, J., et al. 2017. Evaluating the Life Cycle Environmental Benefits and Trade-Offs of Water Reuse Systems for Net-Zero Buildings. *Environmental Science and Technology*, 51 (3):1110–1119.

Henry, C. and Brown, S. 1997. Restoring a Superfund Site with Biosolids and Fly Ash. *Biocycle*, 38 (11): 79–83.

Husnain, T., Mi, B. and Riffat, R. 2015. A Combined Forward Osmosis and Membrane Distillation System for Sidestream Treatment. *Journal of Water Resource and Protection*, 7: 1111–1120.

Maslon, A. 2017. *Analysis of Energy Consumption at the Rzeszów Wastewater Treatment Plant*. E3S Web Conference, 22, 00115.

Mehta, C. M., Khunjar, W.O., Nguyen, V., Tait, S., Batstone, D. J. 2015. Technologies to Recover Nutrients from Waste Streams: A Critical Review. *Critical Reviews in Environmental Science and Technology* 45 (4): 385–427.

Metcalf and Eddy, Tchobanoglous, G., Stensel, H., Tcuchihashi, R., and Burton, F. 2013. *Wastewater Engineering: Treatment and Resource Recovery*. 5th edn. McGraw-Hill, Inc., New York, NY, USA.

Mihelcic, J. R. and Zimmerman, J. B. 2014. *Environmental Engineering: Fundamentals, Sustainability, Design*. 2nd edn., Wiley and Sons, Hoboken, NJ, USA.

Mitchell, C., Fam, D., Abeysuriya, K. 2013. *Transitioning to Sustainable Sanitation – A Transdisciplinary Project of Urine Diversion*. Institute for Sustainable Futures, University of Technology, Sydney, Australia.

NEWater. 2021. http://en.wikipedia.org/wiki/NEWater

Nexus. 2015. Nexus eWater. http://www.nexusewater.com/ (accessed June 16 2017).

Noe-Hayes, A., Nace, K., Patel, N., Lahr, R., Goetsch, H., Mullen, R., et al. 2015. Urine Diversion for Nutrient Recovery and Micropollutant Management: Results from A Regional Urine Recycling Program. *Proceedings of the 88th Annual Conference of Water Environment Federation, WEFTEC 2015*, Chicago, IL, USA.

Nowak, O., Enderle, P., and Varbanov, P. 2015. Ways to Optimize the Energy Balance of Municipal Wastewater Systems: Lessons Learned from Austrian Applications. *Journal of Cleaner Production*, 88:125–131.

NRC. 2012. *Water Reuse: Potential for Expanding the Nation's Water Supply Through Reuse of Municipal Wastewater*. National Research Council. National Academies Press, Washington, DC, USA.

PLWTP. 2021. Point Loma Wastewater Treatment Plant Website, http://www.sandiego.gov/mwwd/facilities/ptloma.shtml (accessed August 11, 2021).

PUB. 2021. NEWater. PUB, Singapore's National Water Agency. https://www.pub.gov.sg/watersupply/fournationaltaps/newater

Rahman, M. M., Salleh, M.A.M., Rashid, U., Ahsan, A., Hossain, M. M., and Ra, C.S. 2014. Production of Slow Release Crystal Fertilizer from Wastewaters Through Struvite Crystallization – A Review. *Arabian Journal of Chemistry* 7 (1): 139–155.

Rodriguez, D. J., Serrano, H. A., Delgado, A., Nolasco, D., and Saltiel, G. 2020. *From Waste to Resource: Shifting Paradigms for Smarter Wastewater Interventions in Latin America and the Caribbean*. World Bank, Washington, DC, USA.

Sheboygan-WWTP. 2021. Sheboygan Regional Wastewater Treatment Facility Website https://sheboygandpw.com/wastewater-treatment (accessed August 17, 2021).

WEF. 2010. *Design of Municipal Wastewater Treatment Plants*, 5th ed. Manual of Practice No. 8, vol 3, Chapter 20–27, Water Environment Federation, Alexandria, VA, USA.

WERF. 2008. *State of the Science Report: Energy and Resource Recovery from Sludge*. Water Environment Research Foundation, Alexandria, VA, USA.

WERF. 2015. *Demonstrated Energy Nuetrality Leadership: A Study of Five Champions of Change*. Water Environment Research Foundation, Alexandria, VA, USA.

WPTP. 2021. West Point Treatment Plant Website, http://www.kingcounty.gov/environment/wtd/About/System/West.aspx (accessed August 11 2021).

Chapter 15

Design examples

15.1 INTRODUCTION

The design principles of various primary, secondary, and advanced processes for the treatment of wastewater have been described in the previous chapters. Using the fundamental principles together with parameters used in practice, the overall design of a wastewater treatment system is explained in this chapter, with the help of example problems. Based on location, such as urban or rural, and other factors, the process design is discussed from conceptualization to selection and design of unit processes. Flow diagrams are used to illustrate the design parameters. The design of a septic tank for a rural household is detailed in an example problem, based on state and local regulatory requirements.

15.2 DESIGN EXAMPLE – CONVENTIONAL WASTEWATER TREATMENT PLANT

Example 15.1: Design a new conventional secondary wastewater treatment plant for an urban community.

Step 1: Investigate the service area

1. Service area and land use
    The wastewater treatment plant will serve an area of approximately 10 sq miles (2500 ha). The service area has 80% (2000 ha) residential (single-family homes, apartments, and mixeduse) facilities and 20% (500 ha) commercial and institutional facilities. There is no industrial development in the service area.

2. Population data
    The historical population data for the service area or that of similar areas in the neighboring city should be considered for population projection. The population data of the service area is presented in Table 15.1. Also, the trends of population growth must be studied carefully to decide whether constant growth, log-growth, percent growth, or ratio method can be used for accurate projection of

DOI: 10.1201/9781003134374-15

Table 15.1 Population data for
the service area

Year	Population
1980	16,300
1990	20,500
2000	25,200
2010	29,300
2020	32,400

population. The population in the service area has been increasing linearly in the last 40 years, thus the constant growth method can be used with significant accuracy.

Step 2: Wastewater flow estimation

1. **Design period**

A typical design period of 20–25 years is considered for the design of major wastewater treatment plants that are difficult or expensive to expand in the future. 10–15 years can also be considered with possible expansion planning in the future (Metcalf and Eddy et al., 2013). A 20-year design period is considered for the plant.

2. **Population projection**

The historical population data will be used to estimate the population for the year 2040 with a 20-year design period.

Using population data of 2010 and 2020,

$$P_t = P_0 + k\Delta t$$

or $P_{2020} = P_{2010} + k(2020 - 2010)$

or $32,400 = 29,300 + k \times 10$

or $k = 310$/year

Population in 2040,

$$\begin{aligned} P_{2040} &= P_{2020} + k(2040 - 2020) \\ &= 32400 + 310 \times 20 \\ &= 38,600 \end{aligned}$$

3. **Residential wastewater flow**

The wastewater flow from the service area is calculated using the projected population and per capita water withdrawal rate. Wastewater production varies largely with different factors, as discussed in Chapter 5. Based on the information provided, select an average daily flow = 378 Lpcd

$$\text{Residential wastewater flow} = 38,600 \times 378$$
$$= 14,590,800 \, \text{L/day}$$
$$= 14,591 \, \text{m}^3/\text{day}$$
$$\approx 14,600 \, \text{m}^3/\text{day}$$

4. **Commercial and Institutional flow**
   A range of 7.5–14 m³/ha.d of wastewater flow can be estimated from commercial and institutional facilities without data from well-defined businesses and institutions. A design value of 10 m³/ha.d is considered here for 500 ha commercial and institutional facilities.

   Commercial and institutional wastewater flow
   $$= 10 \times 500 = 5000 \, \text{m}^3/\text{d}$$

5. **Infiltration and Inflow (I&I)**
   I&I contribution can vary largely based on the area served, new or old sewer lines, or seasonal variations. Metcalf and Eddy (1981) recommend a design value of 4 m³/ha.d for a service area of 2500 ha.

   I & I wastewater flow $= 2500 \times 4 = 10,000 \, \text{m}^3/\text{day}$

6. **Total wastewater flow**
   The total wastewater flow (or design flow) for the wastewater treatment plant design is calculated from residential, commercial, institutional, and infiltration and inflow contribution.

   $$\text{Design flow} = 14,600 + 5000 + 10,000$$
   $$= 29,600 \approx 30,000 \, \text{m}^3/\text{d}$$

Step 3: Wastewater characteristics and effluent standards
Previously in Chapter 5, Table 5.3 presented the typical characteristics of untreated municipal wastewater. The composition of the wastewater can be considered weak, medium, or strong depending on different parameters and their concentrations. A sample of the wastewater collected from the service area (or neighboring areas with similar land use) and various wastewater parameters are measured. The characteristics of the untreated wastewater, considered in this example, are presented in Table 15.2.

The design of the wastewater treatment system largely depends on the level of treatment that must be achieved, which is generally governed by the wastewater treatment standards. The secondary treatment standards for US and EU are presented in Chapter 1. State and local regulations specify the secondary treatment standards, which are presented in Table 15.2.

Table 15.2 Influent characteristics and effluent standard of the wastewater

Constituents	Units	Influent	Effluent standard
Temperature	°C	10–20°C	—
pH	—	6.8	6–9
BOD_5 (20°C)	mg/L	240	20
TSS	mg/L	300	20
DO	mg/L	1.5	5

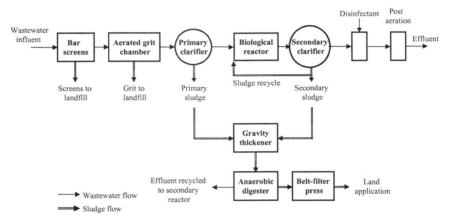

Figure 15.1 Flow diagram of a conventional wastewater treatment plant (for Example 15.1).

Step 4: Treatment plant flow diagram
Based on the influent composition and treatment standards, a flow diagram is developed to identify the processes necessary to achieve the desired removal of contaminants. A preliminary flow diagram for the treatment plant is shown in Figure 15.1.

Step 5: Design of preliminary treatment processes

1. **Headworks**
The headworks include pumping stations and flow measuring devices. A centrifugal pump or screw pumps are common for pumping operations, whereas Parshall flume or magnetic meters are used to measure the incoming flow rate.

2. **Design of mechanically cleaned bar screen**
The design (width and height) of the bar screen channel is based on Manning's equation (Davis, 2020), and a width of 1.0 m and water depth of 0.75 m are selected for the plant.

$$\text{Wastewater flow rate} = 30,000\,\text{m}^3/\text{day} = \frac{30,000}{86,400}$$

$$= 0.35\,\frac{\text{m}^3}{\text{s}}$$

$$\text{Approach velocity} = \frac{0.35 \dfrac{m^3}{s}}{(1.0 \times 0.75) m^2} = 0.47 \frac{m}{s}$$

The approach velocity is greater than the minimum required velocity of 0.4 m/s for a mechanically cleaned bar screen.

Considering 50 mm bar spacing (clear opening between bars) and 10 mm bar width (typical range 5–15 mm), the approximate number of bars is calculated using Equation (6.2),

$$\text{Number of bars, } N_{bar} = \frac{1000 - 50}{15 + 50} = 14.6 \approx 15$$

$$\text{Number of bar spaces} = 15 + 1 = 16$$

$$\text{Area of screen opening} = 16 \times \frac{50}{1000} \times 0.75 = 0.6 \, m^2$$

$$\text{Velocity through bar opening, } V_s = \frac{0.35 \dfrac{m^3}{s}}{0.6 \, m^2} = 0.58 \, m/s$$

The velocity through the bar opening is well below the maximum velocity of 0.9 m/s, to prevent the pass-through of solids.

Calculate the clean water headloss using Equation (6.1) with C_d value of 0.8 (typical value 0.7–0.84):

$$\text{Clean water headloss, } H_L = \frac{1}{0.8} \left(\frac{0.58^2 - 0.47^2}{2 \times 9.81} \right)$$
$$= 0.007 \, m = 7 \, mm$$

Calculate the maximum clogging allowed by considering the 0.9 m/s velocity through the bar opening:

$$\text{Area of screen opening} = \frac{0.35 \dfrac{m^3}{s}}{0.9 \dfrac{m}{s}} = 0.39 \, m^2$$

With $X\%$ clogging of the screen,

$$\text{Area of screen opening} = 0.6 \left(1 - \frac{X}{100} \right) = 0.39 \, m^2$$

Or, $X = 35\%$

Headloss with 35% clogging of the bar screen with $C_d = 0.6$,

$$\text{Clogged Screen Headloss, } H_L = \frac{1}{0.6}\left(\frac{0.9^2 - 0.47^2}{2 \times 9.81}\right)$$
$$= 0.05\,\text{m} = 50\,\text{mm}$$

The bar screen should be mechanically cleaned when the headloss is 50 mm.

3. **Design of aerated grit chamber**

   Aerated grit chamber is designed for peak hourly flow rate. Use two grit chambers in parallel.

   The peak factor is estimated at 2.5 (using Figure 5.1).

   Peak flow rate $= 30,000 \times 2.5 = 75,000\,\text{m}^3/\text{d}$

   $$\text{Flow in each tank at peak flow} = \frac{75,000}{2} = 37,500\,\frac{\text{m}^3}{\text{d}}$$

   Using 3 min detention time (typical value 2–5 min),

   $$\text{Volume of each tank} = \frac{37,500\,\dfrac{\text{m}^3}{\text{d}} \times 3\,\text{min}}{1440\,\dfrac{\text{min}}{\text{d}}} = 78\,\text{m}^3$$

   Assuming a tank depth of 3 m (typical value 2–5 m), and a width-to-depth ratio of 1:1,

   Tank depth $= 3\,\text{m}$

   Tank width $= 3 \times 1 = 3\,\text{m}$

   $$\text{Tank length} = \frac{78\,\text{m}^3}{3\,\text{m} \times 3\,\text{m}} = 8.7\,\text{m} \approx 9\,\text{m}$$

   Tank dimensions $= 9\,\text{m} \times 3\,\text{m} \times 3\,\text{m}$

   Select an air supply rate of 0.3 m³/min.m length of the tank,

   $$\text{Total air required} = 0.3\,\frac{\text{m}^3}{\text{min.m}} \times 9\,\text{m} \times 2\,\text{tanks} = 5.4\,\frac{\text{m}^3}{\text{min}}$$

   Assume a grit quantity of 0.02 m³/1000 m³ of flow

   $$\text{Total grit volume per day} = \frac{0.02\,\text{m}^3}{1000\,\text{m}^3} \times 37,500\,\frac{\text{m}^3}{\text{d}} \times 2\,\text{tanks}$$
   $$= 1.5\,\frac{\text{m}^3}{\text{d}}$$

Step 6: Design of primary treatment
Four circular clarifiers are selected for primary clarification.

$$\text{Flow in each clarifier} = \frac{30{,}000\,\dfrac{m^3}{d}}{4} = 7500\,\frac{m^3}{d}$$

Using a typical surface overflow rate of 40 m³/m².d

$$\text{Surface area} = \frac{7500\,\dfrac{m^3}{d}}{40\,\dfrac{m^3}{m^2 d}} = 187.5\,m^2$$

$$\text{Or } \pi\left(\frac{D^2}{4}\right) = 187.5$$

Or Diameter, $D = 15.45 \approx 15.5\,m$

$$\text{Actual Surface area} = \pi\left(\frac{15.5^2}{4}\right) = 188.7\,m^2$$

Calculate the detention time considering a water depth of 4.0 m (typical value 3–5m),

$$\text{Detention time} = \frac{188.7\,m^2 \times 4\,m}{7500\,\dfrac{m^3}{d}} = 0.1\,d = 2.4\,h$$

The detention time is within the range of 1.5–2.5 h for a clarifier.

$$\text{Weir loading rate} = \frac{7500}{\pi \times 15.5} = 154\,\frac{m^3}{m.d}$$

The weir loading rate is within the range of 125–500 m³/m.d for a clarifier.

The dimension of the primary clarifier is 15.5 m diameter and 4.5 m depth considering 0.5 m freeboard.

Check for peak flow condition:

$$\text{Peak flow in each clarifier} = \frac{30{,}000 \times 2.5}{4} = 18{,}750\,\frac{m^3}{d}$$

$$\text{Overflow rate} = \frac{18{,}750\,\dfrac{m^3}{d}}{188.7\,m^2} = 99.4\,\frac{m^3}{m^2 d}$$

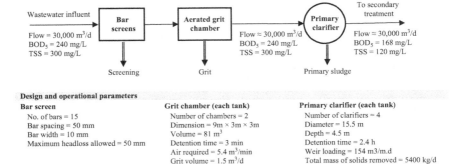

Design and operational parameters

Bar screen	Grit chamber (each tank)	Primary clarifier (each tank)
No. of bars = 15	Number of chambers = 2	Number of clarifiers = 4
Bar spacing = 50 mm	Dimension = 9m × 3m × 3m	Diameter = 15.5 m
Bar width = 10 mm	Volume = 81 m³	Depth = 4.5 m
Maximum headloss allowed = 50 mm	Detention time = 3 min	Detention time = 2.4 h
	Air required = 5.4 m³/min	Weir loading = 154 m3/m.d
	Grit volume = 1.5 m³/d	Total mass of solids removed = 5400 kg/d

Figure 15.2 Design and operational parameters for preliminary and primary treatment processes.

$$\text{Detention time} = \frac{188.7\,\text{m}^2 \times 4\,\text{m}}{18{,}750\dfrac{\text{m}^3}{\text{d}}} = 0.04\,\text{d} = 0.97\,\text{h}$$

$$\text{Weir loading rate} = \frac{18{,}750}{\pi \times 15.5} = 385\,\frac{\text{m}^3}{\text{m}^2\text{d}}$$

The overflow rate and weir loading rate are well within the range for a primary clarifier, whereas the detention time is slightly less than the 1 h minimum value.

The design and operational parameters of the preliminary and primary treatment processes are illustrated in Figure 15.2.

Step 7: Design of secondary treatment – activated sludge process

1. **Design of aeration tank**

   A completely mixed activated sludge process with recycle is considered for the plant. Two tanks will be used in parallel. Mean cell residence time or SRT (θ_c) is a major design and operational parameter for the activated sludge process, which can be accurately determined from different kinetics parameters (i.e. Y, k_d, μ_{max}, and K_s) obtained from lab or pilot plant analysis. In many cases, the kinetic parameters are not readily available, and the sludge age is estimated to maintain a certain MLSS concentration and F/M ratio in the aeration tank.

   $$\text{Flow in each aeration tank} = \frac{30{,}000\dfrac{\text{m}^3}{\text{d}}}{2} = 15{,}000\,\frac{\text{m}^3}{\text{d}}$$

   The primary clarifier typically removes 30% of the raw BOD_5, thus the influent BOD_5 for the aeration tank is

   $$S_o = 240 \times (1 - 0.3) = 168\,\text{mg/L}$$

The effluent BOD_5 based on secondary treatment standards is

$$S = 20 \, mg/L$$

Conventional activated sludge processes are operated at F/M ratio of 0.2–0.4, and MLSS concentration of 1500–4000 mg/L. Select an F/M ratio = 0.3 and MLSS = 2500 mg/L for the plant.

$$F/M \, ratio = \frac{15,000 \times (168 - 20)}{V \times 2500} = 0.3$$

Or $V = 2960 \, m^3$

$$HRT, \theta = \frac{2960}{15,000} = 0.2 \, d = 4.7 \, h$$

Return sludge concentration (X_u) is an operational parameter, generally selected in the range of 10,000–14,000 mg/L for good settling sludge, and 3000–6000 mg/L for poorly settling sludge. The sludge volume index (SVI) experiment can help identify the sludge settling characteristics. Select X_u = 6000 mg/L, which is wasted at 200 m³/d. The SRT is calculated as

$$SRT, \theta_c = \frac{VX}{Q_w X_u} = \frac{2960 \times \dfrac{2500}{1000}}{200 \times \dfrac{6000}{1000}} = 6.2 \, d$$

Return sludge flow rate (Q_R) is calculated using Equation (8.56),

$$Q_r = \frac{15,000 \times \dfrac{2500}{1000} - 200 \times \dfrac{6000}{1000}}{\dfrac{6000}{1000} - \dfrac{2500}{1000}} = 10,370 \, \frac{m^3}{d}$$

$$\text{Recirculation ratio} = \frac{Q_r}{Q} = \frac{10,370}{15,000} = 0.7$$

The SRT, MLSS, HRT, and recirculation ratio are all within the typical design range for a completely mixed activated sludge process.

Two rectangular aeration tanks, each 40 m × 15 m × 5 m, are selected for the plant.

2. **Design of secondary clarifier**

The batch settling column test results with different initial solids concentrations are provided in Table 15.3. The limiting solids flux for 6000 mg/L underflow concentration and the settling velocity for

Table 15.3 Settling column test results

Solids, mg/L	Velocity, m/h	Flux, kg/m².h
500	6.2	3.1
1000	4.5	4.5
1500	2.8	4.2
2000	1.75	3.5
3000	0.8	2.4
4000	0.38	1.52
5000	0.18	0.9
6000	0.1	0.6
7000	0.06	0.42

2500 mg/L MLSS concentration are presented in Figures 15.3 (a) and 15.3 (b), respectively.

Area required for thickening:

Limiting solids flux = 4.5 kg/m². h (determined from Figure 15.3(a))

Flow rate, $Q = 15,000\,\mathrm{m^3/d}$

Recycle flow rate, $Q_r = 10,370\,\mathrm{m^3/d}$

$$\text{Solids loading to clarifier} = \frac{2500}{1000} \times (15,000 + 10,370) \times \frac{1}{24}$$

$$= 2640\,\mathrm{kg/d}$$

$$\text{Surface area required for thickening} = \frac{2640}{4.5} = 587\,\mathrm{m^2}$$

Area required for clarification:

Settling velocity = 1.1 m/h (determined from Figure 15.3(b))

$$\text{Surface area required for clarification} = \frac{15,000 \times \frac{1}{24}}{1.1}$$

$$= 570\,\mathrm{m^2}$$

Design surface area for secondary clarifier = 587 m²

Two circular secondary clarifiers, each with a diameter of 27.5 m and a side water depth of 4 m, are selected for the plant.

$$\text{Underflow rate, } Q_u = Q_r + Q_w$$

$$= 10,370 + 200 = 10,570\,\mathrm{m^3/d}$$

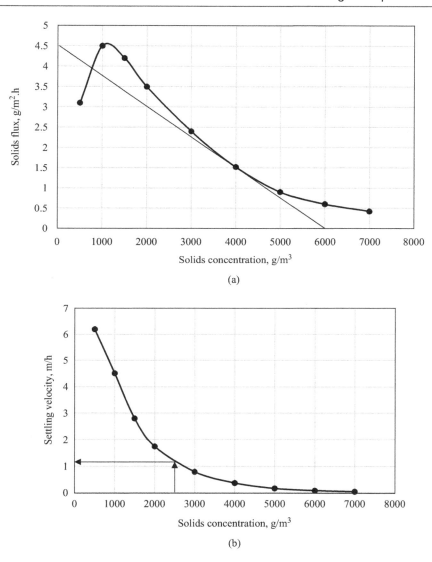

Figure 15.3 (a) Limiting solids flux and (b) settling velocity curve.

$$\text{Underflow velocity} = \frac{4.5}{\dfrac{6000}{1000}} = 0.75 \frac{\text{m}}{\text{h}}$$

Step 8: Disinfection and post-aeration

Chlorination and ultraviolet (UV) disinfection for wastewater treatment are common in the US. The deciding factors for the disinfection process are cost and secondary effluent characteristics. UV disinfection is selected for the plant, with medium-pressure high-intensity lamps provided in two parallel channels. The number, type, and rating of the lamps are selected based on hydraulic loading, aging, and fouling of the lamps, wastewater quality, and discharge standards required in the NPDES permit. The dose maintained after secondary treatment is in the range of 50–140 mJ/cm^2.

The effluent from the secondary clarifier typically has very low dissolved oxygen (DO), thus post-aeration is required to increase the DO level. Cascade aeration is most common, although diffused aerators can also be used when sufficient elevation head is not available. The cascade height is calculated based on the oxygen deficit, cascade step geometry, and temperature.

Considering wastewater temperature of 30°C, and 1.5 mg/L influent DO concentration to the cascade,

Saturation dissolved oxygen at 30°C, C_s = 7.63 mg/L

Effluent DO, C = 5 mg/L

Oxygen deficit ratio, $R = \dfrac{7.63 - 1.5}{7.63 - 5} = 2.33$

Height of cascade, considering steps with broad crested weir

$$H = \frac{2.33 - 1}{0.361 \times 0.8 \times 1 \times (1 + 0.046 \times 30)} = 1.94\,\text{m} \approx 2\,\text{m}$$

Using steps with 200 mm height and 450 mm length,

$$\text{Number of steps} = \frac{2\,\text{m}}{\dfrac{200}{1000}\,\text{m}} = 10$$

A 2 m cascade height is provided with 10 steps with a broad-crested weir to achieve a discharge DO concentration of 5 mg/L.

The design and operational parameters of secondary treatment, disinfection, and post-aeration processes are illustrated in Figure 15.4.

Step 9: Solids processing and disposal

1. **Design of gravity thickener**

   Assume 60% removal of solids in the primary clarifier.

$$\text{Mass of the primary sludge} = 300\,\frac{\text{mg}}{\text{L}} \times 0.60 \times 30{,}000\,\frac{\text{m}^3}{\text{d}}$$
$$\times \frac{1000\,\text{L}}{\text{m}^3} \times \frac{1\,\text{kg}}{10^6\,\text{mg}}$$

$$= 5400\,\text{kg/d}$$

Although the specific gravity of the primary and secondary sludge is slightly higher than water, they can be assumed to be the same as water.

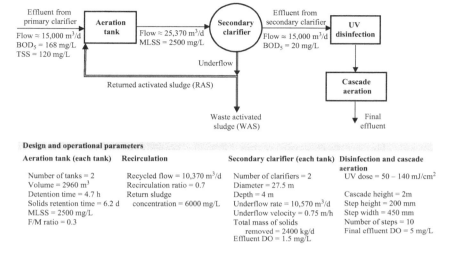

Figure 15.4 Design and operational parameters for secondary treatment, disinfection and post-aeration processes.

Volume of primary sludge with 5% solids:

$$\text{Volume of primary sludge} = \frac{5400}{1000 \times 1 \times 0.05} = 108 \frac{m^3}{d}$$

$$\text{Mass of secondary sludge} = 2 \times 200 \times \frac{6000}{1000} = 2400 \frac{kg}{d}$$

Volume of the secondary sludge with 4% solids:

$$\text{Volume of secondary sludge} = \frac{2400}{1000 \times 1 \times 0.04} = 60 \frac{m^3}{d}$$

Total mass of sludge = 5400 + 2400 = 7800 kg/d

The design of the gravity thickener is similar to the design of a secondary clarifier using the solids flux or state point analysis.

Four gravity thickeners, each with a diameter of 6 m and a side water depth of 5 m are selected, with 2 m of sludge blanket at the bottom.

$$\text{Area of each thickener} = \frac{\pi \times 6^2}{4} = 28.30 \, m^2 s$$

Inflow from the primary and secondary clarifier,

$$Q_{in} = \frac{(108 + 60)}{4} = 42.0 \frac{m^3}{d}$$

An additional 200 m³/d of treated wastewater is applied to each thickener for improved overflow rate, and thickened sludge is withdrawn at 15 m³/d at 7% solids content:

$$Q_{effluent} = (42 + 200 - 15) = 227 \, m^3/d$$

$$\text{Overflow rate} = \frac{227}{28.3} = 8.02 \, \frac{m^3}{m^2 d}$$

The overflow rate is within the range of 6–12 m³/m².d for combined primary and waste-activated sludge. The amount of additional treated wastewater can be varied to maintain the overflow rate within range.

$$\text{Solids loading} = \frac{(5400 + 2400)}{4 \times 28.3} = 68.90 \, \frac{kg}{m^2 d}$$

The solids loading is within the range of 25–80 kg/m².d. The number of gravity thickeners can be modified to keep the value within the specified range.

With 2 m sludge blanket,

$$\text{Sludge volume ratio} = \frac{28.3 \times 2}{15}$$
$$= 3.8 d \quad (\text{Typical range} \, 0.5 - 20 \, d)$$

$$\%\text{solids capture} = \frac{(15 \times 0.07 \times 1000)}{\dfrac{(5400 + 2400)}{4}} \times 100 = 53.85\%$$

2. **Design of anaerobic digester**

   A single-stage mesophilic anaerobic digester is considered for the plant. The influent characteristics of the anaerobic digester are determined from the effluent concentration of the gravity thickener. There are four gravity thickeners, and thickened sludge is withdrawn at 15 m³/d from each thickener at 7% solids content.

   $$\text{Raw sludge loading rate}, V_1 = 4 \times 15 = 60 \, m^3/d$$

   $$\text{Mass of solids to digester} = 4 \times (15 \times 0.07 \times 1000)$$
   $$= 4200 \, kg/d$$

   The total solids analysis of the thickener sludge can be used to identify the percentage of organic (volatile solids) and inorganic (fixed solids) fractions. Assume 70% of the solids are organic.

Organic fraction = $4200 \times 0.7 = 2940 \, kg/d$

Inorganic fraction = $4200 \times (1 - 0.7) = 1260 \, kg/d$

The percent reduction of organic matter (volatile solids) in the digester can be estimated from pilot studies. Volatile solids destruction of 55–65% can be achieved at SRT of 15–30 d. Assume 60% destruction of volatile solids in the digester at 20 d SRT.

$$Organic \; fraction \; remaining = 2940 \times (1 - 0.60)$$
$$= 1176 \, kg/d$$

Total mass remaining = $1176 + 1260 = 2436 \, kg/d$

Assume 6% solids content in the digested sludge,

$$Digested \; sludge \; accumulation \; rate, \; V_2 = \frac{2436}{1000 \times 1 \times 0.06}$$
$$= 40.6 \frac{m^3}{d} s$$

Select 50 d sludge storage time. The single-stage digester volume is calculated using Equation (12.11),

$$V_s = \frac{(60 + 40.6)}{2} \times 20 + 40.6 \times 50 = 3036 \, m^3$$

Cylindrical digesters are common in the US with a diameter ranging from 6 to 40 m, and height less than 14 m, with at least 7.5 m water depth (Metcalf and Eddy et al., 2013). Two mesophilic (35°C) digesters, each with 15 m diameter, and 9 m height are selected. The digesters are placed 6 m below ground to reduce heat loss.

The digested sludge is withdrawn at 65 m³/d from each digester at 6% solids content and sent to the dewatering process. The effluent from the digester, which has a very high concentration of nutrients (nitrogen and phosphorus) are recycled to the secondary reactor (aeration tank) for additional treatment.

Heating requirement for each digester:

Heating requirements mostly depend on the weather (summer, winter, etc.), surrounding soil condition (soil temperature, dry/wet), and construction materials (concrete, steel, etc.). The maximum heat required is calculated for cold weather conditions.

Digesters considered for the plant are designed to have a 6-mm-thick steel cover, and 300 mm concrete floor and insulated wall with moist surrounding soil. On a typical day with an air temperature of 12°C, soil temperature of 15°C, and feed sludge temperature of 20°C, the total heat required is

$$q_r = \frac{4200}{2} \times 4.186 \times (35-20) = 0.132 \times 10^6 \text{ kJ/d}$$

Heat loss from the floor,

$$q_{\text{floor}} = 2.85 \times \left(\frac{\pi}{4} \times 15^2 \right)(35-15) \times \frac{86,400}{1000}$$
$$= 0.870 \times 10^6 \text{ kJ/d}$$

Heat loss from the cover,

$$q_{\text{cover}} = 5.0 \times \left(\frac{\pi}{4} \times 15^2 \right)(35-12) \times \frac{86,400}{1000}$$
$$= 1.756 \times 10^6 \text{ kJ/d}$$

Heat loss from the wall,

$$q_{\text{wall}} = \left[0.7 \times (\pi \times 15 \times 3)(35-12) + 1.2 \times (\pi \times 15 \times 6)(35-15) \right]$$
$$\times \frac{86,400}{1000}$$

$$= 0.783 \times 10^6 \text{ kJ/d}$$

$$\text{Total heat required} = (0.132 + 0.870 + 1.756 + 0.685) \times 10^6$$
$$= 3.541 \times 10^6 \text{ kJ/d}$$

3. **Design of belt-filter press for dewatering**
   The design sludge flow rate for the belt-press filter is calculated from the digested sludge flow collected from sludge outlets of the anaerobic digesters. The digested sludge is withdrawn at 65 m³/d from each digester at 6% solids content, thus the design sludge flow for the belt-press filter is

$$\text{Design sludge flow} = 2 \times 65 = 130 \text{ m}^3/\text{d}$$

The belt-filter press is operated on a one-shift basis (8 h/d),

$$\text{Hydraulic loading} = \frac{130}{8} = 16.25 \frac{\text{m}^3}{\text{h}}$$

$$\text{Solids loading} = 16.25\frac{m^3}{h} \times 1000\frac{kg}{m^3} \times 1 \times 0.06 = 975\,kg/h$$

The belt width is selected based on the hydraulic loading and solids loading data, and compared with manufacturers' data. A 2 m belt width would be sufficient for the calculated hydraulic and solids loading rates, which is commonly used in municipal sludge dewatering operations (Davis, 2020).

$$\text{Solids loading rate} = \frac{975}{2} = 488\frac{kg}{m.h} \text{ is within the range of } 90-680\,kg/m.h.$$

4. **Biosolids disposal**

    Dewatered cake solids from the belt-filter press contain 12–32% solids, and the amount of cake produced per day is used to calculate the truckload and transportation of the sludge (biosolids) for land application.Sludge produced from the dewatering process is tested for pathogens, metals, VSS reduction, and vector attraction reduction requirement, and classified as Class A or Class B biosolids. For this plant, the treated sludge is characterized as Class B biosolids; thus, land application with site restrictions with respect to crop harvesting, animal grazing, and public access are placed.

    The design and operational parameters of solids processing and disposal are illustrated in Figure 15.5.

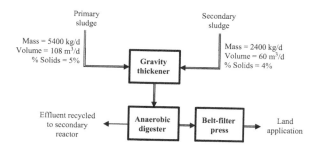

Design and operational parameters

Gravity thickener (each)
Number of thickener = 4
Diameter = 6 m
Depth = 5 m (with 2m sludge blanket)
Inflow = 42 m³/d (plus 200 m³/d wastewater)
Overflow rate = 7.9 m³/m².d
Solid loading = 68.9 kg/m².d
Sludge volume ratio = 3.8 d
% solids capture = 54%
Sludge withdrawal = 15 m³/d at 7% solids

Anaerobic digester (each)
Number of digesters = 2
Diameter = 15 m
Height = 9 m (6 m underground)
SRT = 20 d
Storage period = 50 d
Sludge withdrawal = 65 m³/d at 6% solids
Heat required = 3.54 × 10⁶ kJ/d

Belt-filter press
Operating shift = 8 h/d
Hydraulic loading = 16.25 m³/h
Solids loading = 975 kg/h
Belt width = 2 m
Solids loading rate = 488 kg/m.h

Figure 15.5 Design and operational parameters for solids processing.

15.3 DESIGN EXAMPLE – DECENTRALIZED WASTEWATER TREATMENT IN A RURAL LOCATION

Example 15.2: Design a decentralized wastewater treatment system for a 4-bedroom household situated in a rural community in upstate New York.

SOLUTION

A septic tank will be used for the treatment of the wastewater generated from the 4-bedroom house. The design of the septic tank is based on New York Department of Health (2010) guidelines for wastewater treatment standards for individual household systems.

1. **Tank capacity**
   Tank capacity is calculated based on the amount of wastewater produced each day or based on the number of bedrooms in the household. For a 4-bedroom household,

   $$\text{Minimum tank capacity required} = 1250 \, \text{gal} \approx 167 \, \text{ft}^3$$

   $$\text{Minimum liquid surface area} = 34 \, \text{ft}^2$$

   Use a tank with a capacity of 1250 gal.

2. **Tank geometry**
   The minimum liquid depth of 30 in is recommended, with a maximum allowable depth of 60 in. Considering 48 in (4 ft) of liquid depth and a length-to-width ratio of 2.5:1 (typical $L:W = 2:1$ to 4:1),

   $$\text{Liquid surface area} = \frac{167 \, \text{ft}^3}{4 \, \text{ft}} = 41.75 \, \text{ft}^2$$

   $$\text{or } L \times W = 41.75 \, \text{ft}^2$$

   $$\text{or } 2.5W \times W = 41.75 \, \text{ft}^2$$

   $$\text{Width,} \, W = 4.08 \, \text{ft} \approx 4 \, \text{ft}$$

   $$\text{Length} = 2.5 \times 4 = 10 \, \text{ft} \quad (\text{greater than 6 ft minimum})$$

   $$\text{Effective surface area} = 10 \times 4 = 40 \, \text{ft}^2 \quad (\text{greater than 34 ft}^2 \, \text{minimum})$$

   Use tank dimension of 10 ft length × 4 ft width × 5 ft depth with 1 ft freeboard.

Dual compartments are recommended for all tanks with length of 10 ft or more. Use two compartments in this design.

Length of first compartment = 7 ft

Length of second compartment = 3 ft

The first compartment accounts for 70% of the design volume (Typical range 60–75%).

The two-compartment septic tank is illustrated in Figure 15.6 (a).

3. Construction materials

The tank can be made of concrete, fiberglass, polyethylene, or steel, based on cost, availability, and groundwater level. Due to the high groundwater level in the service area, a polyethylene tank is used.

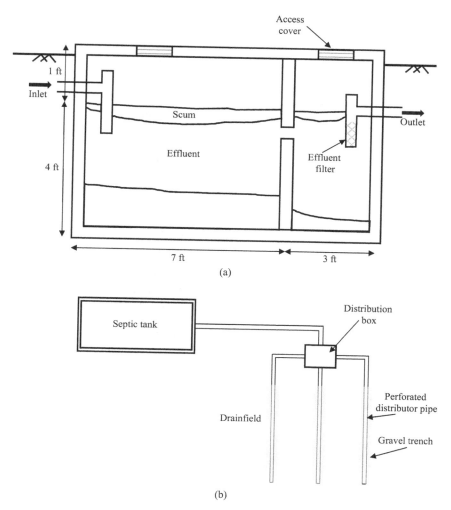

Figure 15.6 (a) Septic tank dimensions and (b) septic tank system – plan view.

4. **Distribution box**

The drop between inlet and outlet inverts shall be at least 2 in. All outlets are placed at the same level for even distribution of flow. The box can be constructed in place or purchased prefabricated.

5. **Effluent treatment system**

All wastewater effluent from the septic tank shall be discharged to a subsurface treatment system. A conventional absorption trench system is considered here.

1. The minimum distance must be maintained for all the components of the treatment system from the nearby wells, streams, lakes, wetlands, or other site features. The local and state guidelines must be followed.
2. The length of the trench is determined based on the percolation test result and soil evaluation. The maximum depth and width of 30 in and 24 in, respectively, are selected for the trench.
3. Perforated distributor pipes are selected and placed in the trenches. Rigid or corrugated plastic pipes are common in the septic system.
4. Gravel or crushed stone of ¾ to 1 ½ in diameter are used to cover the perforated pipes.
5. The pipes are sloped at 1/16 in to 1/32 in per ft for gravity distribution.

The distribution box and drainfield for effluent discharge are illustrated in Figure 15.6 (b).

REFERENCES

Davis, M. 2020. *Water and Wastewater Engineering: Design Principles and Practice.* 2nd edn. McGraw-Hill, Inc., New York, NY, USA.

Metcalf and Eddy, Inc. 1981. *Wastewater Engineering: Collection and Pumping Wastewater*, McGraw-Hill, Inc., New York, NY, USA.

Metcalf and Eddy, Tchobanoglous, G., Stensel, H., Tcuchihashi, R., and Burton, F. 2013. *Wastewater Engineering: Treatment and Resource Recovery.* 5th edn. McGraw-Hill, Inc., New York, NY, USA.

New York State Department of Health. 2010. *Wastewater Treatment Standards - Individual Household Systems*, Appendix 75-A, New York State Department of Health, New York, NY, USA.

Appendix A

Table A-1 Physical properties of water (SI units)

Temperature (°C)	Specific weight γ (kN/m³)	Density ρ (kg/m³)	Modulus of elasticity* $E \times 10^{-6}$ (kN/m²)	Dynamic viscosity $\mu \times 10^3$ (N.s/m²)	Kinematic viscosity $\nu \times 10^6$ (m²/s)	Surface tension** σ (N/m)	Vapor pressure p_v (kN/m²)
0	9.805	999.8	1.98	1.781	1.785	0.0765	0.61
5	9.807	1000.0	2.05	1.518	1.519	0.0749	0.87
10	9.804	999.7	2.10	1.307	1.306	0.0742	1.23
15	9.798	999.7	2.15	1.139	1.139	0.0735	1.70
20	9.789	998.2	2.17	1.002	1.003	0.0728	2.34
25	9.777	997.0	2.22	0.890	0.893	0.0720	3.17
30	9.764	995.7	2.25	0.798	0.800	0.0712	4.24
40	9.730	992.2	2.28	0.653	0.658	0.0696	7.38
50	9.689	988.0	2.29	0.547	0.553	0.0679	12.33
60	9.642	983.2	2.28	0.466	0.474	0.0662	19.92
70	9.589	977.8	2.25	0.404	0.413	0.0644	31.16
80	9.530	971.8	2.20	0.354	0.364	0.0626	47.34
90	9.466	965.3	2.14	0.315	0.326	0.0608	70.10
100	9.399	958.4	2.07	0.282	0.294	0.0589	101.33

*At atmospheric pressure; **In contact with air.

Table A-2 Equilibrium concentrations (mg/L) of dissolved oxygen* as a function of temperature and chloride concentration

Temperature (°C)	Chloride concentration (mg/L)				
	0	5000	10,000	15,000	20,000
0	14.62	13.79	12.97	12.14	11.32
1	14.23	13.41	12.61	11.82	11.03
2	13.84	13.05	12.28	11.52	10.76
3	13.48	12.72	11.98	11.24	10.50
4	13.13	12.41	11.69	10.97	10.25
5	12.80	12.09	11.39	10.70	10.01
6	12.48	11.79	11.12	10.45	9.78
7	12.17	11.51	10.85	10.21	9.57
8	11.87	11.24	10.61	9.98	9.36
9	11.59	10.978	10.36	9.76	9.17
10	11.33	10.73	10.13	9.55	8.98
11	11.08	10.49	9.92	9.35	8.80
12	10.83	10.28	9.72	9.17	8.62
13	10.60	10.05	9.52	8.98	8.46
14	10.37	9.85	9.32	8.80	8.30
15	10.15	9.65	9.14	8.63	8.14
16	9.95	9.46	8.96	8.47	7.99
17	9.74	9.26	8.78	8.30	7.84
18	9.54	9.07	8.62	8.15	7.70
19	9.35	8.89	8.45	8.00	7.56
20	9.17	8.73	8.30	7.86	7.42
21	8.99	8.57	8.14	7.71	7.28
22	8.83	8.42	7.99	7.57	7.14
23	8.68	8.27	7.85	7.43	7.00
24	8.53	8.12	7.71	7.30	6.87
25	8.38	7.96	7.56	7.15	6.74
26	8.22	7.81	7.42	7.02	6.61
27	8.07	7.67	7.28	6.88	6.49
28	8.92	7.53	7.14	6.75	6.37
29	7.77	7.39	7.00	6.62	6.25
30	7.63	7.25	6.86	6.49	6.13

Source: G.C.Whipple and M.C.Whipple: Solubility of Oxygen in Sea Water, *J.Am. Chem. Soc.*, vol. 33, p. 362, 1911. Calculated using data developed by C.J.J. Fox: On the Coefficients of Absorption of Nitrogen and Oxygen in Distilled Water and Sea Water and Atmospheric Carbonic Acid in Sea Water, *Trean. Faraday Soc.*, vol. 5, p. 68, 1909.

*Saturation values of dissolved oxygen in freshwater and seawater exposed to dry air containing 20.90% oxygen by volume under a total pressure of 760 mm of mercury.

Index